Molecular Phylogenetics and Mitochondrial Evolution

Molecular Phylogenetics and Mitochondrial Evolution

Editors

Andrea Luchetti
Federico Plazzi

MDPI • Basel • Beijing • Wuhan • Barcelona • Belgrade • Manchester • Tokyo • Cluj • Tianjin

Editors
Andrea Luchetti
Department of Biological,
Geological and Environmental
Sciences
University of Bologna
Bologna
Italy

Federico Plazzi
Department of Biological,
Geological and Environmental
Sciences
University of Bologna
Bologna
Italy

Editorial Office
MDPI
St. Alban-Anlage 66
4052 Basel, Switzerland

This is a reprint of articles from the Special Issue published online in the open access journal *Life* (ISSN 2075-1729) (available at: www.mdpi.com/journal/life/special_issues/Molecular_Evolution).

For citation purposes, cite each article independently as indicated on the article page online and as indicated below:

LastName, A.A.; LastName, B.B.; LastName, C.C. Article Title. *Journal Name* **Year**, *Volume Number*, Page Range.

ISBN 978-3-0365-3306-3 (Hbk)
ISBN 978-3-0365-3305-6 (PDF)

© 2022 by the authors. Articles in this book are Open Access and distributed under the Creative Commons Attribution (CC BY) license, which allows users to download, copy and build upon published articles, as long as the author and publisher are properly credited, which ensures maximum dissemination and a wider impact of our publications.

The book as a whole is distributed by MDPI under the terms and conditions of the Creative Commons license CC BY-NC-ND.

Contents

About the Editors vii

Preface to "Molecular Phylogenetics and Mitochondrial Evolution" ix

Andrea Luchetti and Federico Plazzi
Molecular Phylogenetics and Mitochondrial Evolution
Reprinted from: *Life* **2021**, *12*, 4, doi:10.3390/life12010004 1

Alessandro Formaggioni, Andrea Luchetti and Federico Plazzi
Mitochondrial Genomic Landscape: A Portrait of the Mitochondrial Genome 40 Years after the First Complete Sequence
Reprinted from: *Life* **2021**, *11*, 663, doi:10.3390/life11070663 5

Giulia Furfaro and Paolo Mariottini
Looking at the Nudibranch Family Myrrhinidae (Gastropoda, Heterobranchia) from a Mitochondrial '2D Folding Structure' Point of View
Reprinted from: *Life* **2021**, *11*, 583, doi:10.3390/life11060583 25

Giulia Furfaro and Paolo Mariottini
Nemesignis, a Replacement Name for *Nemesis* Furfaro amp; Mariottini, 2021 (Mollusca, Gastropoda, Myrrhinidae), Preoccupied by *Nemesis* Risso, 1826 (Crustacea, Copepoda)
Reprinted from: *Life* **2021**, *11*, 809, doi:10.3390/life11080809 41

Arsen V. Dotsev, Elisabeth Kunz, Veronika R. Kharzinova, Innokentiy M. Okhlopkov, Feng-Hua Lv and Meng-Hua Li et al.
Mitochondrial DNA Analysis Clarifies Taxonomic Status of the Northernmost Snow Sheep (*Ovis nivicola*) Population
Reprinted from: *Life* **2021**, *11*, 252, doi:10.3390/life11030252 43

Micky M. Mwamuye, Isaiah Obara, Khawla Elati, David Odongo, Mohammed A. Bakheit and Frans Jongejan et al.
Unique Mitochondrial Single Nucleotide Polymorphisms Demonstrate Resolution Potential to Discriminate *Theileria parva* Vaccine and Buffalo-Derived Strains
Reprinted from: *Life* **2020**, *10*, 334, doi:10.3390/life10120334 55

Adrianna Kilikowska, Monika Mioduchowska, Anna Wysocka, Agnieszka Kaczmarczyk-Ziemba, Joanna Rychlińska and Katarzyna Zajac et al.
The Patterns and Puzzles of Genetic Diversity of Endangered Freshwater Mussel *Unio crassus* Philipsson, 1788 Populations from Vistula and Neman Drainages (Eastern Central Europe)
Reprinted from: *Life* **2020**, *10*, 119, doi:10.3390/life10070119 71

Stefania Chiesa, Livia Lucentini, Paula Chainho, Federico Plazzi, Maria Manuel Angélico and Francisco Ruano et al.
One in a Million: Genetic Diversity and Conservation of the Reference *Crassostrea angulata* Population in Europe from the Sado Estuary (Portugal)
Reprinted from: *Life* **2021**, *11*, 1173, doi:10.3390/life11111173 95

Pedro María Alarcón-Elbal, Ricardo García-Jiménez, María Luisa Peláez, Jose Luis Horreo and Antonio G. Valdecasas
Molecular Correlation between Larval, Deutonymph and Adult Stages of the Water Mite *Arrenurus (Micruracarus) Novus*
Reprinted from: *Life* **2020**, *10*, 108, doi:10.3390/life10070108 107

Dan Zhao, Zhaozhe Xin, Hongxia Hou, Yi Zhou, Jianxia Wang and Jinhua Xiao et al.
Inferring the Phylogenetic Positions of Two Fig Wasp Subfamilies of Epichrysomallinae and Sycophaginae Using Transcriptomes and Mitochondrial Data
Reprinted from: *Life* **2021**, *11*, 40, doi:10.3390/life11010040 121

Nicola Zadra, Annapaola Rizzoli and Omar Rota-Stabelli
Chronological Incongruences between Mitochondrial and Nuclear Phylogenies of *Aedes* Mosquitoes
Reprinted from: *Life* **2021**, *11*, 181, doi:10.3390/life11030181 133

Xuhua Xia
Improving Phylogenetic Signals of Mitochondrial Genes Using a New Method of Codon Degeneration
Reprinted from: *Life* **2020**, *10*, 171, doi:10.3390/life10090171 151

Juliana E. Arcila-Galvis, Rafael E. Arango, Javier M. Torres-Bonilla and Tatiana Arias
The Mitochondrial Genome of a Plant Fungal Pathogen *Pseudocercospora fijiensis* (Mycosphaerellaceae), Comparative Analysis and Diversification Times of the Sigatoka Disease Complex Using Fossil Calibrated Phylogenies
Reprinted from: *Life* **2021**, *11*, 215, doi:10.3390/life11030215 167

Steinar Daae Johansen, Sylvia I. Chi, Arseny Dubin and Tor Erik Jørgensen
The Mitochondrial Genome of the Sea Anemone *Stichodactyla haddoni* Reveals Catalytic Introns, Insertion-Like Element, and Unexpected Phylogeny
Reprinted from: *Life* **2021**, *11*, 402, doi:10.3390/life11050402 187

Maria-Eleni Parakatselaki and Emmanuel D. Ladoukakis
mtDNA Heteroplasmy: Origin, Detection, Significance, and Evolutionary Consequences
Reprinted from: *Life* **2021**, *11*, 633, doi:10.3390/life11070633 201

About the Editors

Andrea Luchetti

The main research interest of Andrea Luchetti is the study of molecular evolution in different Arthropoda with special regard to the study of the evolution of repetitive DNA: tandem repeats (satellite DNA, ribosomal DNA) and interspersed repeats (transposable elements). Since the evolutionary dynamics of these sequences is strictly linked to the taxonomy and the biology of analyzed animal systems, he also investigated the phylogeny, phylogeography, molecular systematics and the reproductive biology of several insect and crustacean species.

The main research lines are:

–Molecular phylogenetics, systematics and reproductive biology of Insecta, particularly Isoptera and Phasmida; Crustacea, Branchiopoda: Choncostraca and Notostraca.

–Evolution of eusociality, colony structure and breeding system in termites (Reticulitermes spp., Rhintoermitidae; Kalotermes spp., Kalotermitidae).

–Evolutionary dynamics of repetitive DNA (mainly satellite DNA and retrotransposons) in the genome of arthropods with non-canonical reproductive strategies (parthenogenesis; ermaphroditism) or non-panmictic (eusociality).

Federico Plazzi

Federico Plazzi is Adjunct Professor of Applied Statistics and Evolutionary Zoology at the University of Bologna since 2015. While dealing above all with bivalve molluscs, he also worked and still works on other groups, such as hexapods and bony fishes. His research interests focus on phylogenetics, phylogenomics, and mitogenomics, as well as on the impact of small regulatory elements (like retroposons and small noncoding RNAs) on eukaryotic genomes. To this extent, he also aims to develop statistical and computational methods and tools.

Preface to "Molecular Phylogenetics and Mitochondrial Evolution"

The aim of the present Special Issue is to address the state-of-art of mitochondrial genomics and phylogenomics. Mitochondrial markers are widespread in phylogenetics; however, it is becoming increasingly clear that (i) many discordance issues arise with respect to nuclear markers and (ii) many features that are normally considered 'typical' for the mitochondrial genome are indeed highly unstable and unconserved.

The Special Issue is addressed to molecular and evolutionary biologists and involves papers from authors all over the world. We are deeply indebted to Donna Zheng, who conceived the Special Issue, and the whole *Life* Editorial Office staff, who followed the project from the beginning up to the present book.

Andrea Luchetti, Federico Plazzi
Editors

Editorial

Molecular Phylogenetics and Mitochondrial Evolution

Andrea Luchetti and Federico Plazzi *

Department of Biological, Geological and Environmental Sciences, University of Bologna, Via Selmi, 3, 40126 Bologna, BO, Italy; andrea.luchetti@unibo.it
* Correspondence: federico.plazzi@unibo.it; Tel.: +39-051-2094-172

Citation: Luchetti, A.; Plazzi, F. Molecular Phylogenetics and Mitochondrial Evolution. *Life* 2022, 12, 4. https://doi.org/10.3390/life12010004

Received: 13 December 2021
Accepted: 16 December 2021
Published: 21 December 2021

Publisher's Note: MDPI stays neutral with regard to jurisdictional claims in published maps and institutional affiliations.

Copyright: © 2021 by the authors. Licensee MDPI, Basel, Switzerland. This article is an open access article distributed under the terms and conditions of the Creative Commons Attribution (CC BY) license (https://creativecommons.org/licenses/by/4.0/).

The myth of a "typical" mitochondrial genome (mtDNA) is a rock-hard belief in the field of genetics, at least for the animal kingdom [1]. The first complete mitochondrial genomes were published in the 1980s [2–5]; since then, thousands of mtDNAs have been studied, and it is now well evident that only a few features (if any) are conserved among eukaryotic mtDNAs. Nonetheless, mtDNA has demonstrated, and it is everyday demonstrating, its suitability as a phylogenetic marker, ranging from population to phylum scale. In many cases, however, incompatibility issues arose between mitochondrial and nuclear phylogenies, and they have to be reconciled case by case.

The present Special Issue is an attempt to set and assess the state-of-art of mitochondrial genomics, with special reference to the suitability of mtDNAs for phylogenetic inference. The work by Formaggioni and colleagues [6] opens the Special Issue with a broad-scale analysis of the current knowledge on mitochondrial genomics, presenting the largest-to-date database of mtDNA features, which was made freely available to the scientific community. Similarly, the review by Parakatselaki and Ladoukakis [7], which closes the Special Issue, is specifically focused on mitochondrial heteroplasmy, a well-known phenomenon which is in fact neglected among the dogmatic "widespread" features of mitochondrial genomics, which include the strict maternal inheritance. Within this frame, several examples are provided of the usefulness of mtDNA for taxonomy, population genetics, and phylogenomics.

Furfaro and Mariottini [8,9] demonstrate the potential of mitochondrial rDNA secondary structures to unravel the taxonomy of a controversial gastropod family, Myrrhinidae, with the original description of a new genus; moreover, Dotsev and colleagues [10] address the taxonomic status of the Northernmost Snow Sheep (*Ovis nivicola*) using the well-exploited mitochondrial gene cytochrome *b*.

Several example of mtDNA-based population genetics are provided. Different strains of the cattle protozoan parasite *Theileria parva* are identified using mitochondrial single nucleotide polymorphisms in the work by Mwamuye and colleagues [11]; the population structure of the endangered freshwater mussel *Unio crassus* from Eastern Europe is unraveled by Kilikowska and colleagues [12] using one nuclear and two mitochondrial markers; similarly, the population genetics of the Portuguese oyster *Crassostrea angulata* is addressed by Chiesa and colleagues [13]. Conversely, Alarcón-Elbal and colleagues [14] provide a high-quality example of tracking the complete life cycle the water mite *Arrenurus (Micruracarus) novus* by means of morphology interwoven with the sequence of mitochondrial genes *cox1* and *cytb* at different life stages.

Mitochondrial phylogeny and comparison with nuclear data is the main purpose of the work by Zhao and colleagues [15] on two subfamilies of fig wasps, Epichrysomallinae and Sycophaginae, which are investigated using a large cluster of ortholog nuclear genes, as well as complete mitochondrial genomes; by Zadra and colleagues [16] on the genus *Aedes*, who found a strong and consistent incongruence between nuclear and mitochondrial phylogenetic inference (with special reference to dating), whose disentanglement is thoroughly discussed. Finally, Xia [17] reports on proper codon degeneration techniques to avoid phylogenetic artifacts when a mitochondrial phylogeny is inferred–a new method is implemented and applied to mammalian and avian lineages.

Furthermore, new complete mtDNAs are hereby reported. Arcila-Galvis and colleagues [18] present the complete mitochondrial genome of a plant fungal pathogen, *Pseudocercospora fijiensis* (Ascomycota: Pezizomycotina), as well as the phylogenetic reconstruction of the family Mycosphaerellaceae, providing clues for multiple invasions on introns within mitochondrial genes. Conversely, Johansen and colleagues [19] present the complete mitochondrial genome of a sea anemone, *Stichodactyla haddoni*, that shows group I introns harboring expressed open reading frames and supernumerary genes, and challenges current views of the Actiniidae taxonomy.

Concluding, the present Special Issue highlights the effectiveness of mitochondrial-based analyses in different fields of evolutionary biology, on one hand, while highlighting future, promising perspectives of research in the field of mitochondrial genetics, on the other hand, a field where unexpected is expected and the exception is the rule.

Conflicts of Interest: The authors declare no conflict of interest.

References

1. Lavrov, D.V.; Pett, W. Animal Mitochondrial DNA as We Do Not Know It: Mt-Genome Organization and Evolution in Nonbilaterian Lineages. *Genome Biol. Evol.* **2016**, *8*, 2896–2913. [CrossRef] [PubMed]
2. Anderson, S.; Bankier, A.T.; Barrell, B.G.; de Bruijn, M.H.L.; Coulson, A.R.; Drouin, J.; Eperon, I.C.; Nierlich, D.P.; Roe, B.A.; Sanger, F.; et al. Sequence and organization of the human mitochondrial genome. *Nature* **1981**, *290*, 457–465. [CrossRef] [PubMed]
3. Bibb, M.J.; Van Etten, R.A.; Wright, C.T.; Walberg, M.W.; Clayton, D.A. Sequence and gene organization of mouse mitochondrial DNA. *Cell* **1981**, *26*, 167–180. [CrossRef]
4. Anderson, S.; de Bruijn, M.H.; Coulson, A.R.; Eperon, I.C.; Sanger, F.; Young, I.G. Complete sequence of bovine mitochondrial DNA. Conserved features of the mammalian mitochondrial genome. *J. Mol. Biol.* **1982**, *156*, 683–717. [CrossRef]
5. Clary, D.O.; Wolstenholme, D.R. The *Drosophila* mitochondrial genome. *Oxf. Surv. Eukaryot Genes* **1984**, *1*, 1–35. [PubMed]
6. Formaggioni, A.; Luchetti, A.; Plazzi, F. Mitochondrial Genomic Landscape: A Portrait of the Mitochondrial Genome 40 Years after the First Complete Sequence. *Life* **2021**, *11*, 663. [CrossRef] [PubMed]
7. Parakatselaki, M.; Ladoukakis, E. mtDNA Heteroplasmy: Origin, Detection, Significance, and Evolutionary Consequences. *Life* **2021**, *11*, 633. [CrossRef] [PubMed]
8. Furfaro, G.; Mariottini, P. Looking at the Nudibranch Family Myrrhinidae (Gastropoda, Heterobranchia) from a Mitochondrial '2D Folding Structure' Point of View. *Life* **2021**, *11*, 583. [CrossRef] [PubMed]
9. Furfaro, G.; Mariottini, P. *Nemesignis*, a Replacement Name for *Nemesis* Furfaro & Mariottini, 2021 (Mollusca, Gastropoda, Myrrhinidae), Preoccupied by *Nemesis* Risso, 1826 (Crustacea, Copepoda). *Life* **2021**, *11*, 809. [CrossRef] [PubMed]
10. Dotsev, A.; Kunz, E.; Kharzinova, V.; Okhlopkov, I.; Lv, F.; Li, M.; Rodionov, A.; Shakhin, A.; Sipko, T.; Medvedev, D.; et al. Mitochondrial DNA Analysis Clarifies Taxonomic Status of the Northernmost Snow Sheep (*Ovis nivicola*) Population. *Life* **2021**, *11*, 252. [CrossRef] [PubMed]
11. Mwamuye, M.; Obara, I.; Elati, K.; Odongo, D.; Bakheit, M.; Jongejan, F.; Nijhof, A. Unique Mitochondrial Single Nucleotide Polymorphisms Demonstrate Resolution Potential to Discriminate *Theileria parva* Vaccine and Buffalo-Derived Strains. *Life* **2020**, *10*, 334. [CrossRef] [PubMed]
12. Kilikowska, A.; Mioduchowska, M.; Wysocka, A.; Kaczmarczyk-Ziemba, A.; Rychlińska, J.; Zając, K.; Zając, T.; Ivinskis, P.; Sell, J. The Patterns and Puzzles of Genetic Diversity of Endangered Freshwater Mussel *Unio crassus* Philipsson, 1788 Populations from Vistula and Neman Drainages (Eastern Central Europe). *Life* **2020**, *10*, 119. [CrossRef] [PubMed]
13. Chiesa, S.; Lucentini, L.; Chainho, P.; Plazzi, F.; Angélico, M.M.; Ruano, F.; Freitas, R.; Costa, J.L. One in a Million: Genetic Diversity and Conservation of the Reference *Crassostrea angulata* Population in Europe from the Sado Estuary (Portugal). *Life* **2021**, *11*, 1173. [CrossRef] [PubMed]
14. Alarcón-Elbal, P.; García-Jiménez, R.; Peláez, M.; Horreo, J.; Valdecasas, A. Molecular Correlation between Larval, Deutonymph and Adult Stages of the Water Mite *Arrenurus* (*Micruracarus*) *Novus*. *Life* **2020**, *10*, 108. [CrossRef] [PubMed]
15. Zhao, D.; Xin, Z.; Hou, H.; Zhou, Y.; Wang, J.; Xiao, J.; Huang, D. Inferring the Phylogenetic Positions of Two Fig Wasp Subfamilies of Epichrysomallinae and Sycophaginae Using Transcriptomes and Mitochondrial Data. *Life* **2021**, *11*, 40. [CrossRef] [PubMed]
16. Zadra, N.; Rizzoli, A.; Rota-Stabelli, O. Chronological Incongruences between Mitochondrial and Nuclear Phylogenies of *Aedes* Mosquitoes. *Life* **2021**, *11*, 181. [CrossRef] [PubMed]
17. Xia, X. Improving Phylogenetic Signals of Mitochondrial Genes Using a New Method of Codon Degeneration. *Life* **2020**, *10*, 171. [CrossRef] [PubMed]

18. Arcila-Galvis, J.; Arango, R.; Torres-Bonilla, J.; Arias, T. The Mitochondrial Genome of a Plant Fungal Pathogen *Pseudocercospora fijiensis* (Mycosphaerellaceae), Comparative Analysis and Diversification Times of the Sigatoka Disease Complex Using Fossil Calibrated Phylogenies. *Life* **2021**, *11*, 215. [CrossRef] [PubMed]
19. Johansen, S.D.; Chi, S.; Dubin, A.; Jørgensen, T. The Mitochondrial Genome of the Sea Anemone *Stichodactyla haddoni* Reveals Catalytic Introns, Insertion-Like Element, and Unexpected Phylogeny. *Life* **2021**, *11*, 402. [CrossRef] [PubMed]

Article

Mitochondrial Genomic Landscape: A Portrait of the Mitochondrial Genome 40 Years after the First Complete Sequence

Alessandro Formaggioni, Andrea Luchetti and Federico Plazzi *

Department of Biological, Geological and Environmental Sciences, University of Bologna, Via Selmi, 3, 40126 Bologna, BO, Italy; alessand.formaggioni@studio.unibo.it (A.F.); andrea.luchetti@unibo.it (A.L.)
* Correspondence: federico.plazzi@unibo.it; Tel.: +39-051-2094-172

Abstract: Notwithstanding the initial claims of general conservation, mitochondrial genomes are a largely heterogeneous set of organellar chromosomes which displays a bewildering diversity in terms of structure, architecture, gene content, and functionality. The mitochondrial genome is typically described as a single chromosome, yet many examples of multipartite genomes have been found (for example, among sponges and diplonemeans); the mitochondrial genome is typically depicted as circular, yet many linear genomes are known (for example, among jellyfish, alveolates, and apicomplexans); the chromosome is normally said to be "small", yet there is a huge variation between the smallest and the largest known genomes (found, for example, in ctenophores and vascular plants, respectively); even the gene content is highly unconserved, ranging from the 13 oxidative phosphorylation-related enzymatic subunits encoded by animal mitochondria to the wider set of mitochondrial genes found in jakobids. In the present paper, we compile and describe a large database of 27,873 mitochondrial genomes currently available in GenBank, encompassing the whole eukaryotic domain. We discuss the major features of mitochondrial molecular diversity, with special reference to nucleotide composition and compositional biases; moreover, the database is made publicly available for future analyses on the MoZoo Lab GitHub page.

Keywords: mitochondrial genome; mtDNA architecture; mtDNA structure; nucleotide composition; compositional bias; strand asymmetry; Eukaryota; mtDNA expansion

1. Introduction

Few myths in molecular biology are as stubbornly long-lived as the stability and conservation of mitochondrial genome (mtDNA) among animals (and eukaryotes), be it in terms of content, structure, or architecture. The first evidence that some animals harbor a covalently-closed mtDNA was provided in 1966 for chickens, cows, and mice [1,2]; in the very same years, a comparably small size was reported from a handful of animal groups [3]. As discussed in Williamson [4], this became the first, indisputable evidence for the intriguing hypothesis that was initially put forward by Altmann [5] about 80 years before: mitochondria are endosymbionts with a prokaryotic descent.

In this context, when linear DNA molecules were reported from unicellular eukaryotes [6], it was tempting to classify them as exceptions, and the "broken-circle theory" [7,8] was proposed for yeast mtDNA: if any linear mtDNA is observed in yeast, it ought to be a broken circle (see [4]). Moreover, this claim extended the supposed conservation of mtDNA to a different eukaryotic realm. The complete sequences of mtDNA from humans [9], mice [10], and cattle [11] were soon followed by the complete sequence from *Drosophila yakuba* [12]: genomes that were found to be (i) single, (ii) closed circles of (iii) comparable size, with (iv) a conserved genetic content. The myth of a "typical" mtDNA was born, at least for Metazoa [13].

Nevertheless, 40 years after the first complete mtDNAs, thousands of mtDNAs have been completely or partially sequenced, annotated, and compared, and it has become increasingly clear that these features are hardly conserved (if at all) among eukaryotes. In extreme cases, some eukaryotes did even lose mtDNA (or even organelles themselves; [13–17]).

(i) Multipartite genomes. Multipartite mtDNAs are, in fact, widespread among eukaryotes [15,18]. The mtDNA of *Trypanosoma brucei* (Euglenozoa: Kinetoplastea) is organized as a kinetoplast, a compact network of maxicircles (~25 kb) and thousands of minicircles (~1 kb), where mitochondrial genes and regulatory small RNAs are located, respectively ([18–21] and references therein). The mtDNA in other euglenozoans, Diplonemea, is composed of dozens of circular chromosomes; they can be subdivided into two size classes, with chromosomes of the same class sharing approximately 95% of the sequence. The remainder constitutes the only coding region of the chromosome, where one or more exons are located, ranging from 40 to 540 bp in length and relying on a complex trans-splicing and post-transcriptional machinery [19,22]. Mitochondrial genomes from Alveolata, and specifically of dinoflagellates, are also highly fragmented and possibly constitute the most divergent mitochondrial genomes among eukaryotes along with diplonemeans [23,24].

The structure of plant mtDNA is better understood as an entangled pattern where alternative molecules can coexist and recombine [7,15,25–36], while some mitochondria may contain only partial or no genome at all [37]. Occasionally, however, the mtDNA appears to be organized into stable, autonomous circles (e.g., [38,39]).

Among Opisthokonta, multiple mitochondrial chromosomes have been reported from Calcarea [13,40,41]; Hydrozoa and Cubozoa [42,43]; *Dicyema* [44]; Syndermata *sensu* Witek et al. [45–47]; Nematoda [48,49]; Hexapoda [50–54], where mtDNA fragmentation was indeed suggested as an autapomorphy for the clade Mitodivisia [55]; Ichthyosporea [56,57]; and Saccharomycotina (e.g., [4,58]).

(ii) Chromosome architecture. Many examples are currently known of linear mtDNA [18,59]. Moreover, mtDNA is not always organized as a single chromosome; many species with multipartite mitochondrial genomes have been identified. Among Metazoans, linear chromosomes are known to be present in mammals with a wide array of concatenated forms ([60] and references therein); all medusozoans (cnidarians, excluding Anthozoa) analyzed so far show linear mtDNAs, which are further subdivided into multiple chromosomes in Hydrozoa and Cubozoa [40,42,43,61–63]. Linear, multipartite mtDNAs are also known to exist in calcareous sponges [13,40,41].

Among Fungi, the "broken-circle theory" has now been discontinued and the existence of polydisperse, linear mtDNAs in the brewer's yeast *Saccharomyces cerevisiae* and in other yeasts is currently accepted [4,58,64–67]. It appears that linear mtDNA forms evolved from circular chromosomes in yeasts, but the shape of the genome also depends on the life stage of the yeast cell, with linear concatenamers dominating in mature bud cells [58,65,68]. More generally, in yeasts and land plants, mitochondria are best described as concatenated, linear-branched structures [15,26,29].

Among Alveolata, the ciliates *Paramecium* and *Tetrahymena* have been known, since 1968, to possess linear mtDNA [6,69–71]; the apicomplexan *Plasmodium* has a small 6 kb-long linear mtDNA with only three protein coding genes [72,73]. Additionally, mitochondria of *Amoebidium parasiticum* (Opisthokonta: Ichthyosporea) harbor several hundreds of small linear chromosomes [57].

Finally, besides the core mitochondrial genome, many land plant species and fungi harbor linear mitochondrial plasmids (e.g., [26,74,75] and references therein), which were reported from ciliates as well [76].

(iii) Genome size. Genome size is highly variable among eukaryotes, ranging from 6 kb in apicomplexan [23,24,72,73] and <13 kb in some green algae [77], ctenophores [78,79], and some fungi [80]; through 43 kb in placozoans [78,81] and >70 kb in choanoflagellates and ciliates [57,76]; up to >200 kb in other green algae and fungi [82–84], and

11 Mb in flowering plants [38]. Moreover, phenomena of the punctuated expansion of mtDNA have been reported within clades with generally reduced genome size (e.g., frogs [85], ark shells [86,87]). In most cases, this variability is not related to gene content; rather, the expansion and reduction of the intergenic region appear to be the main drivers of genome size among eukaryotes (e.g., [38,56,58,82,83,88,89]).

(iv) Gene content. Only three genes are located in the mtDNA of apicomplexans and their relatives [90–92], as well as in dinoflagellates [23,24]; only a dozen genes are encoded in euglenozoans' mtDNAs [19], but up to ~100 have been identified in jakobids. The order Jakobida is included in the eukaryotic supergroup Discoba (see [93–95] and references therein); jakobids have been found to have up to ~100 mitochondrially-encoded genes [96], and, to the best of our knowledge, *Andalucia godoyi* has the most gene-rich mtDNA [97]. The choanoflagellate *Monosiga brevicollis* has an intermediate gene complement of 55 genes [57], while there are 47 for the ichthyosporean parasite *Sphaerothecum destruens* [98]. Conversely, a relatively constant gene content is known to be present in fungi and animals [56].

Chytridiomycetes typically harbor circular mtDNAs coding for the full complement of genes inferred from the opisthokont common ancestor, including tRNAs; mtDNAs from other Fungi appear to have lost many genes [56]. All yeast mtDNAs encode for three subunits of complex V (*atp6*, *atp8*, and *atp9*), for apocytochrome b (*cytb*), and for three subunits of complex IV (*cox1*, *cox2*, *cox3*) [99–102]. Additionally, seven subunits of complex I (*nad1-6* and *nad4L*) and two additional genes (*var1* and *rpm1*) may be present in fungal mtDNAs. However, complex I subunits were lost in the *Saccharomyces* group, while the *var1* gene was lost in the *Candida* group ([58] and references therein).

In bilaterian animals, the gene content encompasses two subunits of complex V (*atp6* and *atp8*), three subunits of complex IV (*cox1*, *cox2*, *cox3*), apocytochrome b (*cytb*), and seven subunits of complex I (*nad1-6* and *nad4L*) (e.g., [40,103,104]). Nonetheless, several exceptions have been observed. For instance, the *atp8* gene is often very divergent (e.g., [89,105,106]) and in some cases it has been claimed to be completely absent ([78,107–113]; also see [88,114,115]). Furthermore, many Open Reading Frames (ORFs) with no clear homology have been detected in many bilaterian lineages (e.g., [89,103,116–121]).

However, the picture of mtDNAs gets more confused among non-bilaterian animals, and many other ORFs have been identified (reviewed in [40]). Placozoans are considered to likely possess the mtDNA that is more similar to that of the metazoan common ancestor [81], which is a large, circular molecule with a full complement of tRNAs [122]. The number of tRNAs is variable among sponges, from 2 to 27, and *tatC* and *atp9* genes may be found [40]. A handful of tRNAs have been identified in cnidarians (e.g., [78,123]), where additional genes are present ([124] and references therein); similarly, many genes that are usually found in animal mtDNA are missing from that of ctenophores (tRNAs, *atp6*, *atp8*; [79,81,116,125]).

This summary of mitochondrial molecular structures and architectures certainly gives an idea of the stunning variability of these organellar genomes, which largely surpasses that of plastid trans-splicing phenomena [19,78,81,126]; the use of different genetic codes [13,127,128]; bewildering gene rearrangement [81–88,105,106,129–131]; and biparental and doubly uniparental inheritance [28,30,58,132–136].

In the present paper, we obtained from GenBank all the available complete mitochondrial genomes and used a slightly modified version of a recently published tool [137] to analyze the dataset. Exactly 40 years after the first complete mitochondrial sequence, we present a general description of our results; we also identify mitochondrial features typical of different taxa, aiming to provide a global overview of mitochondrial molecular diversity.

2. Materials and Methods

Mitochondrial genomes were mined from NCBI GenBank database (accessed on February 2021) using two different queries: "mitochondrion(title) AND complete(title) AND genome(title)" and "mitochondrial(title) AND DNA(title) AND complete(title) AND genome(title)". In order to avoid unnecessary network load to the database (and machine time for subsequent analyses), overrepresented species were manually identified and relative records were excluded. Only one representative—or a few of them, depending on whether different populations were available—for each of the excluded species was manually selected and added to the automatically generated list (Supplementary Table S1).

A customized version of the HERMES tool [137] was used to analyze the dataset. The method involves the computation of several variables from annotated complete mitochondrial genomes. Variables are associated to gene content, nucleotide composition, phylogeny, and more. In fact, a HERMES analysis is typically carried out in a phylogenetic framework, which must be separately assessed. These metrics are usually summarized in a single number, the HERMES index, by means of a maximum likelihood factor analysis. For the present purpose, though, the HERMES index itself and variables stemming from a phylogenetic tree—such as AMIGA [137], root-to-tip distance, and maximum likelihood distance)—which were obviously not available in this context, were excluded. The following 11 variables are, therefore, considered and computed for each entry: length of the mtDNA, topology (linear or circular), number of annotated genes, absolute value of the Strand Usage skew (SU-skew; see [137] for definition), A+T content, AT-skew, GC-skew, CAI, percentage of Unassigned Regions (URs), UR-based A+T content, and UR-based median length.

Many NCBI hits were discarded as unsuitable for further analysis due to annotation errors/flaws or unsupported format. Some examples are entries with no annotation (raw sequences) or those with no annotated Coding Domain Sequences (CDSs). In six cases, it was possible to edit the minor details of annotations to include sequences that would have been otherwise excluded (Supplementary Table S2). Unfortunately, the HERMES approach has different constraints on a mtDNA annotation when carrying out the analyses. For example, at least one gene must be annotated to compute the UR proportion, and at least one CDS must be annotated to compute CAI. Consequently, our pipeline is blind to mitochondrial chromosomes where only tRNAs are annotated (or even no genes at all), which is sometimes the case for multipartite mtDNAs. It is also blind to unannotated entries resulting, for example, from studies on mitochondrial variation and displaying only mutations with respect to a reference sequence, and to entries with no sequence and linked to assembly data are also not detected.

The taxonomic information was retrieved from the NCBI page of each entry. We used the python class NCBITaxa from the package ETE Toolkit [138] to assign each lineage name to its proper taxonomic rank. All the analyses were carried out using custom-tailored Pyhton3 and R [139] scripts (available from F.P. and A.F. upon reasonable request). Plots were displayed using the "ggplot2" R package [140]. We calculated the Spearman's rank correlation coefficient between pairwise groups of variables through the function rcorr from the "Hmisc" R package [141] and displayed the results through correlograms using the "corrplot" R package [142].

We used the database WoRMS [143] to collect ecological information such as feeding type and functional group. These pieces of information are annotated with a three-rank quality score; data marked with the lowest rank ("unreviewed") were discarded from our analysis. Each piece of information was recorded along with the respective life stage. Ecological data were recorded for three taxonomic ranks: species, genus, and family. They were then applied to all matching entries using the package "worrms" [144].

3. Results and Discussion

3.1. Dataset Composition

The two queries combined, filtered for overrepresented taxa, returned 31,065 entries. Out of these entries, we were unable to compute the variables of 3192 entries because of poor annotation or unsupported format. Overall, we discarded 10.50% of the Metazoa entries (2965 entries), 11.90% of the Fungi entries (171 entries), 4.69% of the Viridiplantae entries (29 entries), and 3.56% of the remaining entries (27 entries). All entries were correctly assigned to one of the major eukaryotic subdivisions ([145]; Figure 1a): Diaphoretickes, comprised by Archaeplastida (including Viridiplantae), Excavata, Haptista, and SAR clade (Stramenopiles, Alveolata, Rhizaria); Amorphea, comprised by Amoebozoa and Opisthokonta (including Fungi and Metazoa); and CRuMs (Collodictyonidae, Rigifilida, Mantamonas). Only five GenBank entries were not correctly placed, four of which were eukaryotes *incertae sedis* (GenBank Accession Numbers NC_034794, NC_036491, MN082145, MG202007), while the last one was a dsDNA virus (GenBank Accession Number BK012062) that was removed from subsequent analyses.

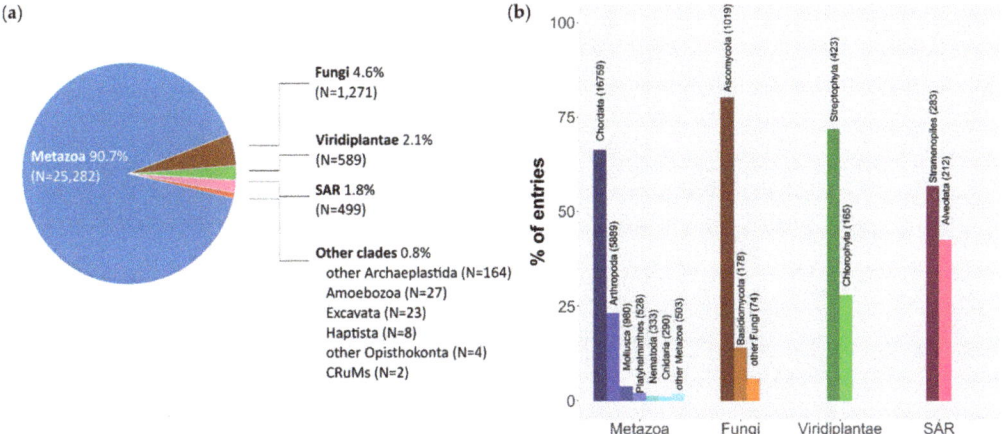

Figure 1. Database composition. (**a**) Major eukaryote subdivisions; (**b**) composition within the three major kingdoms (Metazoa, Fungi, and Viridiplantae) and the SAR clade.

The current version of the database was made publicly available as a CSV-formatted plain text file along with most R functions used for the present work on the MoZoo GitHub page, at the URL https://github.com/mozoo/almighto (accessed on: 7 July 2021). The final dataset is a 27873 × 80 matrix, where each row contains a taxonomic entry and columns are as follows:

1–2: accession number and definition of the genome on the NCBI page.
3–13: the 11 variables described above, which were obtained by the modified version of HERMES.
14–49: taxonomic ranks retrieved by NCBITaxa. The intersection of a row and a column is the entry's lineage name for that taxonomic rank or NA if information is missing. In the 49th column, which is named "Eu_divisions", each entry is placed in one of the major eukaryotic subdivisions described above.
50: the mitochondrial genetic code, which was retrieved from the relevant NCBI Taxonomy page.
51–80: ecological data. Each column is a different ecological feature. For the "functional group" columns, the value in each cell can be "FuncAdult" or "FuncLarva", depending on which life stage the feature is at, or NA if information is missing. The same organization was used for the "feeding type" columns using the values "FeedAdult" and "FeedLarva".

As expected, most of the retrieved entries come from the kingdom Metazoa (Figure 1a). Most of these entries belong to the phyla Chordata (66.2%) and Arthropoda (23.3%) (Figure 1b). On average, we obtained 2.4 mitogenomes per (available) species in Chordata, while the mean value for metazoans was 1.98 (1.44 for Arthropoda, 1.59 for Mollusca, 1.51 for Nematoda). In Fungi, most of the entries belong to the Ascomycota phylum (80.2%), which is mainly grouped in two classes, the Sordariomycetes and the Saccharomycetes, 41.1% and 36.1% of the Ascomycota entries, respectively (Figure 1b). On average, 3.13 entries correspond to each Sordariomycetes species, and 2.61 entries correspond to each Saccharomycetes species. Among Viridiplantae, the richest phylum is the Streptophyta, which is mainly represented by the classes Magnoliopsida (72.8%) and Bryopsida (11.3%) (Figure 1b). On average, 1.56 entries correspond to each Magnoliopsida species, whereas 1.28 entries correspond to each other Viridiplantae species.

The SAR clade is the third biggest clade in the dataset, after Opisthokonta and Archaeplastida (Figure 1a). It is divided into three main clades: Alveolata (42.5% of SAR entries), Stramenopiles (56.7% of SAR entries) and Rhizaria (4 entries) (Figure 1b). On average, 2.06 entries correspond to each Alveolata species; 1.73 entries correspond to each Stramenopiles species.

3.2. Mitogenome Reduction and Expansion

Excluding multipartite mtDNAs, the shortest complete mitogenomes in our dataset belong to three different Chinese isolates of the genus *Babesia*, an apicomplexan taxon that causes babesiosis, a tick-transmitted disease. Their linear mitogenome ranges from 5767 bp to 5790 bp and it encodes for nine genes: three protein coding genes and six rRNA genes [146]. The shortest metazoan mitogenome was the Ctenophora *Mnemiopsis leidyi*, which resulted in only 10326 pb long: this is mostly due to the absence of tRNA genes, to the scarcity of intergenic nucleotides, and to the relocation of atp6 to the nuclear genome [79].

On the contrary, the longest mitogenomes found belong to *Corchorus capsularis* and *Corchorus olitorius*, 1999 kbp and 1829 kbp, respectively (GenBank Accession Numbers NC_031359 and NC_031360, respectively). The latter species are commonly named jute and belong to the Malvaeae family. In Metazoa, the longest mitogenomes belong to ark shells of the genus *Anadara* (GenBank Accession Numbers NC_020787, NC_024927, KF750628); the mtDNA encodes for 42s tRNA and for a total of 56 genes, which constitute the largest number of tRNAs and genes encoded by a metazoan mtDNA. However, its length is mostly due to URs, which represent 67.7% of the entire sequence. The mtDNA was found to be 47–50 kbp long, depending upon the number of repeats in the URs [87]. However, if considering mitogenomes composed of several chromosomes, then the longest metazoan mitogenome belongs to the calcareous sponge *Clathrina clathrus* (six chromosomes, for a total length of 51 kb; [13]).

Viridiplantae show the highest median in length, URs number, and URs median length among eukaryotes, as well as a high variability inside the clade (Figure 2a,c,d). Globally, the mtDNA length seems to increase with the UR content (Figure 3; Supplementary Figure S1). It is worth recalling that metazoans comprise the largest part of our dataset and may consistently drive the observed pattern; nonetheless, the correlation between mtDNA length and UR content was also observed in the isolated groups—Metazoa (Figure 3b), Fungi (Figure 3c), Viridiplantae (Figure 3d) and Stramenopiles (Figure 3f). However, in Alveolata (which includes apicomplexans), the length is negatively correlated with URs, but positively correlated with the number of genes (Figure 3e). Therefore, although metazoans share a reduced mtDNA with alveolates, in the latter group this reduction appears to be associated with gene loss rather than to URs reduction. Indeed, the Alveolata show the lowest median in length and genes, even if they show the third richest mtDNA in terms of URs (Figure 2a–c).

It has been shown that the expansion of the mitochondrial genome is mostly associated with the expansion of non-coding or unassigned regions [56]. Among Viridiplantae, the mitogenome expansion is concurrent with the transition from water to land and it accelerated after the appearance of vascular plants [147–149]. Indeed, Chlorophyta shows the smallest mtDNA in terms of length and URs content. It is followed by the freshwater green algae in the Streptophyta (named non-embryophytes Streptophyta in Figure 4), the non-vascular Embryophyta (mosses, liverworts, and hornworts), and the Tracheophyta, which shows the longest and UR-richest mtDNA (Figure 4). This data underly an evolutionary pattern from the (hypothetical) ancestral mtDNA of Viridiplantae to the more derived and longer one of vascular plants. Conversely, among animals, an opposite autapomorphy seems to have arisen in Bilateria: mitogenomes from Porifera, Cnidaria, and Placozoa are generally regarded as more similar to the metazoan common ancestor, and on average, they are larger and harbor more unassigned regions (Figure 4; [150]).

The gene content is highly variable in eukaryotes, and during the evolution the mtDNA underwent losses and relocations of genes to the nucleus. Species in the Jakobida clade are considered the eukaryotes with the mtDNA most similar to the ancestral state, since they show a high gene content and some unique mitochondrial genes, such as the RNA polymerases [93,96]. Indeed, the clade Excavata, which includes the order Jackobida, shows the highest median gene content among eukaryotes (Figure 2b).

The protein-encoding genes are well conserved in the three main kingdoms: 14 in the Fungi, 13 in the Metazoa (excluding non-bilaterians), and 24 in the Viridiplantae [151]. The higher standard deviation of the gene content in Viridiplantae and Fungi (Figure 2b) is mainly due to the homing endonucleases encoded inside the introns and unassigned ORFs [82,152,153].

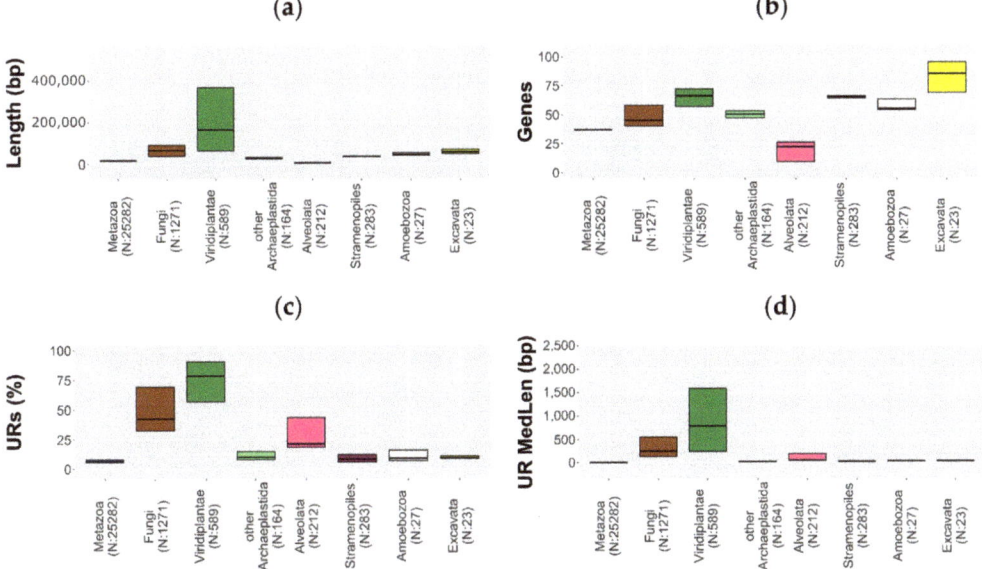

Figure 2. Mitochondrial genome dimension. The thick line depicts the median value; the boxplot ranges from the first to the third quantile. (**a**) mtDNA length (bp); (**b**) number of annotated genes; (**c**) UR content (%); (**d**) UR median length (bp).

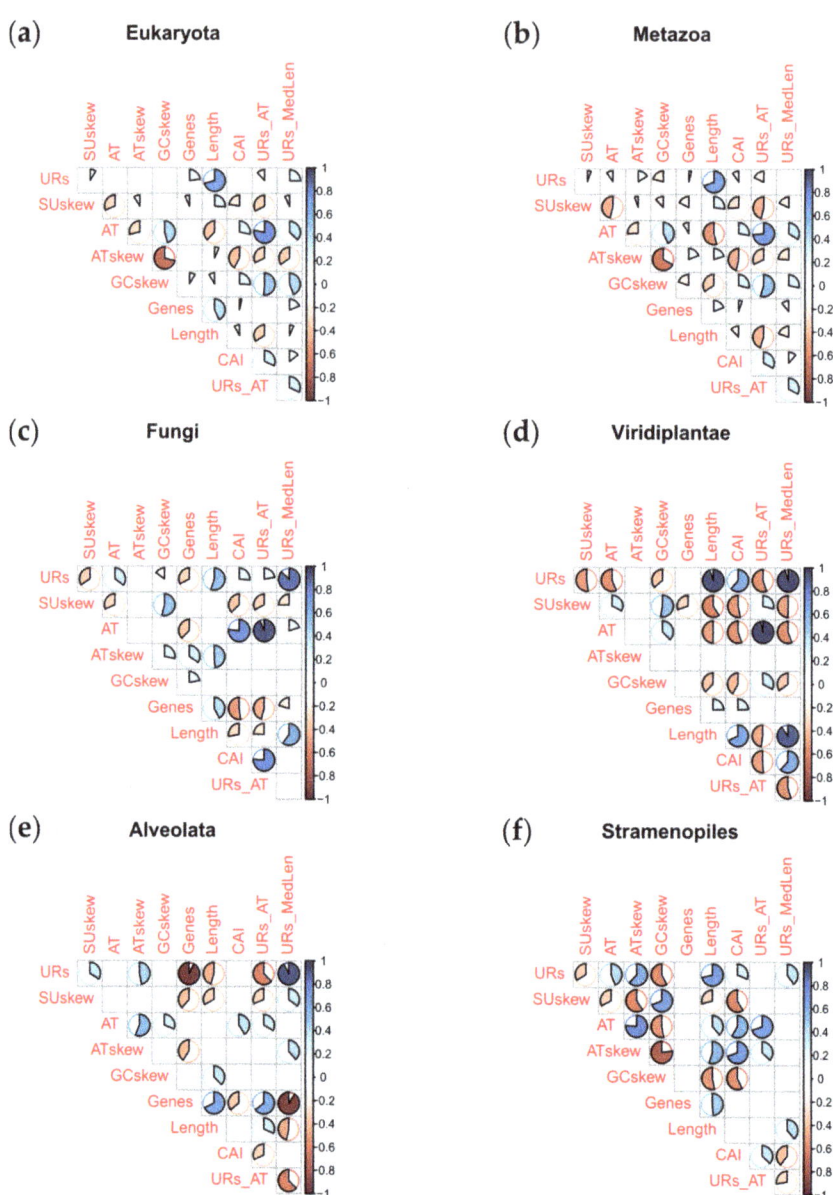

Figure 3. Correlograms for major mtDNA features. Each pie chart represents the value of a significant Spearman's rho; where the pie chart is not shown, the correlation is not significant. A blue pie shows a positive Spearman's rho, increasing clockwise from 0 to 1; a red pie shows a negative Spearman's rho, increasing counterclockwise from 0 to 1. (**a**) Whole database; (**b**) Metazoa; (**c**) Fungi; (**d**) Viridiplantae; (**e**) Alveolata; (**f**) Stramenopiles.

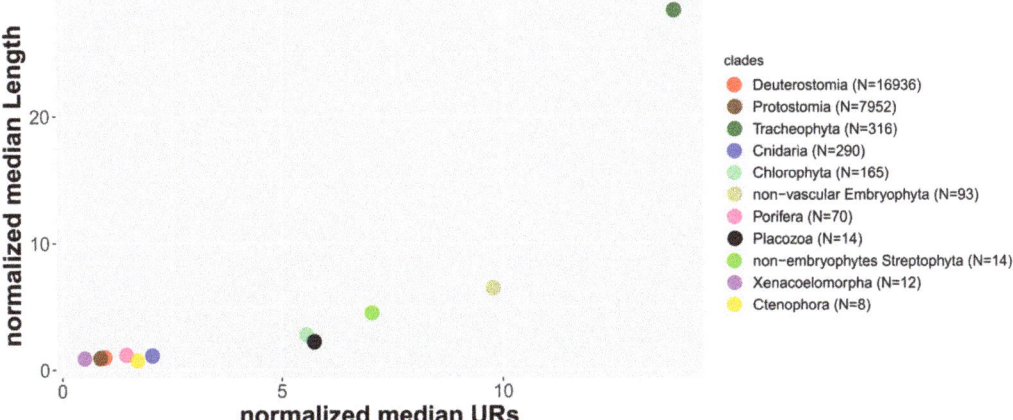

Figure 4. Mitogenomes expansion and contraction. For each group, the median UR content (%) has been normalized on the whole-database median UR content; the median length (bp) has also been normalized on the whole database median length.

3.3. The Strand Asymmetry in Eukaryota

The Metazoa is the only clade showing a negative median for the GC-skew (Supplementary Figure S2), meaning that the cytidines are overabundant on the (putative) plus strand.

Moreover, AT-skew and GC-skew are strongly inversely correlated in Metazoa (Figure 3b), as well as in Stramenopiles (Figure 3f); on the other hand, the two variables are directly correlated in Fungi (Figure 3c), and have no significant correlation in Viridiplantae and Alveolata (Figure 3d,e). Therefore, in Metazoa, the plus strand is rich in A and C, whereas the minus strand is rich in T and G. According to the literature, this feature is due to the unidirectional replication of metazoan mitogenome. The H strand (which most of the times is considered the minus strand) is firstly replicated as single-stranded; during this condition, the deamination phenomenon is more frequent, leading to the mutation of C into U and A into hX (which base pairs with a C on the opposite strand [154]).

Different kind of correlations in the other kingdoms could be due to different replication and repair mechanisms of the mitogenome; in fact, a similar explanation has been proposed for the different evolutionary rates among eukaryote mtDNAs [155]. For instance, in Eubacteria, the GC-skew can be used to determine whether there are multiple origins of replication or not [156]. Although mitogenome replication is poorly understood outside metazoans, evidence suggests that plants, *Plasmodium falciparum*, and yeasts mitogenomes replicate through a rolling circle mechanism [157,158]. A thorough revision of mtDNA replication dynamics is well beyond the purpose of the present paper, and further investigation is needed to fully unveil and understand the different DNA replication mechanisms in different eukaryotic mitochondria, as well as to associate them to precise nucleotide compositional biases.

Notably, there are many exceptions even in Metazoa. In some cases, a reversed strand asymmetry (RSA) can be observed, so that on the plus strand there are more G than C as well as more T than A (resulting in a positive GC-skew and a negative AT-skew). This can be found, for example, in Porifera, Cnidaria, Platyhelminthes, Nemertea, and Nematoda, whereas in other spiralians, Mollusca, and other Ecdysozoa both conditions are present (Figure 5a,b). Moreover, the RSA can also be observed in other phyla at lower taxonomic ranks; examples are reported in the literature for fish [159], echinoderms [160], and arthropods [161,162]. The RSA is often related with the inversion of the control region; an AT-rich region is normally pivotal for the replication and determines which strand is replicated first [163,164]. Therefore, an inversion of the control region could invert the mutational pattern on both strands [162,165]. The localization of this region is not easy, and since it is one of the most variable regions of the mitogenome, it is impossible to align if the phylogenetic distance between the species is high [161]. Therefore, it is still uncertain if the inversion of the control region is the only process that can lead to RSA. Although some examples have been reported, a phylogenetically wider analysis is needed to determine what can affect the strand asymmetry.

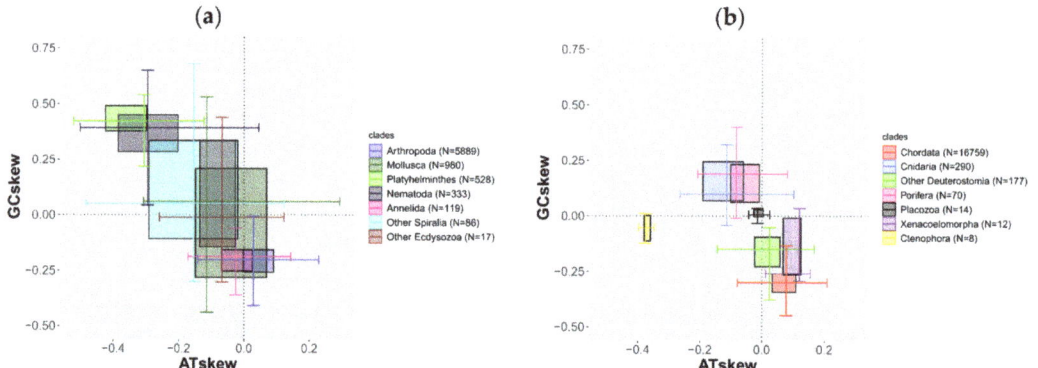

Figure 5. Bivariate boxplots of AT-skew and GC-skew. The lines within the boxplots depict the two median values; the boxplot ranges from the first to the third quantile along both axes; whiskers extend to roughly 95% confidence interval. For the sake of visibility, genomes have been divided into protostomes (**a**) and non-protostomes (**b**).

3.4. Codon Adaptation and A+T Content

Mitogenomes are generally biased toward a high A+T content (which implies a low G+C content). Indeed, all the clades show a median above 50% (Supplementary Figure S2B). As mentioned before, the replication of the mtDNA leads to the deamination of A and C on the H strand. However, the deamination of C is more common; this results in an accumulation of T on the H strand, and of A on the L strand [166]. Therefore, the replication process may explain the biased A+T content as well as the opposite values of AT-skew and GC-skew, which are commonly observed, at least among Metazoa. However, as detailed above, the correlation between the skews is not consistent outside metazoans. Nonetheless, the A+T content is generally biased, with most clades showing an A+T content even higher than Metazoa (Supplementary Figure S2b). This data would imply that either the replication mechanisms outside metazoans lead to the increase in the A+T content without affecting the strand asymmetry or the nucleotide biases are affected by different causes, which are predominant in some clades and negligible in others. Viridiplantae showed the lowest A+T median content (55.8%), being even remarkably lower than that showed by the other Archaeplastida (70.4%) (Supplementary Figure S2b). It has been suggested that the A+T content decrease is concomitant with the land plants' mtDNA genome expansion. The accumulation of URs can lead to a higher recombination frequency, which eventually raises the G+C content, thus decreasing the A+T content [147]. This mutational pattern would contrast with that observed in the Metazoa, where the replication leads to an accumulation of A and T, as reported above. The clades with a higher median URs and length show a lower A+T content (Table 1, Figure 4). The data agree with the hypothesis of Pedrola-Monfort and colleagues [147]; moreover, in Viridiplantae, URs and A+T content are negatively correlated, (Figure 3d). However, only non-vascular Embryophyta show a significant negative correlation (Table 1).

Table 1. Mitochondrial genome expansion of land plants. The three columns present the A+T content percentage, the Spearman's rho coefficient of correlation between A+T content and UR percentage on the mtDNA, and the *p*-value associated with the Spearman's rho.

	A+T Content	Spearman's Rho	*p*-Value
Chlorophyta	62.8%	ρ = 0.15	*p* = 0.0506
Non-embryophytes Streptophyta	60.3%	ρ = −0.51	*p* = 0.065
Non-vascular Embryophyta	58.9%	ρ = −0.64	*p* = 0
Tracheophyta	55.0%	ρ = 0.07	*p* = 0.2

In Metazoa, the median A+T content is 60.3%; moreover, they show a high variability (Supplementary Figure S2B). Arthropoda and Nematoda show the highest median A+T content among metazoans at 76.4% and 74.1%, respectively. On the other hand, the lowest A+T content can be observed in the phylum Chordata (57.6%).

The CAI statistic is a measure of how unbalanced the use of codons is in the same codon family. Their biased use can be explained in several ways. Specific codons can be selected to enhance translation efficiency; for instance, a correlation between the most used codons and the most abundant tRNA has been reported in bacteria [167,168], with the accepted idea being that the tRNA bias affects codon usage.

However, this does not seem the case for mtDNA, where codons seem to be biased according to the mitochondrial mutational pattern. More specifically, vertebrates A and C are found in the most frequent nucleotides at the third codon position [169], and in this clade most of the genes are located on the L strand, the one that in fact mostly accumulates A and C. In Bivalvia, where the strand asymmetry is reversed, the most used codons end with T and G [170]. Several works report that in Arthropoda and Nematoda mitogenomes, the most used codons end with A and T [171–173]. As detailed above, these phyla show the highest A+T content among Eukaryota; the neutral accumulation of A and T is confirmed by the highest frequency at the third codon position compared to the first and the second one [174]. Moreover, the accumulation of A and T is highly correlated with a biased use of codons inside Arthropoda and Nematoda (ρ = 0.79 and *p*-value = 0, ρ = 0.58 and *p*-value = 0, respectively). Indeed, comparisons between arthropod mtDNAs confirmed that mitogenomes richer in AT use NNT and NNA codons more frequently [175].

Interestingly, Actinopterygii show one of the lowest A+T content in Eukaryota (55.4%). In some entries, the AT-skew and the GC-skew show the same sign; moreover, in the cyprinid *Opsariichthys bidens*, it has been proven that C is the most used nucleotide at the third codon-position, while G-ending codons are more frequent than in other Chordata, suggesting that this feature is a result of a more efficient repair reaction to deamination [159]. This would also explain the low A+T content and the same sign of AT-skew and GC-skew in some Actinopterygii entries.

4. Conclusions and Final Remarks

Unfortunately, it was revealed that inconsistent annotation conventions led to systematic biases in data. Differences in the strand asymmetry can sometimes be related to different annotation decisions. Indeed, the signs of AT-skew and GC-skew depend on how the authors decide which strand is the plus strand, as a positive AT-skew on the L strand is obviously associated with an opposite negative AT-skew on the H strand. However, the localization of the L strand is not obvious for the clades where genes are located on both strands.

For example, we discovered that the order Unionida is split into two groups: 42 entries show RSA (as the other Bivalvia orders), while 119 entries show normal strand asymmetry (Supplementary Table S3). To prove that these groups are due to annotation issues, we calculated the number of genes on the plus strand for each entry, the number of genes on the minus strand, and on which strand the *cox1* gene is located, since its position is often taken as a reference to decide on the location of the plus strand. In the first group, most of the genes are located on the minus strand for all the entries, but the *cox1* gene is located on the plus strand; in the second group, most of the genes are located on the plus strand for all the entries, but the *cox1* gene is located on the minus strand (Supplementary Table S3). Although different gene orders have been detected in the order Unionida, the *cox1* gene in each group is located on the strand that harbors less genes [176]. Therefore, the authors applied different procedures to determine the plus strand: it is likely that the first group selected the strand with the *cox1* gene as the plus strand, whereas the second group selected the strand with more genes than the plus strand. This specific annotation issue has already been discussed [177]. Once again, we underline the necessity to determine and adopt conventions for the annotation of mitogenomes.

Moreover, when the two strands are correctly annotated, it is possible to unravel many phylogenetic artifacts. It has been proven that the RSA bias can affect the phylogenetic signal, also at the ammino acid level, thus leading to the clustering of clades that acquired RSA independently [161,178,179]. Therefore, before a phylomitogenomic analysis starts, it is pivotal to determine the strand asymmetry of each marker to exclude wrong phylogenetic signals.

To our knowledge, the database set up for the present paper is the first attempt to present most of the available mitochondrial genomes together. It has the potential to elucidate several molecular patterns underlying the figure of mitochondrial evolution across the eukaryotic domain. Further research is requested to overcome annotation errors and issues, and a release of the database including multipartite mtDNAs, as well as other GenBank entries, is currently under preparation in our laboratory. Recalling the vagaries of mitochondrial evolutionary history, it is clear that only an adequate sampling of eukaryotic biodiversity and the analysis of a huge number of genomes can shed light on the structure, architecture, and the role of the mitochondrial genome in the eukaryotic cell.

Supplementary Materials: The following are available online at https://www.mdpi.com/article/10.3390/life11070663/s1, Figure S1: Correlograms for major mtDNA features, Figure S2: Nucleotide composition and strand asymmetry, Table S1: Manually excluded species and relative selected GenBank Accession Numbers, Table S2: Manually edited annotations, Table S3: Annotation bias in Unionida.

Author Contributions: A.L. and F.P. conceived and supervised the study during all its stages. A.L. and F.P. prepared the dataset. F.P. modified the original HERMES script and parsed data from GenBank. A.F. analyzed the results, wrote the R and Python3 scripts that were used for the present analyses, and drafted the manuscript. A.L. and F.P. revised the manuscript. All authors have read and approved the final version of the manuscript.

Funding: This research was funded by the "Canziani Bequest" fund.

Institutional Review Board Statement: Not applicable.

Informed Consent Statement: Not applicable.

Data Availability Statement: The database which is presented in the present manuscript is publicly available on the MoZoo GitHub page, at the URL https://github.com/mozoo/almighto (accessed on: 7 July 2021).

Acknowledgments: Thanks are due to two anonymous reviewers whose comments and suggestions improved the original version of the present paper.

Conflicts of Interest: The authors declare no conflict of interest.

References

1. Sinclair, J.H.; Stevens, B.J. Circular DNA filaments from mouse mitochondria. *Proc. Natl. Acad. Sci. USA* **1966**, *56*, 508–514. [CrossRef]
2. Van Bruggen, E.F.J.; Borst, P.; Ruttenberg, G.J.C.M.; Gruber, M.; Kroon, A.M. Circular mitochondrial DNA. *Biochim. Biophys. Acta* **1966**, *119*, 437–439. [CrossRef]
3. Borst, P.; Kroon, A. Mitochondrial DNA: Physicochemical properties, replication, and genetic function. *Int. Rev. Cytol.* **1969**, *26*, 107–190.
4. Williamson, D. The curious history of yeast mitochondrial DNA. *Nat. Rev. Genet.* **2002**, *3*, 1–7. [CrossRef]

5. Altmann, R. *Die Elementarorganismen und ihre Beziehungen zu den Zellen*; Veit: Leipzig, Germany, 1890.
6. Suyama, Y.; Miura, K. Size and structural variations of mitochondrial DNA. *Proc. Natl. Acad. Sci. USA* **1968**, *60*, 235–242. [CrossRef] [PubMed]
7. Bendich, A.J. Reaching for the ring: The study of mitochondrial genome structure. *Curr. Genet.* **1993**, *24*, 279–290. [CrossRef]
8. Hollenberg, C.P.; Borst, P.; Thuring, R.W.J.; Van Bruggen, E.F.J. Size, structure and genetic complexity of yeast mitochondrial DNA. *Biochim. Biophys. Acta* **1969**, *186*, 417–419. [CrossRef]
9. Anderson, S.; Bankier, A.T.; Barrell, B.G.; de Bruijn, M.H.L.; Coulson, A.R.; Drouin, J.; Eperon, I.C.; Nierlich, D.P.; Roe, B.A.; Sanger, F.; et al. Sequence and organization of the human mitochondrial genome. *Nature* **1981**, *290*, 457–465. [CrossRef] [PubMed]
10. Bibb, M.J.; Van Etten, R.A.; Wright, C.T.; Walberg, M.W.; Clayton, D.A. Sequence and gene organization of mouse mitochondrial DNA. *Cell* **1981**, *26*, 167–180. [CrossRef]
11. Anderson, S.; de Bruijn, M.H.; Coulson, A.R.; Eperon, I.C.; Sanger, F.; Young, I.G. Complete sequence of bovine mitochondrial DNA. Conserved features of the mammalian mitochondrial genome. *J. Mol. Biol.* **1982**, *156*, 683–717. [CrossRef]
12. Clary, D.O.; Wolstenholme, D.R. The Drosophila mitochondrial genome. *Oxf. Surv. Eukaryot. Genes* **1984**, *1*, 1–35.
13. Lavrov, D.V.; Pett, W.; Voigt, O.; Wörheide, G.; Forget, L.; Lang, B.F.; Kayal, E. Mitochondrial DNA of *Clathrina clathrus* (Calcarea, Calcinea): Six linear chromosomes, fragmented rRNAs, tRNA editing, and a novel genetic code. *Mol. Biol. Evol.* **2013**, *30*, 865–880. [CrossRef] [PubMed]
14. Karnkowska, A.; Vacek, V.; Zubáčová, Z.; Treitli, S.C.; Petrželková, R.; Eme, L.; Novák, L.; Žárský, V.; Barlow, L.D.; Herman, E.K.; et al. A Eukaryote without a mitochondrial organelle. *Curr. Biol.* **2016**, *26*, 1274–1284. [CrossRef] [PubMed]
15. Smith, D.R.; Keeling, P.J. Mitochondrial and plastid genome architecture: Reoccurring themes, but significant differences at the extremes. *Proc. Natl. Acad. Sci. USA* **2015**, *112*, 10177–10184. [CrossRef]
16. Müller, M.; Mentel, M.; van Hellemond, J.J.; Henze, K.; Woehle, C.; Gould, S.B.; Yu, R.Y.; van der Giezen, M.; Tielens, A.G.M.; Martin, W.F. Biochemistry and evolution of anaerobic energy metabolism in eukaryotes. *Microbiol. Mol. Biol. Rev.* **2012**, *76*, 444–495. [CrossRef]
17. Hjort, K.; Goldberg, A.V.; Tsaousis, A.D.; Hirt, R.P.; Embley, T.M. Diversity and reductive evolution of mitochondria among microbial eukaryotes. *Philos. Trans. R Soc. Lond. B Biol. Sci.* **2010**, *365*, 713–727. [CrossRef] [PubMed]
18. Nosek, J.; Tomáska, L. Mitochondrial genome diversity: Evolution of the molecular architecture and replication strategy. *Curr. Genet.* **2003**, *44*, 73–84. [CrossRef] [PubMed]
19. Burger, G.; Valach, M. Perfection of Eccentricity: Mitochondrial Genomes of Diplonemids. *IUBMB Life* **2018**, *70*, 1197–1206. [CrossRef]
20. Lukeš, J.; Wheeler, R.; Jirsová, D.; David, V.; Archibald, J.M. Massive mitochondrial DNA content in diplonemid and kinetoplastid protists. *IUBMB Life* **2018**, *70*, 1267–1274. [CrossRef]
21. Jensen, R.E.; Englund, P.T. Network news: The replication of kinetoplast DNA. *Annu Rev. Microbiol.* **2012**, *66*, 473–491. [CrossRef]
22. Kaur, B.; Záhonová, K.; Valach, M.; Faktorová, D.; Prokopchuk, G.; Burger, G.; Lukeš, J. Gene fragmentation and RNA editing without borders: Eccentric mitochondrial genomes of diplonemids. *Nucleic Acids Res.* **2020**, *48*, 2694–2708. [CrossRef] [PubMed]
23. Waller, R.F.; Jackson, C.J. Dinoflagellate mitochondrial genomes: Stretching the rules of molecular biology. *BioEssays* **2009**, *31*, 237–245. [CrossRef]
24. Slamovits, C.H.; Saldarriaga, J.F.; Larocque, A.; Keeling, P.J. The highly reduced and fragmented mitochondrial genome of the early branching dinoflagellate *Oxyrrhis marina* shares characteristics with both apicomplexan and dinoflagellate mitochondrial genomes. *J. Mol. Biol.* **2007**, *372*, 356–368. [CrossRef] [PubMed]
25. Wu, Z.; Waneka, G.; Sloan, D.B. The Tempo and Mode of Angiosperm Mitochondrial Genome Divergence Inferred from Intraspecific Variation in *Arabidopsis thaliana*. *G3* **2020**, *10*, 1077–1086. [CrossRef]
26. Kozik, A.; Rowan, B.A.; Lavelle, D.; Berke, L.; Schranz, M.E.; Michelmore, R.W.; Christensen, A.C. The alternative reality of plant mitochondrial DNA: One ring does not rule them all. *PLoS Genet.* **2019**, *15*, e1008373. [CrossRef]
27. Gualberto, J.M.; Newton, K.J. Plant mitochondrial genomes: Dynamics and mechanisms of mutation. *Annu Rev. Plant. Biol.* **2017**, *68*, 225–252. [CrossRef]
28. McCauley, D.E. Paternal leakage, heteroplasmy, and the evolution of plant mitochondrial genomes. *New Phytol.* **2013**, *200*, 966–977. [CrossRef] [PubMed]
29. Bendich, A.J. The size and form of chromosomes are constant in the nucleus, but highly variable in bacteria, mitochondria and chloroplasts. *BioEssays* **2007**, *29*, 474–483. [CrossRef]
30. Barr, C.M.; Neiman, M.; Taylor, D.R. Inheritance and recombination of mitochondrial genomes in plants, fungi and animals. *New Phytol.* **2005**, *168*, 39–50. [CrossRef]
31. Abdelnoor, R.V.; Yule, R.; Elo, A.; Christensen, A.C.; Meyer-Gauen, G.; Mackenzie, S.A. Substoichiometric shifting in the plant mitochondrial genome is influenced by a gene homologous to *MutS*. *Proc. Natl. Acad. Sci. USA* **2003**, *100*, 5968–5973. [CrossRef]
32. Sloan, D.B. One ring to rule them all? Genome sequencing provides new insights into the 'master circle' model of plant mitochondrial DNA structure. *New Phytol.* **2013**, *200*, 978–985. [CrossRef] [PubMed]
33. Mower, J.P.; Case, A.L.; Floro, E.R.; Willis, J.H. Evidence against equimolarity of large repeat arrangements and a predominant master circle structure of the mitochondrial genome from a monkeyflower (*Mimulus guttatus*) lineage with cryptic CMS. *Genome Biol. Evol.* **2012**, *4*, 670–686. [CrossRef]

34. Backert, S.; Borner, T. Phage T4-like intermediates of DNA replication and recombination in the mitochondria of the higher plant *Chenopodium album* (L.). *Curr. Genet.* **2000**, *37*, 304–314. [CrossRef]
35. Bendich, A.J. Structural analysis of mitochondrial DNA molecules from fungi and plants using moving pictures and pulsedfield gel electrophoresis. *J. Mol. Biol.* **1996**, *255*, 564–588. [CrossRef] [PubMed]
36. Palmer, J.D.; Shields, C.R. Tripartite structure of the Brassica campestris mitochondrial genome. *Nature* **1984**, *307*, 437–440. [CrossRef]
37. Preuten, T.; Cincu, E.; Fuchs, J.; Zoschke, R.; Liere, K.; Börner, T. Fewer genes than organelles: Extremely low and variable gene copy numbers in mitochondria of somatic plant cells. *Plant. J.* **2010**, *64*, 948–959. [CrossRef]
38. Sloan, D.B.; Alverson, A.J.; Chuckalovcak, J.P.; Wu, M.; McCauley, D.E.; Palmer, J.D.; Taylor, D.R. Rapid Evolution of Enormous, Multichromosomal Genomes in Flowering Plant Mitochondria with Exceptionally High Mutation Rates. *PLoS Biol* **2012**, *10*, e1001241. [CrossRef]
39. Alverson, A.J.; Rice, D.W.; Dickinson, S.; Barry, K.; Palmer, J.D. Origins and recombination of the bacterial-sized multichromosomal mitochondrial genome of cucumber. *Plant. Cell* **2011**, *23*, 2499–2513. [CrossRef]
40. Lavrov, D.V.; Pett, W. Animal Mitochondrial DNA as We Do Not Know It: Mt-Genome Organization and Evolution in Nonbilaterian Lineages. *Genome Biol. Evol.* **2016**, *8*, 2896–2913. [CrossRef]
41. Lavrov, D.; Adamski, M.; Chevaldonné, P.; Adamska, M. Extensive mitochondrial mRNA editing and unusual mitochondrial genome organization in calcaronean sponges. *Curr. Biol.* **2016**, *26*, 86–92. [CrossRef] [PubMed]
42. Voigt, O.; Erpenbeck, D.; Wörheide, G. A fragmented metazoan organellar genome: The two mitochondrial chromosomes of *Hydra magnipapillata*. *BMC Genom.* **2008**, *9*, 350. [CrossRef]
43. Kayal, E.; Bentlage, B.; Collins, A.G.; Kayal, M.; Pirro, S.; Lavrov, D.V. Evolution of linear mitochondrial genomes in medusozoan cnidarians. *Genome Biol. Evol.* **2012**, *4*, 1–12. [CrossRef] [PubMed]
44. Watanabe, K.; Bessho, Y.; Kawasaki, M.; Hori, H. Mitochondrial genes are found on minicircle DNA molecules in the mesozoan animal *Dicyema*. *J. Mol. Biol.* **1999**, *286*, 645–650. [CrossRef]
45. Witek, A.; Herlyn, H.; Ebersberger, I.; Mark Welch, D.B.; Hankeln, T. Support for the monophyletic origin of Gnathifera from phylogenomics. *Mol. Phylogenet. Evol.* **2009**, *53*, 1037–1041. [CrossRef] [PubMed]
46. Suga, K.; Mark Welch, D.B.; Tanaka, Y.; Sakakura, Y.; Hagiwara, A. Two circular chromosomes of unequal copy number make up the mitochondrial genome of the rotifer *Brachionus plicatilis*. *Mol. Biol. Evol.* **2008**, *25*, 1129–1137. [CrossRef]
47. Hwang, D.S.; Suga, K.; Sakakura, Y.; Park, H.G.; Hagiwara, A.; Rhee, J.S.; Lee, J.S. Complete mitochondrial genome of the monogonont rotifer, *Brachionus koreanus* (Rotifera, Brachionidae). *Mitochondrial DNA* **2014**, *25*, 29–30. [CrossRef]
48. Gibson, T.; Blok, V.C.; Dowton, M. Sequence and characterization of six mitochondrial subgenomes from *Globodera rostochiensis*: Multipartite structure is conserved among close nematode relatives. *J. Mol. Evol.* **2007**, *65*, 308315. [CrossRef] [PubMed]
49. Gibson, T.; Blok, V.C.; Phillips, M.S.; Hong, G.; Kumarasinghe, D.; Riley, I.T.; Dowton, M. The mitochondrial subgenomes of the nematode *Globodera pallida* are mosaics: Evidence of recombination in an animal mitochondrial genome. *J. Mol. Evol.* **2007**, *64*, 463471. [CrossRef]
50. Sweet, A.D.; Johnson, K.P.; Cameron, S.L. Mitochondrial genomes of *Columbicola* feather lice are highly fragmented, indicating repeated evolution of minicircle-type genomes in parasitic lice. *PeerJ* **2020**, *8*, e8759. [CrossRef]
51. Johnson, K.P.; Nguyen, N.; Sweet, A.D.; Boyd, B.M.; Warnow, T.; Allen, J.M. Simultaneous radiation of bird and mammal lice following the K-Pg boundary. *Biol. Lett.* **2018**, *14*, 20180141. [CrossRef] [PubMed]
52. Shi, Y.; Chu, Q.; Wei, D.D.; Qui, Y.J.; Shang, F.; Dou, W.; Wang, J.J. The mitochondrial genome of booklouse, *Liposcelis sculptilis* (Psocoptera: Liposcelididae) and the evolutionary timescale of Liposcelis. *Sci. Rep.* **2016**, *6*, 30660. [CrossRef]
53. Dickey, A.M.; Kumar, V.; Morgan, J.K.; Jara-Cavieres, A.; Shatters, R.G., Jr.; McKenzie, C.L.; Osborne, L.S. A novel mitochondrial genome architecture in thrips (Insecta: Thysanoptera): Extreme size asymmetry among chromosomes and possible recent control region duplication. *BMC Genomics* **2015**, *16*, 439. [CrossRef]
54. Wei, D.D.; Shao, R.; Yuan, M.L.; Dou, W.; Barker, S.C.; Wang, J.J. The multipartite mitochondrial genome of *Liposcelis bostrychophila*: Insights into the evolution of mitochondrial genomes in bilateral animals. *PLoS ONE* **2012**, *7*, e33973. [CrossRef]
55. Song, F.; Li, H.; Lio, G.H.; Wang, W.; James, P.; Colwell, D.D.; Tran, A.; Gong, S.; Cai, W.; Shao, R. Mitochondrial genome fragmentation unites the parasitic lice of eutherian mammals. *Syst. Biol.* **2019**, *68*, 430440. [CrossRef]
56. Bullerwell, C.E.; Gray, M.W. Evolution of the mitochondrial genome: Protist connections to animals, fungi and plants. *Curr. Opin. Microbiol* **2004**, *7*, 528–534. [CrossRef]
57. Burger, G.; Forget, L.; Zhu, Y.; Gray, M.W.; Lang, B.F. Unique mitochondrial genome architecture in unicellular relatives of animals. *Proc. Natl. Acad. Sci. USA* **2003**, *100*, 892–897. [CrossRef]
58. Freel, K.C.; Friedrich, A.; Schacherer, J. Mitochondrial genome evolution in yeasts: An all-encompassing view. *FEMS Yeast Res.* **2015**, *15*, fov023. [CrossRef] [PubMed]
59. Pohjoismäki, J.L.O.; Goffart, S. Of circles, forks and humanity: Topological organisation and replication of mammalian mitochondrial DNA. *BioEssays* **2011**, *33*, 290–299. [CrossRef] [PubMed]
60. Smith, D.R.; Keeling, P.J. Gene conversion shapes linear mitochondrial genome architecture. *Genome Biol. Evol.* **2013**, *5*, 905–912. [CrossRef] [PubMed]
61. Ender, A.; Schierwater, B. Placozoa are not derived cnidarians: Evidence from molecular morphology. *Mol. Biol. Evol.* **2003**, *20*, 130–134. [CrossRef] [PubMed]

62. Bridge, D.; Cunningham, C.W.; Schierwater, B.; Desalle, R.; Buss, L.W. Class-level relationships in the phylum Cnidaria: Evidence from mitochondrial genome structure. *Proc. Natl. Acad. Sci. USA* **1992**, *89*, 8750–8753. [CrossRef]
63. Warrior, R.; Gall, J. The mitochondrial DNA of *Hydra attenuata* and *Hydra littoralis* consists of two linear molecules. *Arch. Sci. Geneve* **1985**, *38*, 439–445.
64. Dujon, B. Mitochondrial genetics revisited. *Yeast* **2020**, *37*, 191–205. [CrossRef] [PubMed]
65. Valach, M.; Farkas, Z.; Fricova, D.; Kovac, J.; Brejova, B.; Vinar, T.; Pfeiffer, I.; Kucsera, J.; Tomaska, L.; Lang, B.F.; et al. Evolution of linear chromosomes and multipartite genomes in yeast mitochondria. *Nucleic Acids Res.* **2011**, *39*, 4202–4219. [CrossRef] [PubMed]
66. Gerhold, J.M.; Aun, A.; Sedman, T.; Jõers, P.; Sedman, J. Strand invasion structures in the inverted repeat of *Candida albicans* mitochondrial DNA reveal a role for homologous recombination in replication. *Mol. Cell* **2010**, *39*, 851–861. [CrossRef]
67. Ling, F.; Shibata, T. Recombination-dependent mtDNA partitioning. In vivo role of Mhr1p to promote pairing of homologous DNA. *EMBO J* **2002**, *21*, 4730–4740. [CrossRef] [PubMed]
68. Kosa, P.; Valach, M.; Tomaska, L.; Wolfe, K.H.; Nosek, J. Complete DNA sequences of the mitochondrial genomes of the pathogenic yeasts *Candida orthopsilosis* and *Candida metapsilosis*: Insight into the evolution of linear DNA genomes from mitochondrial telomere mutants. *Nucleic Acids Res.* **2006**, *34*, 2472–2481. [CrossRef]
69. Morin, G.B.; Cech, T.R. Mitochondrial telomeres: Surprising diversity of repeated telomeric DNA sequences among six species of *Tetrahymena*. *Cell* **1988**, *52*, 367–374. [CrossRef]
70. Morin, G.B.; Cech, T.R. The telomeres of the linear mitochondrial DNA of *Tetrahymena thermophila* consist of 53 bp tandem repeats. *Cell* **1986**, *46*, 873–883. [CrossRef]
71. Goddard, J.M.; Cummings, D.J. Mitochondrial DNA replication in *Paramecium aurelia*. Cross-linking of the initiation end. *J. Mol. Biol.* **1977**, *109*, 327–344. [CrossRef]
72. Feagin, J.E.; Mericle, B.L.; Werner, E.; Morris, M. Identification of additional rRNA fragments encoded by the Plasmodium falciparum 6 kb element. *Nucl Acids Res.* **1997**, *25*, 438–446. [CrossRef] [PubMed]
73. Feagin, J.E. The 6-kb element of *Plasmodium falciparum* encodes mitochondrial cytochrome genes. *Mol. Biochem. Parasitol.* **1992**, *52*, 145–148. [CrossRef]
74. Warren, J.M.; Simmons, M.P.; Wu, Z.; Sloan, D.B. Linear Plasmids and the Rate of Sequence Evolution in Plant Mitochondrial Genomes. *Genome Biol. Evol.* **2016**, *8*, 364–374. [CrossRef]
75. Handa, H. Linear plasmids in plant mitochondria: Peaceful coexistences or malicious invasions? *Mitochondrion* **2008**, *8*, 15–25. [CrossRef]
76. Swart, E.C.; Nowacki, M.; Shum, J.; Stiles, H.; Higgins, B.P.; Doak, T.G.; Schotanus, K.; Magrini, V.J.; Minx, P.; Mardis, E.R.; et al. The *Oxytricha trifallax* mitochondrial genome. *Genome Biol. Evol.* **2012**, *4*, 136–154. [CrossRef]
77. Smith, D.R.; Hua, J.; Lee, R.W. Evolution of linear mitochondrial DNA in three known lineages of *Polytomella*. *Curr. Genet.* **2010**, *56*, 427–438. [CrossRef]
78. Osigus, H.J.; Eitel, M.; Bernt, M.; Donath, A.; Schierwater, B. Mitogenomics at the base of Metazoa. *Mol. Phylogenet. Evol.* **2013**, *69*, 339–351. [CrossRef]
79. Pett, W.; Ryan, J.F.; Pang, K.; Mullikin, J.C.; Martindale, M.Q.; Baxevanis, A.D.; Lavrov, D.V. Extreme mitochondrial evolution in the ctenophore *Mnemiopsis leidyi*: Insight from mtDNA and the nuclear genome. *Mitochondrial DNA* **2011**, *22*, 130–142. [CrossRef]
80. Pramateftaki, P.V.; Kouvelis, V.N.; Lanaridis, P.; Typas, M.A. The mitochondrial genome of the wine yeast *Hanseniaspora uvarum*: A unique genome organization among yeast/fungal counterparts. *FEMS Yeast Res.* **2006**, *6*, 77–90. [CrossRef] [PubMed]
81. Bernt, M.; Braband, A.; Schierwater, B.; Stadler, P.F. Genetic aspects of mitochondrial genome evolution. *Mol. Phylogenet. Evol.* **2013**, *69*, 328–338. [CrossRef]
82. Araújo, D.S.; De-Paula, R.B.; Tomé, L.M.R.; Quintanilha-Peixoto, G.; Salvador-Montoya, C.A.; Del-Bem, L.-E.; Badotti, F.; Azevedo, V.A.C.; Brenig, B.; Aguiar, E.R.G.R.; et al. Comparative mitogenomics of Agaricomycetes: Diversity, abundance, impact and coding potential of putative open-reading frames. *Mitochondrion* **2021**, *58*, 1–13. [CrossRef]
83. Repetti, S.I.; Jackson, C.J.; Judd, L.M.; Wick, R.R.; Holt, K.E.; Verbruggen, H. The inflated mitochondrial genomes of siphonous green algae reflect processes driving expansion of noncoding DNA and proliferation of introns. *PeerJ* **2020**, *8*, e8273. [CrossRef]
84. Losada, L.; Pakala, S.B.; Fedorova, N.D.; Joardar, V.; Shabalina, S.A.; Hostetler, J.; Pakala, S.M.; Zafar, N.; Thomas, E.; Rodriguez-Carres, M.; et al. Mobile elements and mitochondrial genome expansion in the soil fungus and potato pathogen *Rhizoctonia solani* AG-3. *FEMS Microbiol. Lett.* **2020**, *352*, 165–173. [CrossRef]
85. Hemmi, K.; Kakehashi, R.; Kambayashi, C.; Du Preez, L.; Minter, L.; Furuno, N.; Kurabayashi, A. Exceptional Enlargement of the Mitochondrial Genome Results from Distinct Causes in Different Rain Frogs (Anura: Brevicipitidae: *Breviceps*). *Int J. Genomics* **2020**, *2020*, 6540343. [CrossRef]
86. Pu, L.; Liu, H.; Wang, G.; Li, B.; Xia, G.; Shen, M.; Yang, M. Complete mitochondrial genome of the cockle *Anadara antiquata* (Linnaeus, 1758). *Mitochondrial DNA Part. B* **2019**, *4*, 2293–2294. [CrossRef] [PubMed]
87. Liu, Y.; Kurokawa, T.; Sekino, M.; Tanabe, T.; Watanabe, K. Complete mitochondrial DNA sequence of the ark shell *Scapharca broughtonii*: An ultra-large metazoan mitochondrial genome. *Comp. Biochem. Physiol. Part. D Genomics Proteomics* **2013**, *8*, 72–81. [CrossRef] [PubMed]
88. Ghiselli, F.; Gomes-dos-Santos, A.; Adema, C.M.; Lopes-Lima, M.; Sharbrough, J.; Boore, J.L. Molluscan mitochondrial genomes break the rules. *Phil. Trans. R Soc. B* **2021**, *376*, 20200159. [CrossRef]

89. Plazzi, F.; Puccio, G.; Passamonti, M. Comparative Large-Scale Mitogenomics Evidences Clade-Specific Evolutionary Trends in Mitochondrial DNAs of Bivalvia. *Genome Biol. Evol.* **2016**, *8*, 2544–2564. [CrossRef] [PubMed]
90. Flegontov, P.; Michálek, J.; Janouškovec, J.; La, D.H.; Jirk, M.; Hajdušková, E.; Tomčala, A.; Otto, T.D.; Keeling, P.J.; Pain, A.; et al. Divergent mitochondrial ararespiratory chains in phototrophic relatives of apicomplexan parasites. *Mol. Biol. Evol.* **2015**, *32*, 1115–1131. [CrossRef]
91. Rehkopf, D.H.; Gillespie, D.E.; Harrell, M.I.; Feagin, J.E. Transcriptional mapping and RNA processing of the *Plasmodium falciparum* mitochondrial mRNAs. *Mol. Biochem. Parasitol.* **2000**, *105*, 91–103. [CrossRef]
92. Feagin, J.E. The extrachromosomal DNAs of apicomplexan parasites. *Annu Rev. Microbiol.* **1994**, *48*, 81–104. [CrossRef]
93. Gray, M.W.; Burger, G.; Derelle, R.; Klimeš, V.; Leger, M.M.; Sarrasin, M.; Vlček, Č.; Roger, A.J.; Eliáš, M.; Lang, B.F. The draft nuclear genome sequence and predicted mitochondrial proteome of *Andalucia godoyi*, a protist with the most gene-rich and bacteria-like mitochondrial genome. *BMC Biol.* **2020**, *18*, 22. [CrossRef] [PubMed]
94. Derelle, R.; Torruella, G.; Klimeš, V.; Brinkmann, H.; Kim, E.; Vlček, Č.; Lang, B.F.; Eliáš, M. Bacterial proteins pinpoint a single eukaryotic root. *Proc. Natl Acad. Sci. USA* **2015**, *112*, E693–E699. [CrossRef]
95. Hampl, V.; Hug, L.; Leigh, J.W.; Dacks, J.B.; Lang, B.F.; Simpson, A.G.B.; Roger, A.J. Phylogenomic analyses support the monophyly of Excavata and resolve relationships among eukaryotic "supergroups". *Proc. Natl. Acad. Sci. USA* **2009**, *106*, 3859–3864. [CrossRef]
96. Lang, B.F.; Burger, G.; O'Kelly, C.J.; Cedergren, R.; Golding, G.B.; Lemieux, C.; Sankoff, D.; Turmel, M.; Gray, M.W. An ancestral mitochondrial DNA resembling a eubacterial genome in miniature. *Nature* **1997**, *387*, 493–497. [CrossRef]
97. Burger, G.; Gray, M.W.; Forget, L.; Lang, B.F. Strikingly bacteria-like and gene-rich mitochondrial genomes throughout jakobid protists. *Genome Biol Evol* **2013**, *5*, 418–438. [CrossRef]
98. Sana, S.; Hardouin, E.A.; Paley, R.; Zhang, T.; Andreou, D. The complete mitochondrial genome of a parasite at the animal-fungal boundary. *Parasites Vectors* **2020**, *13*, 81. [CrossRef] [PubMed]
99. Solieri, L. Mitochondrial inheritance in budding yeasts: Towards an integrated understanding. *Trends Microbiol* **2010**, *18*, 521–530. [CrossRef]
100. Foury, F.; Roganti, T.; Lecrenier, N.; Purnelle, B. The complete sequence of the mitochondrial genome of *Saccharomyces cerevisiae*. *FEBS Lett* **1998**, *32*, 325–331. [CrossRef]
101. Wolf, K.; Del Giudice, L. The variable mitochondrial genome of ascomycetes: Organization, mutations, alterations, and expression. *Adv. Genet.* **1988**, *25*, 185–308.
102. Zamaroczy, M.; Bernardi, G. The primary structure of the mitochondrial genome of *Saccharomyces cerevisiae*—A review. *Gene* **1986**, *47*, 155–177. [CrossRef]
103. Breton, S.; Milani, L.; Ghiselli, F.; Guerra, D.; Stewart, D.T.; Passamonti, M. A resourceful genome: Updating the functional repertoire and evolutionary role of animal mitochondrial DNAs. *Trends Genet.* **2014**, *30*, 555–564. [CrossRef] [PubMed]
104. Boore, J.L. Animal mitochondrial genomes. *Nucleic Acids Res.* **1999**, *27*, 1767–1780. [CrossRef]
105. Monnens, M.; Thijs, S.; Briscoe, A.G.; Clark, M.; Frost, E.J.; Littlewood, D.T.J.; Sewell, M.; Smeets, K.; Artois, T.; Vanhove, M.O.M. The first mitochondrial genomes of endosymbiotic rhabdocoels illustrate evolutionary relaxation of atp8 and genome plasticity in flatworms. *Int J. Biol Macromol* **2020**, *162*, 454–469. [CrossRef] [PubMed]
106. Trindade Rosa, M.; Oliveira, D.S.; Loreto, E.L.S. Characterization of the first mitochondrial genome of a catenulid flatworm: *Stenostomum leucops* (Platyhelminthes). *J. Zool Syst. Evol. Res.* **2017**, *55*, 98–105. [CrossRef]
107. Solà, E.; Álvarez-Presas, M.; Frías-López, C.; Littlewood, D.T.J.; Rozas, J.; Riutort, M. Evolutionary Analysis of Mitogenomes from Parasitic and Free-Living Flatworms. *PLoS ONE* **2015**, *10*, e0120081. [CrossRef] [PubMed]
108. Sultana, T.; Kim, J.; Lee, S.H.; Han, H.; Kim, S.; Min, G.S.; Nadler, S.A.; Park, J.K. Comparative analysis of complete mitochondrial genome sequences confirms independent origins of plant-parasitic nematodes. *BMC Evol. Biol.* **2013**, *13*, 12. [CrossRef]
109. Helfenbein, K.; Fourcade, H.; Vanjani, R.; Boore, J. The mitochondrial genome of *Paraspadella gotoi* is highly reduced and reveals that chaetognaths are a sister group to protostomes. *Proc. Natl. Acad. Sci. USA* **2004**, *101*, 10639–10643. [CrossRef]
110. Papillon, D.; Perez, Y.; Caubit, X.; Le Parco, Y. Identification of chaetognaths as protostomes is supported by the analysis of their mitochondrial genome. *Mol. Biol. Evol.* **2004**, *21*, 2122–2129. [CrossRef]
111. Von Nickisch-Rosenegk, M.; Brown, W.M.; Boore, J.L. Complete sequence of the mitochondrial genome of the tapeworm *Hymenolepis diminuta*: Gene arrangements indicate that Platyhelminths are Eutrochozoans. *Mol. Biol. Evol.* **2001**, *18*, 721–730. [CrossRef]
112. Le, T.H.; Blair, D.; Agatsuma, T.; Humair, P.F.; Campbell, N.J.H.; Iwagami, M.; Littlewood, D.T.J.; Peacock, B.; Johnston, D.A.; Bartley, J.; et al. Phylogenies inferred from mitochondrial gene orders—a cautionary tale from the parasitic flatworms. *Mol. Biol. Evol.* **2000**, *17*, 1123–1125. [CrossRef]
113. Okimoto, R.; Macfarlane, J.; Clary, D.; Wolstenholme, D. The mitochondrial genomes of two nematodes, *Caenorhabditis elegans* and *Ascaris suum*. *Genetics* **1992**, *130*, 471–498. [CrossRef] [PubMed]
114. Barthelemy, R.; Seligmann, H. Cryptic tRNAs in chaetognath mitochondrial genomes. *Comput. Biol. Chem.* **2016**, *62*, 119–132. [CrossRef] [PubMed]
115. Lavrov, D.V.; Brown, W.M. *Trichinella spiralis* mtDNA: A nematode mitochondrial genome that encodes a putative ATP8, normally-structured tRNAs, and has a gene arrangement relatable to those of coelomate metazoans. *Genetics* **2001**, *157*, 621–637. [CrossRef] [PubMed]

116. Schultz, D.T.; Eizenga, J.M.; Corbett-Detig, R.B.; Francis, W.R.; Christianson, L.M.; Haddock, S.H.D. Conserved novel ORFs in the mitochondrial genome of the ctenophore *Beroe forskalii*. *PeerJ* **2020**, *8*, e8356. [CrossRef]
117. Arafat, H.; Alamaru, A.; Gissi, C.; Huchon, D. Extensive mitochondrial gene rearrangements in Ctenophora: Insights from benthic platyctenida. *BMC Evol. Biol.* **2018**, *18*, 65. [CrossRef]
118. Cohen, P. New role for the mitochondrial peptide humanin: Protective agent against chemotherapy-induced side effects. *J. Natl. Cancer Inst.* **2014**, *106*, dju006. [CrossRef]
119. Lee, C.; Yen, K.; Cohen, P. Humanin: A harbinger of mitochondrial-derived peptides? *Trends Endocrinol Metab* **2013**, *24*, 222–228. [CrossRef]
120. Plazzi, F.; Ribani, A.; Passamonti, M. The complete mitochondrial genome of *Solemya velum* (Mollusca: Bivalvia) and its relationships with Conchifera. *BMC Genomics* **2013**, *14*, 409. [CrossRef] [PubMed]
121. Endo, K.; Noguchi, Y.; Ueshima, R.; Jacobs, H.T. Novel repetitive structures, deviant protein-encoding sequences and unidentified ORFs in the mitochondrial genome of the brachiopod *Lingula anatina*. *J. Mol. Evol.* **2005**, *61*, 36–53. [CrossRef]
122. Signorovitch, A.; Buss, L.; Dellaporta, S. Comparative genomics of large mitochondria in Placozoans. *PLoS Genet.* **2007**, *3*, e13. [CrossRef]
123. Zhang, B.; Zhang, Y.H.; Wang, X.; Zhang, H.X.; Lin, Q. The mitochondrial genome of a sea anemone *Bolocera sp.* exhibits novel genetic structures potentially involved in adaptation to the deep-sea environment. *Ecol. Evol.* **2017**, *7*, 4951–4962. [CrossRef] [PubMed]
124. McFadden, C.S.; France, S.C.; Sánchez, J.A.; Alderslade, P. A molecular phylogenetic analysis of the Octocorallia (Cnidaria:Anthozoa) based on mitochondrial protein-coding sequences. *Mol. Phylogenet Evol.* **2006**, *41*, 513–527. [CrossRef]
125. Kohn, A.B.; Citarella, M.R.; Kocot, K.M.; Bobkova, Y.V.; Halanych, K.M.; Moroz, L.L. Rapid evolution of the compact and unusual mitochondrial genome in the ctenophore, *Pleurobrachia bachei*. *Mol. Phylogenet. Evol.* **2012**, *63*, 203–207. [CrossRef]
126. Osigus, H.J.; Eitel, M.; Schierwater, B. Deep RNA sequencing reveals the smallest known mitochondrial micro exon in animals: The placozoan *cox1* single base pair exon. *PLoS ONE* **2017**, *12*, e0177959. [CrossRef] [PubMed]
127. Žihala, D.; Eliáš, M. Evolution and Unprecedented Variants of the Mitochondrial Genetic Code in a Lineage of Green Algae. *Genome Biol. Evol.* **2019**, *11*, 2992–3007. [CrossRef]
128. Li, Y.; Kocot, K.M.; Tassia, M.G.; Cannon, J.T.; Bernt, M.; Halanych, K.M. Mitogenomics Reveals a Novel Genetic Code in Hemichordata. *Genome Biol. Evol.* **2018**, *11*, 29–40. [CrossRef]
129. Kutyumov, V.A.; Predeus, A.V.; Starunov, V.V.; Maltseva, A.L.; Ostrovsky, A.N. Mitochondrial gene order of the freshwater bryozoan *Cristatella mucedo* retains ancestral lophotrochozoan features. *Mitochondrion* **2021**, *59*, 96–104. [CrossRef]
130. Zhang, J.; Kan, X.; Miao, G.; Hu, S.; Sun, Q.; Tian, W. qMGR: A new approach for quantifying mitochondrial genome rearrangement. *Mitochondrion* **2020**, *52*, 20–23. [CrossRef] [PubMed]
131. Bernt, M.; Middendorf, M. A method for computing an inventory of metazoan mitochondrial gene order rearrangements. *BMC Bioinform.* **2011**, *12* (Suppl. 9), S6. [CrossRef]
132. Zouros, E.; Rodakis, G.C. Doubly Uniparental Inheritance of mtDNA: An Unappreciated Defiance of a General Rule. *Adv. Anat Embryol. Cell Biol.* **2019**, *231*, 25–49. [PubMed]
133. Gusman, A.; Lecomte, S.; Stewart, D.T.; Passamonti, M.; Breton, S. Pursuing the quest for better understanding the taxonomic distribution of the system of doubly uniparental inheritance of mtDNA. *PeerJ* **2016**, *4*, e2760. [CrossRef] [PubMed]
134. Zouros, E. Biparental inheritance through uniparental transmission: The doubly uniparental inheritance (DUI) of mitochondrial DNA. *Evol. Biol.* **2013**, *40*, 1–31. [CrossRef]
135. Passamonti, M.; Ghiselli, F. Doubly Uniparental Inheritance: Two mitochondrial genomes, one precious model for organelle DNA inheritance and evolution. *DNA Cell Biol.* **2009**, *28*, 79–89. [CrossRef]
136. Breton, S.; Doucet-Beaupré, H.; Stewart, D.T.; Hoeh, W.R.; Blier, P.U. The unusual system of doubly uniparental inheritance of mtDNA: Isn't one enough? *Trends Genet.* **2007**, *23*, 465–474. [CrossRef]
137. Plazzi, F.; Puccio, G.; Passamonti, M. HERMES: An improved method to test mitochondrial genome molecular synapomorphies among clades. *Mitochondrion* **2021**, *58*, 285–295. [CrossRef] [PubMed]
138. Huerta-Cepas, J.; Serra, F.; Bork, P. ETE 3: Reconstruction, analysis and visualization of phylogenomic data. *Mol. Biol Evol.* **2016**, *33*, 1635–1638. [CrossRef]
139. R Development Core Team. *R: A Language and Environment for Statistical Computing*; R Foundation for Statistical Computing: Vienna, Austria, 2008.
140. Wickham, H. *ggplot2: Elegant Graphics for Data Analysis*; Springer: New York, NY, USA, 2016.
141. Harrell, F.E., Jr. Hmisc: Harrell Miscellaneous. R Package Version 4.4-1. 2020. Available online: https://CRAN.R-project.org/package=Hmisc (accessed on 30 April 2021).
142. Wei, T.; Simko, V. R Package "corrplot": Visualization of a Correlation Matrix (Version 0.84). 2017. Available online: https://github.com/taiyun/corrplot (accessed on 30 April 2021).
143. WoRMS Editorial Board. World Register of Marine Species. Available online: https://www.marinespecies.org (accessed on 30 April 2021).
144. Chamberlain, S. Worrms: World Register of Marine Specie (WoRMS) Client. R package version 0.4.2. 2020. Available online: https://CRAN.R-project.org/package=worrms (accessed on 30 April 2021).

145. Adl, S.M.; Bass, D.; Lane, C.E.; Lukeš, J.; Schoch, C.L.; Smirnov, A.; Agatha, S.; Berney, C.; Brown, M.W.; Burki, F.; et al. Revisions to the Classification, Nomenclature, and Diversity of Eukaryotes. *J. Eukaryot. Microbiol.* **2019**, *66*, 4–119. [CrossRef]
146. Wang, X.; Wang, J.; Liu, J.; Liu, A.; He, X.; Xiang, Q.; Li, Y.; Yin, H.; Luo, J.; Guan, G. Insights into the phylogenetic relationships and drug targets of *Babesia* isolates infective to small ruminants from the mitochondrial genomes. *Parasites Vectors* **2020**, *13*, 378. [CrossRef]
147. Pedrola-Monfort, J.; Lázaro-Gimeno, D.; Boluda, C.G.; Pedrola, L.; Garmendia, A.; Soler, C.; Soriano, J.M. Evolutionary Trends in the Mitochondrial Genome of Archaeplastida: How Does the GC Bias Affect the Transition from Water to Land? *Plants* **2020**, *9*, 358. [CrossRef]
148. Liu, Y.; Wang, B.; Li, L.; Qiu, Y.L.; Xue, J. Conservative and Dynamic Evolution of Mitochondrial Genomes in Early Land Plants. In *Genomics of Chloroplasts and Mitochondria. Advances in Photosynthesis and Respiration (including Bioenergy and Related Processes)*; Bock, R., Knoop, V., Eds.; Springer: Dordrecht, The Netherlands, 2012; Volume 35, pp. 159–174.
149. Liu, Y.; Xue, J.Y.; Wang, B.; Li, L.; Qiu, Y.L. The mitochondrial genomes of the early land plants *Treubia lacunosa* and *Anomodon rugelii*: Dynamic and conservative evolution. *PLoS ONE* **2011**, *6*, e25836. [CrossRef]
150. Dellaporta, S.L.; Xu, A.; Sagasser, S.; Wolfgan, J.; Moreno, M.A.; Buss, L.W.; Schierwater, B. Mitochondrial genome of *Trichoplax adhaerens* supports Placozoa as the basal lower metazoan phylum. *Proc. Natl. Acad. Sci. USA* **2006**, *103*, 8751–8756. [CrossRef]
151. Zardoya, R. Recent advances in understanding mitochondrial genome diversity. *F1000Res.* **2020**, *9*, F1000, Faculty Rev–270. [CrossRef] [PubMed]
152. Brown, G.G.; Colas Des Francs-Small, C.; Ostersetzer-Biran, O. Group II intron splicing factors in plant mitochondria. *Front Plant. Sci.* **2014**, *5*, 35. [CrossRef]
153. Férandon, C.; Xu, J.; Barroso, G. The 135 kbp mitochondrial genome of *Agaricus bisporus* is the largest known eukaryotic reservoir of group I introns and plasmid-related sequences. *Fungal. Genet. Biol.* **2013**, *55*, 85–91. [CrossRef]
154. Saccone, C.; De Giorgi, C.; Gissi, C.; Pesole, G.; Reyes, A. Evolutionary genomics in Metazoa: The mitochondrial DNA as a model system. *Gene* **1999**, *238*, 195–209. [CrossRef]
155. Chevigny, N.; Schatz-Daas, D.; Lotfi, F.; Gualberto, J.M. DNA Repair and the Stability of the Plant Mitochondrial Genome. *Int. J. Mol. Sci.* **2020**, *21*, 328. [CrossRef]
156. Xia, X. DNA replication and strand asymmetry in prokaryotic and mitochondrial genomes. *Curr. Genom.* **2012**, *13*, 16–27. [CrossRef] [PubMed]
157. Chen, X.J.; Clark-Walker, G.D. Unveiling the mystery of mitochondrial DNA replication in yeasts. *Mitochondrion* **2018**, *38*, 17–22. [CrossRef]
158. Cupp, J.D.; Nielsen, B.L. Minireview: DNA replication in plant mitochondria. *Mitochondrion* **2014**, *19 Pt. B*, 231–237. [CrossRef]
159. Wang, X.Z.; Wang, J.; He, S.; Mayden, R.L. The complete mitochondrial genome of the Chinese hook snout carp *Opsariichthys bidens* (Actinopterygii: Cypriniformes) and an alternative pattern of mitogenomic evolution invertebrate. *Gene* **2007**, *399*, 11–19. [CrossRef] [PubMed]
160. Scouras, A.; Smith, M.J. The complete mitochondrial genomes of the sea lily *Gymnocrinus richeri* and the feather star *Phanogenia gracilis*: Signature nucleotide bias and unique *nad4L* gene rearrangement within crinoids. *Mol. Phylogenet. Evol.* **2006**, *39*, 323–334. [CrossRef]
161. Hassanin, A. Phylogeny of Arthropoda inferred from mitochondrial sequences: Strategies for limiting the misleading effects of multiple changes in pattern and rates of substitution. *Mol. Phylogenet. Evol.* **2006**, *38*, 100–116. [CrossRef]
162. Hassanin, A.; Léger, N.; Deutsch, J. Evidence for multiple reversals of asymmetric mutational constraints during the evolution of the mitochondrial genome of metazoa, and consequences for phylogenetic inferences. *Syst. Biol.* **2005**, *54*, 277–298. [CrossRef]
163. Jemt, E.; Persson, Ö.; Shi, Y.; Mehmedovic, M.; Uhler, J.P.; Dávila López, M.; Freyer, C.; Gustafsson, C.M.; Samuelsson, T.; Falkenberg, M. Regulation of DNA replication at the end of the mitochondrial D-loop involves the helicase TWINKLE and a conserved sequence element. *Nucleic Acids Res.* **2015**, *43*, 9262–9275. [CrossRef] [PubMed]
164. Zhang, D.; Hewitt, M. Insect mitochondrial control region: A review of its structure, evolution and usefulness in evolutionary studies. *Biochem. Syst. Ecol.* **1997**, *25*, 99–120. [CrossRef]
165. Wei, S.J.; Shi, M.; Sharkey, M.J.; van Achterberg, C.; Chen, X.X. Comparative mitogenomics of Braconidae (Insecta: Hymenoptera) and the phylogenetic utility of mitochondrial genomes with special reference to Holometabolous insects. *BMC Genomics* **2010**, *11*, 371. [CrossRef]
166. Lindahl, T. Instability and decay of the primary structure of DNA. *Nature* **1993**, *362*, 709–715. [CrossRef] [PubMed]
167. Rocha, E.P. Codon usage bias from tRNA's point of view: Redundancy, specialization, and efficient decoding for translation optimization. *Genome Res.* **2004**, *14*, 2279–2286. [CrossRef]
168. Ikemura, T. Correlation between the abundance of Escherichia coli transfer RNAs and the occurrence of the respective codons in its protein genes. *J. Mol. Biol.* **1981**, *146*, 1–21. [CrossRef]
169. Xia, X. Mutation and selection on the anticodon of tRNA genes in vertebrate mitochondrial genomes. *Gene* **2005**, *345*, 13–20. [CrossRef] [PubMed]
170. Yu, H.; Li, Q. Mutation and selection on the wobble nucleotide in tRNA anticodons in marine bivalve mitochondrial genomes. *PLoS ONE* **2011**, *6*, e16147. [CrossRef] [PubMed]
171. Narakusumo, R.P.; Riedel, A.; Pons, J. Mitochondrial genomes of twelve species of hyperdiverse *Trigonopterus* weevils. *PeerJ* **2020**, *8*, e10017. [CrossRef]

172. Mohandas, N.; Pozio, E.; La Rosa, G.; Korhonen, P.K.; Young, N.D.; Koehler, A.V.; Hall, R.S.; Sternberg, P.W.; Boag, P.R.; Jex, A.R.; et al. Mitochondrial genomes of *Trichinella* species and genotypes—A basis for diagnosis, and systematic and epidemiological explorations. *Int. J. Parasitol.* **2014**, *44*, 1073–1080. [CrossRef]
173. Gibson, T.; Farrugia, D.; Barrett, J.; Chitwood, D.J.; Rowe, J.; Subbotin, S.; Dowton, M. The mitochondrial genome of the soybean cyst nematode, *Heterodera glycines*. *Genome* **2011**, *54*, 565–574. [CrossRef] [PubMed]
174. Sun, L.; Zhuo, K.; Lin, B.; Wang, H.; Liao, J. The complete mitochondrial genome of *Meloidogyne graminicola* (Tylenchina): A unique gene arrangement and its phylogenetic implications. *PLoS ONE* **2014**, *9*, e98558. [CrossRef]
175. Chen, S.C.; Wei, D.D.; Shao, R.; Dou, W.; Wang, J.J. The complete mitochondrial genome of the booklouse, *Liposcelis decolor*: Insights into gene arrangement and genome organization within the genus *Liposcelis*. *PLoS ONE* **2014**, *9*, e91902. [CrossRef]
176. Froufe, E.; Bolotov, I.; Aldridge, D.C.; Bogan, A.E.; Breton, S.; Gan, H.M.; Kovitvadhi, U.; Kovitvadhi, S.; Riccardi, N.; Secci-Petretto, G.; et al. Mesozoic mitogenome rearrangements and freshwater mussel (Bivalvia: Unionoidea) macroevolution. *Heredity* **2020**, *124*, 182–196. [CrossRef] [PubMed]
177. Barroso-Lima, N.C.; Prosdocimi, F. The heavy strand dilemma of vertebrate mitochondria on genome sequencing age: Number of encoded genes or G+T content? *Mitochondrial DNA A* **2017**, *27*, 1–6. [CrossRef]
178. Sun, S.; Li, Q.; Kong, L.; Yu, H. Multiple reversals of strand asymmetry in molluscs mitochondrial genomes, and consequences for phylogenetic inferences. *Mol. Phylogenet Evol.* **2018**, *118*, 222–231. [CrossRef]
179. Min, X.J.; Hickey, D.A. DNA Asymmetric Strand Bias Affects the Amino Acid Composition of Mitochondrial Proteins. *DNA Res.* **2007**, *14*, 201–206. [CrossRef] [PubMed]

life

Article

Looking at the Nudibranch Family Myrrhinidae (Gastropoda, Heterobranchia) from a Mitochondrial '2D Folding Structure' Point of View

Giulia Furfaro [1,*] and Paolo Mariottini [2]

1 Department of Biological and Environmental Sciences and Technologies—DiSTeBA, University of Salento, I-73100 Lecce, Italy
2 Department of Science, University of Roma Tre, I-00146 Rome, Italy; paolo.mariottini@uniroma3.it
* Correspondence: giulia.furfaro@unisalento.it; Tel.: +39-0832-29-8660

Citation: Furfaro, G.; Mariottini, P. Looking at the Nudibranch Family Myrrhinidae (Gastropoda, Heterobranchia) from a Mitochondrial '2D Folding Structure' Point of View. *Life* **2021**, *11*, 583. https://doi.org/10.3390/life11060583

Academic Editors: Andrea Luchetti and Federico Plazzi

Received: 16 April 2021
Accepted: 17 June 2021
Published: 18 June 2021

Publisher's Note: MDPI stays neutral with regard to jurisdictional claims in published maps and institutional affiliations.

Copyright: © 2021 by the authors. Licensee MDPI, Basel, Switzerland. This article is an open access article distributed under the terms and conditions of the Creative Commons Attribution (CC BY) license (https://creativecommons.org/licenses/by/4.0/).

Abstract: Integrative taxonomy is an evolving field of multidisciplinary studies often utilised to elucidate phylogenetic reconstructions that were poorly understood in the past. The systematics of many taxa have been resolved by combining data from different research approaches, i.e., molecular, ecological, behavioural, morphological and chemical. Regarding molecular analysis, there is currently a search for new genetic markers that could be diagnostic at different taxonomic levels and that can be added to the canonical ones. In marine Heterobranchia, the most widely used mitochondrial markers, COI and 16S, are usually analysed by comparing the primary sequence. The 16S rRNA molecule can be folded into a 2D secondary structure that has been poorly exploited in the past study of heterobranchs, despite 2D molecular analyses being sources of possible diagnostic characters. Comparison of the results from the phylogenetic analyses of a concatenated (the nuclear H3 and the mitochondrial COI and 16S markers) dataset (including 30 species belonging to eight accepted genera) and from the 2D folding structure analyses of the 16S rRNA from the type species of the genera investigated demonstrated the diagnostic power of this RNA molecule to reveal the systematics of four genera belonging to the family Myrrhinidae (Gastropoda, Heterobranchia). The "molecular morphological" approach to the 16S rRNA revealed to be a powerful tool to delimit at both species and genus taxonomic levels and to be a useful way of recovering information that is usually lost in phylogenetic analyses. While the validity of the genera *Godiva*, *Hermissenda* and *Phyllodesmium* are confirmed, a new genus is necessary and introduced for *Dondice banyulensis*, *Nemesis* gen. nov. and the monospecific genus *Nanuca* is here synonymised with *Dondice*, with *Nanuca sebastiani* transferred into *Dondice* as *Dondice sebastiani* comb. nov.

Keywords: 2D RNA-Barcoding; molecular morphology; Nudibranchia; *Dondice*

1. Introduction

The use of molecular techniques to investigate the evolutionary history of animal groups and to define monophyletic lineages expanded exponentially in the last few decades, becoming one of the essential steps to accomplish a good integrative taxonomy. To increase the robustness of the obtained results, the integration of different kinds of informative characters has led to an exhaustive search for characters that could be diagnostic at different taxonomic levels. Integrative taxonomy applied to marine Heterobranchia is proving to be a fruitful field of study. In fact, in the last few years, many papers have been published, which included information derived from several biological points of view, including morphological, molecular, chemical, ecological and behavioural characters [1–7]. Regarding the molecular approach, there is a continuous search in the genome for new informative DNA coding and non-coding regions, consequentially resulting in molecules such as proteins and RNA [7–9]. Mitochondrial DNA (mtDNA) is currently, in the greatest number of cases, the most powerful genome being exploited to shed light on lower taxonomic levels, such as species and genera, and it remains the most used DNA molecule to delimit species and investigate cryptic diversity and recent speciation events [7,10–12]. In fact, animal mitochondrial DNA is characterised by a high copy number, largely maternal inheritance, lack of recombination, and displays a higher mutation rate than the nuclear DNA. It is affected by some constraints including retention of ancestral polymorphism, male-biased gene flow, selection on any mtDNA nucleotide, introgression following hybridisation and paralogy resulting from transfer of mtDNA gene copies to the nucleus [13–16]. To avoid these pitfalls, nuclear markers are usually added to the molecular analyses. This approach has been applied to marine Heterobranchia (Mollusca, Gastropoda) whose evolutionary history, when studied at family level, is reconstructed using mainly three molecular markers, the two mitochondrial genes, part of Cytochrome oxidase subunit I (COI) and part of the ribosomal subunit 16S (16S), as well as the nuclear gene histone 3 (H3) which is well known to be poorly or not quite informative at lower taxonomic levels [7,17,18]. By means of these markers, the systematics of several heterobranchs families has been clarified and erroneous outcomes derived from previous morphological studies have been resolved and corrected [2,3,5].

However, the potential of some aspects of the known mitochondrial genome are still incompletely exploited today. These include ribosomal RNA genes, which are transcribed into rRNA molecules able to fold into secondary/tertiary structures that are necessary for correct mitoribosome assembly [19]. The primary sequence of some regions of the mitochondrial rRNA is hypervariable and difficult to align among related species. This rRNA hypervariability occurs in the mitochondrial ribosomal 16S RNA (16S rRNA), used as a classical marker, which folds into different domains and stem-loops. When alignments based on the primary sequences are not easy to achieve, it is a common and suggested practice to remove these portions to stabilise the phylogenetic signal (there are specific programs that help to cut these hypervariable regions, for example the GBlocks program [20,21]). The removal of these unreliable alignment regions from the dataset is promoted by several influential works aimed to demonstrate the improvement of statistical support of the phylogenetic analyses when randomness in sequence alignments is cut out [21–23]. However, this practice means that, unavoidably, very informative regions are often not taken into consideration.

In the last few decades, the comparison of the secondary (2D) structure information of RNA molecules, such as the nuclear ribosomal RNA internal transcribed spacer 2 (ITS2 rRNA) in molluscs (bivalves), has been revealed as a promising approach in both phylogenetic reconstruction and species diagnosis [24–27]. In particular, an important character that has proved very useful are Compensatory Base Changes, CBCs, that are defined as two mutations that occur in a paired region of a primary RNA transcript so that pairing itself is maintained (e.g., G-C mutates to A-U) [28]. On the contrary, information derived from the folded structure of the 16S rRNA has not been utilised in molluscan groups, except for two works by Furfaro et al. [18,29], who analysed this barcoding marker in a couple of sympatric sibling sea slugs (gastropods). These authors reported 2D structural diversity of the 16S rRNA, mainly utilising the highly variable L7 and L13 stem-loops [30], demonstrating that secondary RNA structure information is a valuable additional diagnostic tool for integrative taxonomy and species delimitation [18,29].

In this study, we have extended the approach of 2D RNA-barcoding, investigating systematics at a taxonomic level higher than the species level. In particular, it was used as a case study in a polyphyletic clade [31] composed of four genera ranked in the family Myrrhinidae Bergh, 1905: *Dondice* Marcus Er. 1958; *Godiva* Macnae, 1954; *Hermissenda* Bergh, 1879 and *Phyllodesmium* Ehrenberg, 1831. To date, these genera include four accepted species for *Dondice* and *Godiva* respectively, three for *Hermissenda* and 27 accepted species for *Phyllodesmium*, which is the Myrrhinid genus that is richest in species. These Myrrhinid genera were previously considered part of the polyphyletic family Facelinidae, but were recently moved to the family Myrrhinidae by Martynov et al. [32], based on phylogenetic evidence from a systematic study focused on other families (i.e., Tergipedidae).

To date, phylogenetic relationships among these genera and the included species are still incompletely resolved. In fact, there was evidence that the genus *Dondice* as traditionally conceived is not monophyletic and needs an in-depth study based on a wider dataset [31,32]. The family Myrrhinidae includes charismatic species, which is distributed worldwide and interesting from different points of view. *Hermissenda crassicornis* (Eschscholtz, 1831) is a model organism used in various research fields, such as neurology, ecology, ethology, pharmacology and toxicology [33–39]. Species belonging to the genus *Phyllodesmium* are a group of highly specialised nudibranchs with sophisticated species-specific mechanisms of mimicry with external body features indistinguishable from their cnidarian hosts and are animal models for several studies on chemistry and cell biology or focused on the mechanisms of the symbiotic relationship with the zooxanthellae living in their digestive gland cells [40–42]. This family was chosen as a case study to investigate the capability of the 2D folding structure analyses to give useful insight on the systematics of this animal group by looking at alternative morphological, yet molecular, features. This choice was due to the peculiarity of the genera involved, their unresolved phylogenetic histories and a unique ecological defensive strategy shown by most of the species involved (i.e., the capability to autotomise their cerata if disturbed [31]).

Suitable in-group and out-group species were selected, and the definitive dataset defined by being able to describe, with statistically supported analyses, the evolutionary history among representatives of the family Myrrhinidae. Particular attention was given to the inclusion of all the type species of the genera considered in the present study. Phylogenetic analyses were performed, based on an enlarged dataset that includes additional related genera, to investigate the evolutionary history of this interesting group of sea slugs. According to the phylogenetic analysis carried out, the 16S rRNA folding structures of all the type species of the genera belonging to the family Myrrhinidae (*Dondice occidentalis* (Engel, 1925); *Hermissenda opalescens* (J. G. Cooper, 1863); *Godiva quadricolor* (Barnard, 1927); *Phyllodesmium hyalinum* Ehrenberg, 1831] plus *Nanuca sebastiani* Er. Marcus, 1957, currently ascribed to the Family Facelinidae Bergh, 1889 and of *Aeolidiella alderi* (Cocks, 1852), *A. sanguinea* (Norman, 1877) (Aeolidiidae Gray, 1827), *Babakina anadoni* (Ortea, 1979), *B. indopacifica* Gosliner, Gonzalez-Duarte & Cervera, 2007 (Babakinidae Roller, 1973) and *Dicata odhneri* Schmekel, 1967 (Facelinidae Bergh, 1889)) were examined for comparison. After a stem-loop structure analysis of the 16S rRNA molecule, the specific and highly variable L7 stem-loop was chosen as the most divergent and informative for this group and was revealed to be diagnostic to unambiguously discriminate different 16S rRNA structures. This approach, based on the description of the "molecular morphology" of this very variable region of the mitochondrial 16S rRNA [18], can be considered as an additional tool for species delimitation and integrative taxonomy in Heterobranchia and more generally in marine molluscs. Considering all these issues, the aims of the present work were to: (1) provide an updated phylogenetic framework for the genera of the family Myrrhinidae; (2) describe for the first time the consensus 2D secondary structure of the genus *Dondice*; (3) investigate the diagnostic power of the 16S 2D molecular morphology as an additional tool useful for integrative taxonomy and for shedding light on the particularly intricate history of the family Myrrhinidae; (4) introduce *Nemesis* gen. nov. as a new genus of the family Myrrhinidae; (5) rename the taxon *Nanuca sebastiani* to *Dondice sebastiani* comb. nov.

2. Materials and Methods

2.1. Phylogenetic Analyses

The dataset consisted of 47 sequences, retrieved from GenBank, belonging to 30 species, including the out-group (Table 1). The mitochondrial markers COI and 16S and the nuclear marker H3 are commonly used in nudibranch phylogenetics (e.g., [6,17,43]), the mitochondrial ones being highly variable and informative at a shallow level of divergence, and the nuclear one having a slow rate of mutation, and thus, being more suitable to detect deep divergences in the tree, such as those with a distant out-group.

Table 1. Species name, collection localities, Voucher IDs and sequence accession numbers (H3: Histone 3; 16S; COI: Cytochrome Oxidase subunit 1) of the specimens analysed.

SPECIES	LOCALITY	VOUCHER	H3	16S	COI
Aeolidiella alderi (Cocks, 1852)	Italy	ZSMMol20012341	HQ616795	HQ616766	HQ616729
Aeolidiella sanguinea (Norman, 1877)	France (Atlantic Ocean)	MNCN/ADN51932	JX087600	JX087538	JX087466
Babakina anadoni (Ortea, 1979)	Brazil	MNRJ10893	HQ616775	HQ616709	HQ616746
Babakina anadoni (Ortea, 1979)	Galicia, Spain	MNCN15.05/46704	HQ616796	HQ616730	HQ616767
Babakina indopacifica Gosliner, Gonzalez-Duarte & Cervera, 2007	Luzon, Batangas, Philippines	CASIZ177458	HM162587	HM162678	HM162754
Dicata odhneri Schmekel, 1967	Ballanera, Algesiras, Spain	BAU2674	LT596569	LT596549	LT596560
Dicata odhneri Schmekel, 1967	Andalusia, Spain	MNCN15.05/53692		HQ616739	HQ616773
Dondice banyulensis Portmann & Sandmeier, 1960	Djerba, Tunisia	RM3_129	LS483284	LS483274	LS483267
Dondice banyulensis Portmann & Sandmeier, 1960	Argentario, Tuscany, Italy	RM3_356	LS483285	LS483275	LS483268
Dondice banyulensis Portmann & Sandmeier, 1960	Sant'Agostino, Latium, Italy	RM3_290	LS483286	LS483276	LS483269
Dondice banyulensis Portmann & Sandmeier, 1960		Db_60		GQ403751	GQ403773
Dondice occidentalis (Engel, 1925)	Exuma, Bahamas	LACM177715	KC526529	KC526510	
Dondice occidentalis (Engel, 1925)		LACM2003-41.5	JQ699394	JQ699482	JQ699570
Dondice occidentalis (Engel, 1925)	Exuma, Bahamas	D252	KC526527	KC526518	
Dondice occidentalis (Engel, 1925)	Jamaica	JG61	KC526534	KC526512	
Dondice parguerensis Brandon & Cutress, 1985	La Parguera, Puerto Rico	LACM177705	KC526535	KC526520	
Dondice trainitoi Furfaro & Mariottini, 2020	Civitavecchia, Latium, Italy	RM3_425	LS483287	LS483277	LS483270
Dondice trainitoi Furfaro & Mariottini, 2020	Civitavecchia, Latium, Italy	RM3_596	LS483288	LS483278	LS483271
Godiva quadricolor (Barnard, 1927)	Sabaudia, Latium, Italy	RM3_117	LS483289	LS483279	MG546001
Godiva quadricolor (Barnard, 1927)	Sabaudia, Latium, Italy	RM3_153	LS483290	LS483280	MG546002
Godiva quadricolor (Barnard, 1927)	Sabaudia, Latium, Italy	RM3_154	LS483291	LS483281	MG546003
Godiva quadricolor (Barnard, 1927)	Knysna Lagoon, South Africa	CASIZ176385	HM162589	HM162680	HM162756
Hermissenda opalescens	Monterey Bay, CA, USA	isolate_TL270	KU950225	KU950130	KU950196
Hermissenda opalescens	Malibu, CA, USA	isolate_TL275	KU950224	KU950129	KU950195
Hermissenda opalescens	Long Beach, CA, USA	isolate_TL269	KU950222	KU950128	KU950193
Hermissenda emurai	Tateyama-Chiba, Japan	isolate_TL185	KU950215	KU950123	KU950186
Hermissenda emurai	Tateyama-Chiba, Japan	isolate_TL184	KU950214	KU950122	KU950185
Hermissenda crassicornis	Victoria, B.C., Canada	isolate_TL200	KU950210	KU950118	KU950174
Hermissenda crassicornis	Victoria, B.C., Canada	isolate_TL204	KU950212	KU950121	KU950178
Nanuca sebastiani			JQ699469	JQ699557	JQ699633
Phyllodesmium briareum (Bergh, 1896)	Batangas, Philippines	CASIZ 177239	HQ010460	HQ010528	HQ010492
Phyllodesmium colemani Rudman, 1991	Batangas, Philippines	CASIZ 177647	HQ010466	HQ010534	HQ010498
Phyllodesmium crypticum Rudman, 1981	Batangas, Philippines	CASIZ 180381	HQ010477	HQ010543	HQ010507

Table 1. Cont.

SPECIES	LOCALITY	VOUCHER	H3	16S	COI
Phyllodesmium horridum (Macnae, 1954)	Cape Region, South Africa	CASIZ176127	HM162590	HM162681	HM162757
Phyllodesmium hyalinum Ehrenberg, 1831		Phy.orig.		GQ403756	GQ403778
Phyllodesmium jakobsenae Burghardt & Wägele, 2004	Batangas, Philippines	CASIZ 177576	HQ010456	HQ010524	HQ010489
Phyllodesmium karenae Moore & Gosliner, 2009	Batangas, Philippines	CASIZ 180384	HQ010478	HQ010544	HQ010508
Phyllodesmium koehleri Burghardt, Schrödl & Wägele, 2008	Batangas, Philippines	CASIZ 177693	HQ010462	HQ010530	HQ010494
Phyllodesmium lizardensis Burghardt, Schrödl & Wägele, 2008	Batangas, Philippines	CASIZ 180382	HQ010474	HQ010540	HQ010505
Phyllodesmium macphersonae (Burn, 1962)	Batangas, Philippines	CASIZ 177493	HQ010453	HQ010522	HQ010487
Phyllodesmium opalescens Rudman, 1991	Batangas, Philippines	CASIZ 177541	HQ010450	HQ010519	HQ010485
Phyllodesmium parangatum Ortiz & Gosliner, 2003	Batangas, Philippines	CASIZ 180383B	HQ010476	HQ010542	HQ010506
Phyllodesmium poindimiei (Risbec, 1928)	Batangas, Philippines	CASIZ 177783	HQ010463	HQ010531	HQ010495
Phyllodesmium rudmani Burghardt & Gosliner, 2006	Batangas, Philippines	CASIZ 177622	HQ010461	HQ010529	HQ010493
Phyllodesmium tuberculatum Moore & Gosliner, 2009	Batangas, Philippines	CASIZ 177663	HQ010465	HQ010533	HQ010497
Duvaucelia striata Haefelfinger, 1963	Giannutri Is., Tuscany, Italy	BAU2695	LT615407	LT596542	LT596540
Duvaucelia striata Haefelfinger, 1963	Formiche Is., Tuscany, Italy	BAU2696	LT615408	LT596543	LT596541

The selected sequences (Table 1) were aligned using the Muscle algorithm implemented in MEGA 6.0 [44]. The Gblocks 0.91b web server [20,21] was used to remove the hyper-divergent regions of the 16S rDNA alignment, using the less stringent parameter setting. For each gene alignment the best evolutionary model was selected by JModelTest 0.1 [45] according to the Bayesian Information Criterion (BIC).

Downstream phylogenetic analyses were performed on the single 16S dataset and on the concatenated and partitioned dataset. Phylogenetic analyses were performed using Bayesian Inference (BI) and Maximum Likelihood (ML) methods. *Duvaucelia striata* Haefelfinger, 1963 was used as the out-group in BI and ML phylogenetic analyses based on previous tests aimed to find the best in-group and out-group. BI analyses were carried out with MrBayes 3.2.6 [46], implementing the models selected by JModel Test for each gene partition. We ran four Markov-chains of five million generations each, sampled every 1000 generations. Consensus trees were calculated on trees sampled after a burn-in of 25%. MCMC chain convergence was verified by average standard deviation of split frequencies values below 0.006 and confirmed in Tracer 1.7 [47]. Maximum Likelihood analyses were performed in raxmlGUI 1.5b2 [48], a graphical front end for RAxML 8.2.1 [49], with 100 independent ML searches and 1000 bootstrap replicates (command "-f b"), applying the general time-reversible model with a gamma model of rate heterogeneity (GTRGAMMA), with individual gene partitions.

2.2. RNA Secondary Structure Modelling and Compensatory Base Changes (CBCs)

Partial 16S rRNA (sequence region from L7 to L13 stem-loops) secondary structure was obtained using the program "The Mfold web server" [50,51]. The best-supported folding models were predicted combining a thermodynamic approach [52] with a close study of the paired conserved regions and the identification of CBCs and semi-CBCs.

2.3. Morphological Analyses

To provide additional morphological data that were missing in the original description of *Dondice banyulensis*, anatomical analyses of the radula and jaws from two individuals of *D. banyulensis* were carried out. Buccal masses were removed and dissolved in a Proteinase K solution for the extraction of the chitinous structures. Radulae and jaws were rinsed in water, dried and mounted for examination by optical microscopy. The buccal structures were then mounted and gold coated in an Emitech K550 unit for SEM analysis. To obtain high resolution pictures, SEM images at different magnification levels were performed using the JSM-6480LV Scanning Electron Microscope (JEOL Ltd., Tokyo, Japan) at the Laboratorio di Microscopia Elettronica, Department of Mathematics and Physics, University of Salento. Analyses of the reproductive system of *D. banyulensis* were carried out by anatomical dissection under a stereomicroscope.

3. Results

3.1. Phylogenetic Analyses

The definitive dataset consisted of 134 sequences from 47 specimens belonging to 30 different species, including the out-group (Figure 1, Table 1). The single 16S dataset consisted of 350 nucleotides, while the final concatenated alignment was 1232 nucleotides long. The best evolutionary models selected by JModelTest 0.1 according to the Bayesian Information Criterion (BIC) were: TIM1 + I + G, HKY + I + G and TIM2ef + G for the COI, 16S and H3 alignments, respectively. Results from Bayesian and Maximum Likelihood analyses yielded congruent topologies (Figure 1) both grouping all the *Phyllodesmium* species in a single and strongly supported (BI = 1; ML = 100) monophyletic clade, sister to *Godiva quadricolor* (BI = 1; ML = 100) with a low statistical support (BI = 75; ML = 45). These two clades are sister to another monophyletic group (BI = 1; ML = 81), which includes one monophyletic clade (BI = 1; ML = 100) that grouped species such as *Dondice occidentalis* and *D. parguerensis* (BI = 0.99; ML = 100) and *D. trainitoi* (BI = 1; ML = 97) sister to *Nanuca sebastiani* (BI = 1; ML = 100), and another monophyletic clade, which includes all the species belonging to the genus *Hermissenda* (BI = 1; ML = 100). *Dondice banyulensis* is the sister to all this big monophyletic clade (BI = 0.91; ML = 70), with strong statistical support (BI = 1; ML = 100). The two *Aeolidiella* sister species (BI = 1; ML = 100) are sister to all the previously reported clades forming a large monophyletic group (BI = 0.99; ML = 69). The latter group is sister to *Babakina* spp. (BI = 73; ML = 56) with *Dicata odhneri* (BI = 1; ML = 100) basal to all the previously reported clades and with the out-group *Duvaucelia striata* as the basal sister group to all the mentioned species. BI and ML analyses of the single 16S dataset yielded congruent results but with low statistical support (Figure S1).

3.2. 16S RNA Primary and Secondary Structures Analysis

The mitochondrial 16S rRNA multiple sequence alignments were performed simultaneously considering the secondary structure of each sequence to obtain the optimisation of the final alignment. The 16S rRNA single gene alignment consisted of 407 positions and 47 sequences, obtained from 30 species belonging to 9 accepted genera, including the out-group (Table 1). We have analysed the secondary structure (2D) of the 3' half portion of the 16S rRNA gene, corresponding to the V domain and including the L7-L13 stem-loops [30,53] (Figure S3), in order to highlight the different diagnostic nucleotides characterising genus and species [54]. No variability in the folding 2D structures was detected within the same species. As expected, the V domain showed a high conservation of the primary sequence and global folding when compared among the species analysed (Table 1), which conforms to the canonical molluscan architecture [18,24,25,29,30,53]. Among the variable stem-loops, the L7 was the most divergent rRNA sequence containing diagnostic nucleotides and it was taken into consideration as an RNA barcoding region [18,29]. In Figure 2, the folding models of the 16S rRNA V domain are displayed of eight representative species belonging to seven genera. The L7 stem-loops analysed range from 14 (*G. quadricolor*) to 23 (*B. anadoni*) nucleotides (nt) in length and share a high homology of the sequence/2D structure among three *Dondice* spp. (i.e., *D. occidentalis, D. parguerensis* and *D. trainitoi*) and the monotypic *N. sebastiani* (Figure 3). As shown in Figure 3, the stem-loop is composed of 19 nucleotides in these taxa, the stem composed of five base pairs is identical, while the nine nt long loop shows six identical and three variable nucleotides. Unexpectedly, the L7 stem-loop of *Dondice banyulensis* revealed to be not correlated to the one of the congeneric species when considering both the primary sequence and 2D structure, exhibiting no significant folding homology (Figure 3). Moreover, in Figure 3 the L7 stem-loops of *H. opalescens, G. quadricolor, P. hyalinum, A. alderi* and *B. anadoni* are depicted for comparison. The secondary structure of the *H. opalescens* L7 stem-loop, chosen as an example of the genus *Hermissenda* since it is the type species of the genus, reveals a close homology in sequence/folding with the ones of *D. occidentalis* and *N. sebastiani*, which are the type species of *Dondice* and *Nanuca* genera, respectively; the nucleotide sequence of the shorter stem in particular is identical, while the eight nucleotides-long loop shares six identical nucleotides (Figure 3). Interestingly, two groups of nucleotide positions can be observed in the L7 stem-loop of the members grouped in the *Dondice* clade, which is formed by three taxa (Figure 3 and Figure S4): the first, variable and diagnostic for species identification and the second conserved and diagnostic at the genus level. Three diagnostic nucleotide substitutions separate *Dondice* spp. and the closely related *Nanuca* taxon, indicating that the latter monotypic genus is strongly related to *Dondice*. On the other hand, four nucleotide positions are conserved and shared in the clade which includes both genera *Dondice* and *Nanuca*. Taking into account the folding models of the L8-11 stem-loops, the 2D structures of these hairpins revealed a closer homology between *D. occidentalis* and *N. sebastiani* (Figure S3) than with *D. banyulensis*. In Figure S3, the L8-11 stem-loops are circled to show both the 2D structure and the nucleotides differences occurring among the three taxa. In particular, despite the dissimilar loop size of the L10 stem-loop, due to a single mismatch in the stem, *D. occidentalis* and *N. sebastiani* share a higher 2D folding homology than with *D. banyulensis*. Regarding the type species of the *Godiva* genus, the *G. quadricolor* L7 stem-loop is slightly divergent in sequence and 2D structure, but still exhibiting five identical nucleotides in its 10 nucleotides-long loop. The folding homology decreases analysing the 2D structures of the L7 stem-loop of *P. hyalinum*, a species type of the genus *Phyllodesmium* (four identical nucleotide in the 12 nt long stem). Two other examples of the L7 stem-loop refer to *A. alderi* and *B. anadoni*, which both show very divergent 2D structures.

To summarise, the L7 stem-loop of *D. banyulensis* revealed to be the more divergent 2D structure compared to the ones of *Dondice* spp., *N. sebastiani, H. opalescens, G. quadricolor* and *P. hyalinum*, exhibiting no significant folding homology (Figure 3 and Figure S4). According to both phylogenetic and 16S rRNA 2D structure analyses, the monotypic genus *Nanuca* must be synonymised with *Dondice*, and consequently, the taxon *Nanuca sebastiani* can be renamed to *Dondice sebastiani* comb. nov.

3.3. Morphological Analyses

Morphological analyses of the buccal apparatus from two specimens (Vouchers RM3_290 and RM3_311) belonging to *Dondice banyulensis* (Figure 4) are shown in Figure 5. The high-resolution SEM pictures obtained showed that features of the jaws and radula are congruent with drawings reported in the original description [55]. The reproductive system was examined and the resulted drawing congruent with the one reported from the holotype (Figure S2).

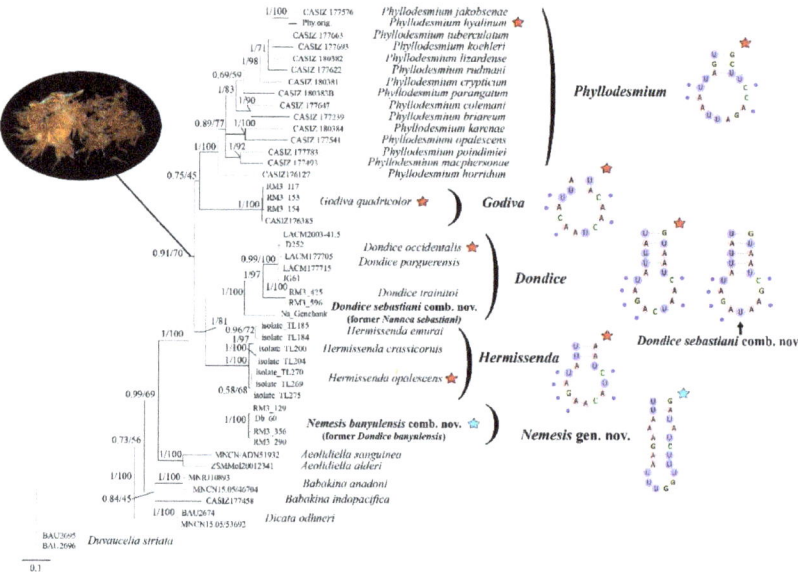

Figure 1. Bayesian phylogenetic tree based on the concatenated dataset (H3, 16S, COI). Bayesian posterior probability (**left**) and Bootstrap (**right**) values are indicated at each node. Red stars highlight the type species of the genera analysed and the relative 16S 2D L7 structures. The blue stars refer to the type species of the *Nemesis* gen. nov. and its 16S rRNA 2D L7 structure. In the black oval at the top left, *G. quadricolor* is shown as well as the autotomy of its cerata as the defensive strategy characteristic and common to the monophyletic group which includes *Dondice*, *Godiva*, *Hermissenda* and *Phyllodesmium*.

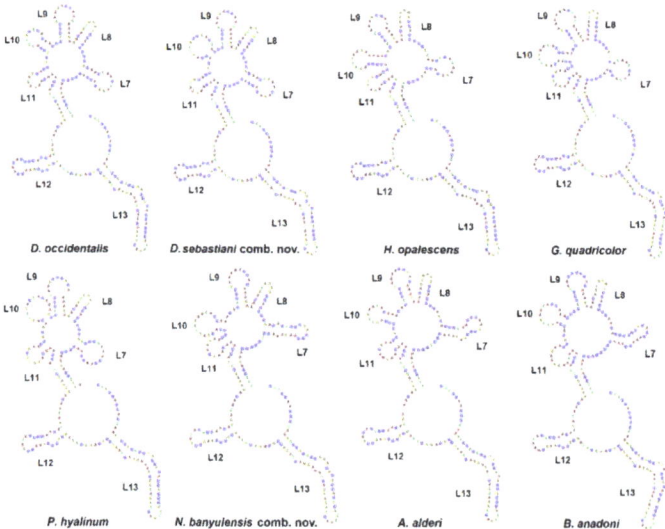

Figure 2. Folding models of the 16S rRNA V domain of the representative species analysed in the present work.

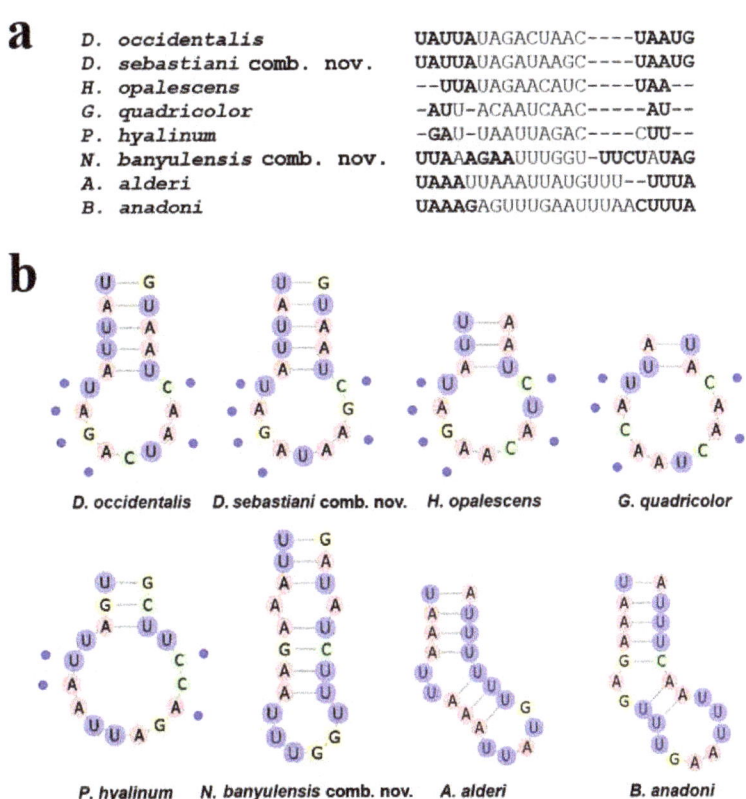

Figure 3. L7 stem-loops of the species here analysed. (**a**) Primary sequence alignment of the L7 stem-loops among the species analysed in the present work. (**b**) Blue circles indicate the conserved nucleotide positions in *Dondice*, *Nanuca*, *Hermissenda*, *Godiva* and *Phyllodesmium* genera.

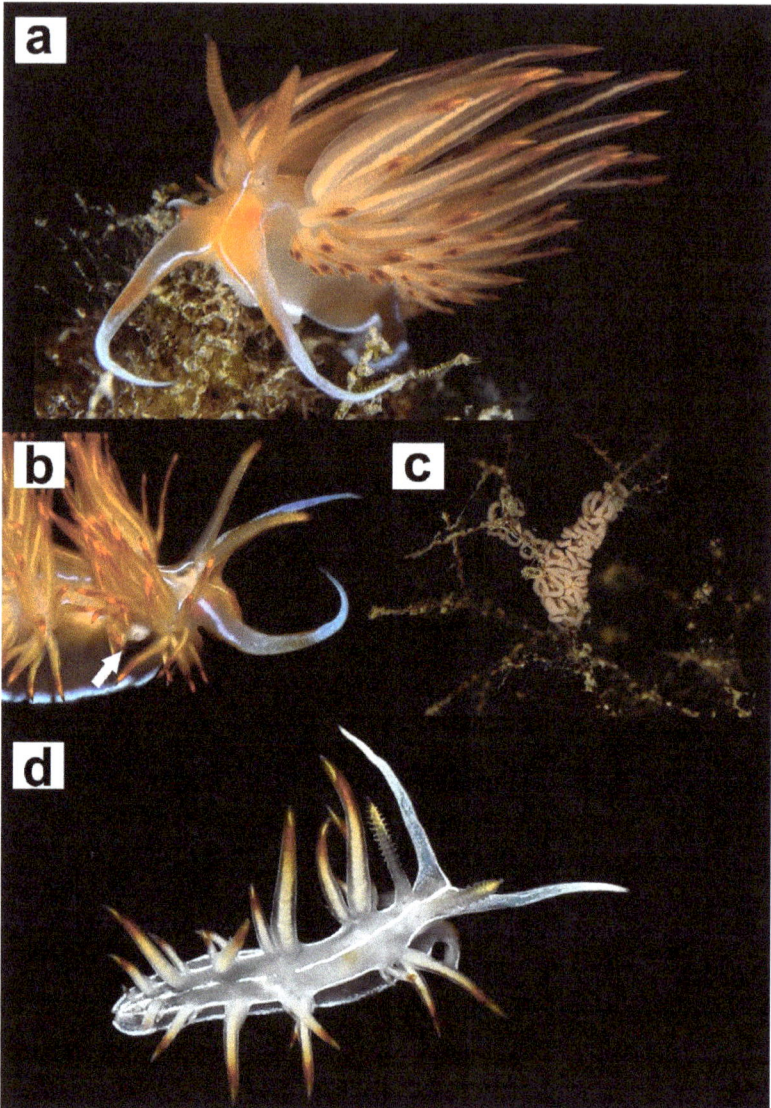

Figure 4. Images of individuals belonging to *Nemesis banyulensis* comb. nov. (original name *Dondice banyulensis*) (**a**,**b**,**d**) and the typical egg ribbon (**c**) of the species. (**a**) Adult specimen from the 'Peschereccio' wreck, Gallipoli, Apulia (Central Tyrrhenian Sea, Mediterranean Sea). (**b**) Particular of the anterior right portion of one specimen from 'Asia' wreck in Civitavecchia (Central Tyrrhenian Sea, Mediterranean Sea) and its reproductive openings highlighted with a white arrow. (**c**) The typical orange to pinkish egg ribbon lead on a hydroid colony in Sant' Agostino, Civitavecchia (Central Tyrrhenian Sea, Mediterranean Sea). (**d**) Picture of a juvenile individual from Scilla in Calabria (South Tyrrhenian Sea, Mediterranean Sea) showing the difference in the chromatic pattern from young to adult individuals.

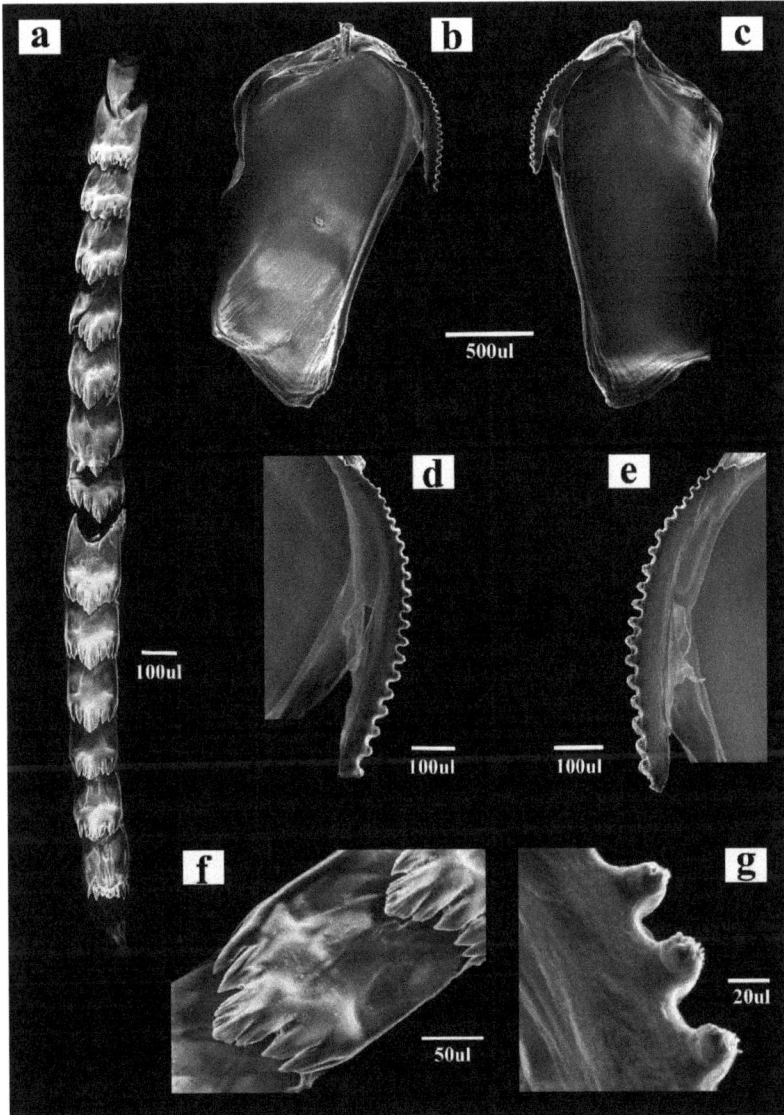

Figure 5. SEM images of the jaws and radula of *Nemesis banyulensis* comb. nov. (former *Dondice banyulensis*) (voucher RM3_290) (**a**), the entire radula of (**b**,**c**) both the left (**b**) and the right (**c**) jaws (**d**,**e**) particular of the masticatory boarders of the jaws. (**f**) Detail of the tooth of the radula with the cusp and the lateral denticles clearly visible. (**g**) The denticulated denticles of the masticatory border of the jaws at a high magnification level.

4. Discussion

Based on these results, *Dondice banyulensis* clearly represents an independent lineage from the species of *Dondice* s.s. and deserves a genus on its own.

Taxonomy
Family Myrrhinidae Bergh, 1905
Nemesis gen. nov.
urn:lsid:zoobank.org:act:8F4E1E1D-6E8A-41CB-81FA-487A58829C65

Figures 4 and 5 and Figure S1

Type species. *Dondice banyulensis* Portmann & Sandmeier, 1960

Etymology. The genus name was chosen to recall the Greek goddess *Nemesis* and her sense of justice to allocate the former *D. banyulensis* in the appropriate genus.

Diagnosis. Myrrhinid with rhinophores annulate and produced anterior foot corners. It has jaws with a single row of denticles that can be denticulated in their apical part. The central cusp of the radular tooth is not marked and is a little longer than lateral denticles. The large stomach, followed by a short smooth intestine, terminates with the anus, which is cleioproct and located inside the second group of cerata on the right side of the body. The long distal and proximal deferent ducts of the male portion of the reproductive system terminates with the unarmed penis. It is not able to autotomise the cerata when stressed by possible predators.

Species included. *Nemesis banyulensis* (Portmann & Sandmeier, 1960), comb. nov. (Original description in [55]; additional data in Figures 4 and 5 and Figure S2).

Remarks. The new genus is diagnosed by a radular tooth characterised by the cuspid that is slightly marked and a little longer than lateral denticles compared to *Dondice* s.s., *Godiva* and *Phyllodesmium* and by the more elongated distal and proximal male deferent ducts of the reproductive system compared to *Dondice* s.s., *Godiva* and *Phyllodesmium*.

Mitochondrial DNA is, to date, the most powerful DNA molecule able to discriminate at species and genus taxonomic levels and is necessary to carry out any integrative systematic study [15,56–60]. In fact, even if today a lot of work is being done to find markers that are alternative to and independent from mitochondrial DNA, such as nuclear markers, it is undeniable that mitochondrial DNA is currently the most informative at lower taxonomic levels. Evidence for this is that nowadays there are two markers that are commonly used in most of the nudibranch systematic studies: the mitochondrial COI and the 16S. While waiting to find useful nuclear alternatives for the study of the evolutionary relationships that occurred between highly related groups, it may be useful to improve and optimise the study of the currently available markers, especially if these have characteristics that are potentially very useful; however, little is known of these characteristics in nudibranchs. One of these features is the ability of the 16S rRNA to fold into a 2D structure, which is mandatory for correct mitoribosome assembly [19]. In the mitochondrial 16S rRNA molecule, the V domain includes the L7-L13 stem-loops [30,53] that can be used to search for diagnostic nucleotides [54]. The 2D structure analysis using the 16S rRNA in molluscs has revealed to be capable to discriminate a couple of sympatric sibling sea slugs [18,29]. In particular, the stem-loop L9 nucleotide analyses confirmed the separation of the species *Diaphorodoris alba* Portmann & Sandmeier, 1960 and *D. luteocincta* (M. Sars, 1870), revealing some diagnostic nucleotides producing a 2D structural diversity [18]. Moreover, the L7 and L13 stem-loops resulted to be diagnostic in separating the two sibling Mediterranean species *Caloria elegans* (Alder and Hancock, 1845) and *Facelina quatrefagesi* (Vayssière, 1888) [29].

The case of the family Myrrhinidae is particularly intriguing due to the importance of the charismatic species included and because the genera belonging to this family were the object of recent integrative taxonomic studies, which, however, could not resolve important aspect of their systematics [31,32]. The latter studies have highlighted the possible polyphyletic nature of the Myrrhinidae genera that invoked more in-depth studies to resolve open questions. For this reason, in the present work, a broad dataset, including more individuals and species, was investigated using another, until now poorly known, source of information: the 2D structure of the 16S rRNA marker. In particular, we analysed the L7 stem-loop folding models since this small region proved to be very variable (Figures 2 and 3, Figures S3 and S4). Two groups of nucleotide positions can be observed in the L7 stem-loop of the members grouped in *Dondice* clade (Figure 1): the first group is more variable and was diagnostic for species identification, while the second one includes nucleotides shared by all the taxa of the clade, and for this reason, could be very useful as a source of landmarks able to delimit the genus (Figure S4). Differences in nucleotides usually occur within the loop region and not in the stem, since the latter is under base-pairing constraints, which is what happens in the L7 stem-loop of *Dondice* spp. where no CBCs nor semiCBCs have been observed. The ability showed by the 2D folding structures of the 16S, and, in particular, by its small L7 loop, to give useful information at both species and genus taxonomic levels is quite important and reported here for the first time also at the genus level. This consideration becomes even more evident if we consider that, to improve results from phylogenetic analyses and from other methods of species delimitation, we are forced to exclude the hypervariable and unreliable regions of the 16S marker from our dataset, as strongly suggested by authoritative works [21–23]. The possibility to recover such a small but very significative region that would be otherwise lost constitutes an opportunity that must be considered.

The results from the phylogenetic approach and 16S rRNA 2D structures analyses are congruent with each other and yield a similar scenario, demonstrating the power of integrating these two methods. According to the results obtained, it is proposed that the species *Nanuca sebastiani*, currently included in the family Facelinidae and in the monospecific genus *Nanuca*, should be included in the genus *Dondice* (and consequently, moved to the family Myrrhinidae) as *Dondice sebastiani* comb. nov., with the genus *Nanuca* regarded as a synonym of *Dondice*. On the contrary, *Dondice banyulensis*, historically included in the genus *Dondice*, must be removed from the latter genus and ascribed as the type species of the new genus *Nemesis* gen. nov. here described, as *Nemesis banyulensis* comb. nov. Some very important ecological and morphological characters strongly support *Nemesis* gen. nov. within the Myrrhinidae. First of all, there is an ethological character (which indeed reflects the structural physiological difference) that appeared in the family after *Nemesis banyulensis* speciated, which became the synapomorphy of all the other apical genera (i.e., *Hermissenda*, *Dondice*, *Godiva* and *Phyllodesmium*): the ability to autotomise single or groups of cerata when under a predator's attack or when in stressed conditions (showed in the black circle in Figure 1). This behaviour was a fundamental turning point in the evolution of these clades, as it constituted a new defensive strategy that was effective in increasing the survival skills of these shell-less sea slugs; in situ and laboratory observations confirm that *Nemesis banyulensis* does not perform cerata autotomy [31]. From a morphological and anatomical points of view, there are two other features that support the validity of *Nemesis* gen. nov.: (i) in the new genus, the radula lacks a prominent cusp, which is, instead, characteristic of the related genera (i.e., *Dondice* (see [31], Figure 4F), *Hermissenda, Godiva* and *Phyllodesmium}*. The median cusp of the radular tooth in *Nemesis* gen. nov. is not accentuated (Figure 5f); (ii) in *Nemesis* gen. nov., the distal and proximal male duct of the reproductive system is very long (Figure S2), which is typically different from other related genera (e.g., *Dondice* and *Godiva* see [31] Figure 5F). The results from morphological, ecological, molecular and 16S 2D structure analyses were congruent with each other, revealing the power of integrating evidence from several methods. In particular, the 'molecular morphology' approach revealed to be effective in this group of eolid nudibranchs, being able to give information at species and genera levels by looking at few diagnostic nucleotides. In fact, the diagnostic nucleotides associated to the 16S rRNA L7 stem-loop can be considered landmarks useful to separate at both genus and species levels in the family Myrrhinidae.

5. Conclusions

We have analysed the secondary structure of the 3' half portion of the 16S rRNA gene in the representative species of five genera, i.e., *Dondice, Nanuca, Godiva, Hermissenda* and *Phyllodesmium*, searching for short variable regions displaying diagnostic nucleotides. Among the variable 16S rRNA stem-loops, the L7 resulted to be the most divergent and it was taken into consideration as a short 2D RNA barcoding region. The 2D folding analysis of this 19 nucleotides-long region produced the same phylogenetic relationship obtained with the Bayesian and Maximum Likelihood analysis on the concatenated H3, 16S and COI alignment (1232 nucleotides), pointing out the resolution power of the "molecular morphology" approach when integrated with the standard use of the primary sequence.

We confirm the validity of the genera *Godiva*, *Hermissenda* and *Phyllodesmium* as traditionally conceived, while we propose to exclude *Dondice banyulensis* from the genus *Dondice* and to assign it as the type species of the new genus *Nemesis* gen. nov. Furthermore, the monospecific genus *Nanuca* is here synonymised with the *Dondice* genus, and consequently, the taxon *Nanuca sebastiani* is renamed to *Dondice sebastiani* comb. nov.

Supplementary Materials: The following are available online at https://www.mdpi.com/article/10.3390/life11060583/s1, Figure S1: Bayesian phylogenetic tree based on the single 16S dataset. Bayesian posterior probability (left) and Bootstrap (right) values are indicated at each node. Figure S2: Drawing of the reproductive system of *Nemesis banyulensis* comb. nov. (Voucher RM3_290) performed using optical microscope. Legend: amp = ampulla, bc = bursa copulatory, dd = deferent duct, fgm = female gland mass, md = male duct, p = penis, rs = *receptaculum seminis*, v = vagina. Figure S3: (a) Folding models of the 16S rRNA V domain of *D. occidentalis*, *D. sebastiani* comb. nov. and *N. banyulensis* comb. nov. analysed in the present work. The L7-11 stem-loops are boxed. (b) The variable L9-11 stem-loops are encircled to show differences occurring among the three taxa. Figure S4: L7 stem-loops of the species here analysed. (a) Primary sequence alignment of the L7 stem-loops among the *Dondice* spp. and *N. banyulensis* comb. nov. (b) Folding models of the L7 stem-loops. Blue circles indicate conserved nucleotide positions within the *Dondice* clade and red circles indicate diagnostic nucleotides.

Author Contributions: Conceptualisation, G.F. and P.M.; Data curation, G.F.; Formal analysis, G.F. and P.M.; Investigation, G.F.; Methodology, P.M.; Resources, G.F.; Supervision, P.M.; Writing—original draft, G.F. and P.M.; Writing—review and editing, G.F. and P.M. Both authors have read and agreed to the published version of the manuscript.

Funding: The APC was funded by the Italian Ministry of Education, grant AIM 1848751-2, Linea 2.

Institutional Review Board Statement: Not applicable.

Informed Consent Statement: Not applicable.

Data Availability Statement: Data available in a publicly accessible repository that does not issue DOIs. Publicly available datasets were analysed in this study. This data can be found here: [https://www.ncbi.nlm.nih.gov/].

Acknowledgments: The authors are indebted to Marcella D'Elia (Department of Mathematics and Physics of the University of Salento, Lecce, Italy) for the SEM pictures of the buccal structures and to Michele Solca (Milan, Italy) for the underwater pictures of living specimens. The authors wish to thank Marco Oliverio (La Sapienza University, Rome, Italy) and Egidio Trainito (Marine Protected Area 'Tavolara Punta Coda Cavallo', Olbia, Italy) for their useful suggestions that improved the quality of the present paper. Bernard Picton (Northern Ireland, UK) is acknowledged for the English language editing of the manuscript. Authors wish to thank the three anonymous reviewers for improving the quality of the manuscript. G.F. and P.M. received financial support from the University of Salento and the University Roma Tre. G.F. is supported by funds from the Italian Ministry of Education, University and Research (MIUR, PON 2014-2020, grant AIM 1848751-2, Linea 2).

Conflicts of Interest: The authors declare no conflict of interest. The funder had no role in the design of the study; in the collection, analyses or interpretation of data; in the writing of the manuscript, or in the decision to publish the results.

References

1. Goodheart, J.A.; Bleidißel, S.; Schillo, D.; Strong, E.; Ayres, D.L.; Preisfeld, A.; Collins, A.G.; Cummings, M.P.; Wägele, H. Comparative morphology and evolution of the cnidosac in Cladobranchia (Gastropoda: Heterobranchia: Nudibranchia). *Front. Zool.* **2018**, *15*, 1–18. [CrossRef]
2. Ekimova, I.; Korshunova, T.; Schepetov, D.; Neretina, T.; Sanamyan, N.; Martynov, A. Integrative systematics of northern and Arctic nudibranchs of the genus *Dendronotus* (Mollusca, Gastropoda), with descriptions of three new species. *Zool. J. Linn. Soc.* **2015**, *173*, 841–886. [CrossRef]
3. Padula, V.; Bahia, J.; Stöger, I.; Camacho-García, Y.; Malaquias, M.A.E.; Cervera, J.L.; Schrödl, M. A test of color-based taxonomy in nudibranchs: Molecular phylogeny and species delimitation of the *Felimida clenchi* (Mollusca: Chromodorididae) species complex. *Mol. Phylogenetics Evol.* **2016**, *103*, 215–229. [CrossRef] [PubMed]

4. Schillo, D.; Wipfler, B.; Undap, N.; Papu, A.; Boehringer, N.; Eisenbarth, J.H.; Waegele, H. Description of a new *Moridilla* species from North Sulawesi, Indonesia (Mollusca: Nudibranchia: Aeolidioidea)—Based on MicroCT, histological and molecular analyses. *Zootaxa* **2019**, *4652*, 265–295. [CrossRef] [PubMed]
5. Fernández-Vilert, R.; Giribet, G.; Salvador, X.; Moles, J. Assessing the systematics of Tylodinidae in the Mediterranean Sea and Eastern Atlantic Ocean: Resurrecting *Tylodina rafinesquii* Philippi, 1836 (Heterobranchia: Umbraculida). *J. Molluscan Stud.* **2021**, *87*, eyaa031. [CrossRef]
6. Furfaro, G.; Salvi, D.; Mancini, E.; Mariottini, P. A multilocus view on Mediterranean aeolid nudibranchs (Mollusca): Systematics and cryptic diversity of Flabellinidae and Piseinotecidae. *Mol. Phylogenetic Evol.* **2018**, *118*, 13–22. [CrossRef]
7. Furfaro, G.; Salvi, D.; Trainito, E.; Vitale, F.; Mariottini, P. When morphology does not match phylogeny: The puzzling case of two sibling nudibranchs (Gastropoda). *Zool. Scr.* **2021**, 1–16. [CrossRef]
8. Dayrat, B. Towards integrative taxonomy. *Biol. J. Linn. Soc.* **2005**, *85*, 407–417. [CrossRef]
9. Will, K.W.; Mishler, B.D.; Wheeler, Q.D. The perils of DNA barcoding and the need for integrative taxonomy. *Syst. Biol.* **2005**, *54*, 844–851. [CrossRef]
10. Pola, M.; Cervera, J.L.; Gosliner, T.M. Phylogenetic relationships of Nembrothinae (Mollusca: Doridacea: Polyceridae) inferred from morphology and mitochondrial DNA. *Mol. Phylogenetics Evol.* **2007**, *43*, 726–742. [CrossRef]
11. Pola, M.; Camacho-Garía, Y.E.; Gosliner, T.M. Molecular data illuminate cryptic nudibranch species: The evolution of the Scyllaeidae (Nudibranchia: Dendronotina) with a revision of *Notobryon*. *Zool. J. Linn. Soc.* **2012**, *165*, 311–336. [CrossRef]
12. Furfaro, G.; Modica, M.V.; Oliverio, M.; Mariottini, P. A DNA-barcoding approach to the phenotypic diversity of Mediterranean species of *Felimare* Ev. Marcus & Er. Marcus, 1967 (Mollusca: Gastropoda), with a preliminary phylogenetic analysis. *Ital. J. Zool.* **2016**, *83*, 195–207. [CrossRef]
13. Bensasson, D.; Zhang, D.X.; Hartl, D.L.; Hewitt, G.M. Mitochondrial pseudogenes: Evolution's misplaced witnesses. *Trends Ecol. Evol.* **2001**, *16*, 314–321. [CrossRef]
14. Ballard, J.W.O.; Whitlock, M.C. The incomplete natural history of mitochondria. *Mol. Ecol.* **2004**, *13*, 729–744. [CrossRef] [PubMed]
15. Hebert, P.D.N.; Stoeckle, M.Y.; Zemlak, T.S.; Francis, C.M. Identification of birds through DNA barcodes. *PLoS Biol.* **2004**, *2*, e312. [CrossRef] [PubMed]
16. Moritz, C.; Cicero, C. DNA Barcoding: Promise and Pitfalls. *PLoS Biol.* **2004**, *2*, e354. [CrossRef]
17. Galià-Camps, C.; Carmona, L.; Cabrito, A.; Ballesteros, M. Double trouble. A cryptic first record of *Berghia marinae* Carmona, Pola, Gosliner, & Cervera 2014 in the Mediterranean Sea. *Mediterr. Mar. Sci.* **2020**, *21*, 191–200. [CrossRef]
18. Furfaro, G.; Picton, B.; Martynov, A.; Mariottini, P. *Diaphorodoris alba* Portmann & Sandmeier, 1960 is a valid species: Molecular and morphological comparison with *D. luteocincta* (M. Sars, 1870). *Zootaxa* **2016**, *4193*, 304–316. [CrossRef]
19. Greber, B.J.; Ban, N. Structure and Function of the Mitochondrial Ribosome. *Annu. Rev. Biochem.* **2016**, *85*, 103–132. [CrossRef]
20. Castresana, J. Selection of conserved blocks from multiple alignments for their use in phylogenetic analysis. *Mol. Biol. Evol.* **2000**, *17*, 540–552. [CrossRef]
21. Talavera, G.; Castresana, J. Improvement of phylogenies after removing divergent and ambiguously aligned blocks from protein sequence alignments. *Syst. Biol.* **2007**, *56*, 564–577. [CrossRef] [PubMed]
22. Privman, E.; Penn, O.; Pupko, T. Improving the performance of positive selection inference by filtering unreliable alignment regions. *Mol. Biol. Evol.* **2012**, *29*, 1–5. [CrossRef] [PubMed]
23. Kück, P.; Meusemann, K.; Dambach, J.; Thormann, B.; von Reumont, B.M.; Wägele, J.W.; Misof, B. Parametric and non-parametric masking of randomness in sequence alignments can be improved and leads to better resolved trees. *Front. Zool.* **2010**, *7*, 1–12. [CrossRef]
24. Salvi, D.; Bellavia, G.; Cervelli, M.; Mariottini, P. The analysis of rRNA Sequence-Structure in phylogenetics: An application to the family Pectinidae (Mollusca, Bivalvia). *Mol. Phylogenetics Evol.* **2010**, *56*, 1059–1067. [CrossRef]
25. Salvi, D.; Macali, A.; Mariottini, P. Molecular phylogenetics and systematics of the bivalve family Ostreidae based on rRNA sequence-structure models and multilocus species tree. *PLoS ONE* **2014**, *9*, e108696. [CrossRef] [PubMed]
26. Salvi, D.; Mariottini, P. Molecular phylogenetics in 2D: ITS2 rRNA evolution and sequence-structure barcode from Veneridae to Bivalvia. *Mol. Phylogenetics Evol.* **2012**, *65*, 792–798. [CrossRef]
27. Salvi, D.; Mariottini, P. Molecular taxonomy in 2D: A novel ITS2 rRNA sequence-structure approach guides the description of the oysters' subfamily Saccostreinae and the genus *Magallana* (Bivalvia: Ostreidae). *Zool. J. Linn. Soc.* **2017**, *179*, 263–276. [CrossRef]
28. Müller, T.; Philippi, N.; Dandekar, T.; Schultz, J.; Wolf, M. Distinguishing species. *RNA* **2007**, *13*, 1469–1472. [CrossRef]
29. Furfaro, G.; Mariottini, P.; Modica, M.V.; Trainito, E.; Doneddu, M.; Oliverio, M. Sympatric sibling species: The case of *Caloria elegans* and *Facelina quatrefagesi* (Gastropoda: Nudibranchia). *Sci. Mar.* **2016**, *80*, 511–520. [CrossRef]
30. Lydeard, C.; Holznagel, W.E.; Schnare, M.N.; Gutell, R.R. Phylogenetic analysis of molluscan mitochondrial LSU rDNA sequences and secondary structures. *Mol. Phylogenetics Evol.* **2000**, *15*, 83–102. [CrossRef]
31. Furfaro, G.; Mariottini, P. A new *Dondice* Marcus Er. 1958 (Gastropoda: Nudibranchia) from the Mediterranean Sea reveals interesting insights into the phylogenetic history of a group of Facelinidae taxa. *Zootaxa* **2020**, *477731*, 1–22. [CrossRef] [PubMed]
32. Martynov, A.; Mehrotra, R.; Chavanich, S.; Nakano, R.; Kashio, S.; Lundin, K.; Korshunova, T. The extraordinary genus *Myja* is not a tergipedid, but related to the Facelinidae s. str. with the addition of two new species from Japan (Mollusca, Nudibranchia). *ZooKeys* **2019**, *818*, 89. [CrossRef]

33. Neary, J.T.; Alkon, D.L. Protein phosphorylation/dephosphorylation and the transient, voltage-dependent potassium conductance in *Hermissenda crassicornis*. *J. Biol. Chem.* **1983**, *258*, 8979–8983. [CrossRef]
34. Croll, R.P. Distribution of monoamines in the central nervous system of the nudibranch gastropod, *Hermissenda crassicornis*. *Brain Res.* **1987**, *405*, 337–347. [CrossRef]
35. Blackwell, K.T.; Alkon, D.L. Ryanodine receptor modulation of in vitro associative learning in *Hermissenda crassicornis*. *Brain Res.* **1999**, *822*, 114–125. [CrossRef]
36. Kasheverov, I.E.; Shelukhina, I.V.; Kudryavtsev, D.S.; Makarieva, T.N.; Spirova, E.N.; Guzii, A.G.; Tsetlin, V.I. 6-Bromohypaphorine from marine nudibranch mollusk *Hermissenda crassicornis* is an agonist of human α7 nicotinic acetylcholine receptor. *Mar. Drugs* **2015**, *13*, 1255–1266. [CrossRef]
37. Tamvacakis, A.N.; Senatore, A.; Katz, P.S. Identification of genes related to learning and memory in the brain transcriptome of the mollusc, *Hermissenda crassicornis*. *Learn. Mem.* **2015**, *22*, 617–621. [CrossRef]
38. Lindsay, T.; Valdés, Á. The Model Organism *Hermissenda crassicornis* (Gastropoda: Heterobranchia) Is a Species Complex. *PLoS ONE* **2016**, *11*, e0154265. [CrossRef]
39. Merlo, E.M.; Milligan, K.A.; Sheets, N.B.; Neufeld, C.J.; Eastham, T.M.; Estores-Pacheco, A.K.A.; Wyeth, R.C. Range extension for the region of sympatry between the nudibranchs *Hermissenda opalescens* and *Hermissenda crassicornis* in the northeastern Pacific. *Facets* **2018**, *3*, 764–776. [CrossRef]
40. Wagner, D.; Kahng, S.E.; Toonen, R.J. Observations on the life history and feeding ecology of a specialized nudibranch predator (*Phyllodesmium poindimiei*), with implications for biocontrol of an invasive octocoral (*Carijoa riisei*) in Hawaii. *J. Exp. Mar. Biol. Ecol.* **2009**, *372*, 64–74. [CrossRef]
41. Mao, S.C.; Gavagnin, M.; Mollo, E.; Guo, Y.W. A new rare asteriscane sesquiterpene and other related derivatives from the Hainan aeolid nudibranch *Phyllodesmium magnum*. *Biochem. Syst. Ecol.* **2011**, *39*, 408–411. [CrossRef]
42. Bogdanov, A.; Hertzer, C.; Kehraus, S.; Nietzer, S.; Rohde, S.; Schupp, P.J.; Wägele, H.; König, G.M. Secondary metabolome and its defensive role in the aeolidoidean *Phyllodesmium longicirrum*, (Gastropoda, Heterobranchia, Nudibranchia). *Beilstein J. Org. Chem.* **2017**, *13*, 502–519. [CrossRef]
43. Ekimova, I.A.; Antokhina, T.I.; Schepetov, D.M. Molecular data and updated morphological description of *Flabellina rubrolineata* (Nudibranchia: Flabellinidae) from the Red and Arabian seas. *Ruthenica* **2020**, *30*, 183–194.
44. Tamura, K.; Stecher, G.; Peterson, D.; Filipski, A.; Kumar, S. MEGA6: Molecular evolutionary genetics analysis version 6.0. *Mol. Biol. Evol.* **2013**, *30*, 2725–2729. [CrossRef] [PubMed]
45. Posada, D. jModelTest: Phylogenetic Model Averaging. *Mol. Biol. Evol.* **2008**, *7*, 1253–1256. [CrossRef]
46. Ronquist, F.; Teslenko, M.; Van Der Mark, P.; Ayres, D.; Darling, A.; Höhna, S.; Larget, B.; Liu, L.; Suchard, M.A.; Huelsenbeck, J.P. MrBayes 3.2: Efficient Bayesian phylogenetic inference and model choice across a large model space. *Syst. Biol.* **2011**, *61*, 539–542. [CrossRef] [PubMed]
47. Rambaut, A.; Drummond, A.J.; Xie, D.; Baele, G.; Suchard, M.A. Posterior summarization in Bayesian phylogenetics using Tracer 1.7. *Syst. Biol.* **2018**, *67*, 901. [CrossRef] [PubMed]
48. Silvestro, D.; Michalak, I. raxmlGUI: A graphical front-end for RAxML. *Org. Divers. Evol.* **2012**, *12*, 335–337. [CrossRef]
49. Stamatakis, A. RAxML version 8: A tool for phylogenetic analysis and post-analysis of large phylogenies. *Bioinformatics* **2014**, *30*, 1312–1313. [CrossRef]
50. Zuker, M.; Jacobson, A.B. Using reliability information to annotate RNA secondary structures. *RNA* **1998**, *4*, 669–679. [CrossRef]
51. Zuker, M. Mfold web server for nucleic acid folding and hybridization prediction. *Nucleic Acids Res.* **2003**, *31*, 3406–3415. [CrossRef] [PubMed]
52. Mathews, D.H.; Sabina, J.; Zuker, M.; Turner, D.H. Expanded sequence dependence of thermodynamic parameters improves prediction of RNA secondary structure. *J. Mol. Biol.* **1999**, *288*, 911–940. [CrossRef] [PubMed]
53. Horovitz, I.; Meyer, A. Systematics of New World monkeys (Platyrrhini, Primates) based on 16S mitochondrial DNA sequences: A comparative analysis of different weighting methods in cladistic analysis. *Mol. Phylogenetics Evol.* **1995**, *4*, 448–456. [CrossRef]
54. Jörger, K.M.; Norenburg, J.L.; Wilson, N.G.; Schrödl, M. Barcoding against a paradox? Combined molecular species delineations reveal multiple cryptic lineages in elusive meiofaunal sea slugs. *BMC Evol. Biol.* **2012**, *12*, 245. [CrossRef] [PubMed]
55. Portmann, A.; Sandmeier, E. *Dondice banyulensis* sp. nov. un Eolidien nouveau de la Méditerranée. *Rev. Suisse Zool.* **1960**, *67*, 158–168. [CrossRef]
56. Caterino, M.S.; Cho, S.; Sperling, F.A. The current state of insect molecular systematics: A thriving Tower of Babel. *Annu. Rev. Entomol.* **2000**, *45*, 1–54. [CrossRef]
57. Hebert, P.D.N.; Cywinska, A.; Ball, S.L.; de Waard, J.R. Biological identifications through DNA barcodes. *Proc. R. Soc. Lond. Ser. B* **2003**, *270*, 313–321. [CrossRef]
58. Hebert, P.D.N.; Ratnasingham, S.; de Waard, J.R. Barcoding animal life: Cytochrome c oxidase subunit 1 di-vergencesamong closely related species. *R. Soc. Lond. Ser. B* **2003**, *270*, S96–S99.
59. Stoeckle, M. Taxonomy, DNA, and the bar code of life. *BioScience* **2003**, *53*, 2–3. [CrossRef]
60. Avise, J.C. *Molecular Markers, Natural History and Evolution*, 2nd ed.; Sinauer: Sunderland, UK, 2004.

Communication

Nemesignis, a Replacement Name for *Nemesis* Furfaro & Mariottini, 2021 (Mollusca, Gastropoda, Myrrhinidae), Preoccupied by *Nemesis* Risso, 1826 (Crustacea, Copepoda)

Giulia Furfaro [1,*] and Paolo Mariottini [2]

1 Department of Biological and Environmental Sciences and Technologies—DiSTeBA, University of Salento, I-73100 Lecce, Italy
2 Department of Science, University of Roma Tre, I-00146 Rome, Italy; paolo.mariottini@uniroma3.it
* Correspondence: giulia.furfaro@unisalento.it; Tel.: +39-0832-29-8660

Abstract: The genus *Nemesis* Furfaro & Mariottini, 2021, was recently introduced for an independent lineage of aeolid nudibranchs, and *Dondice banyulensis* Portmann & Sandmeier, 1960, established as its type species. Anyway, the presence of a senior homonym, *Nemesis* Risso, 1826, was evidently missed. In fact, in 1826, Risso established this genus for a group of Copepoda (Arthropoda, Crustacea) and according to the Principle of Priority (ICZN) only the senior homonym may be used as a valid name. Therefore, a new replacement name is here proposed. Furthermore, the genus name *Nanuca* Er. Marcus, 1957, has priority over *Dondice* Er. Marcus, 1958 and consequently, the species in this clade should be classified under *Nanuca*, mostly as new combinations.

Keywords: ICZN; *Dondice*; homonym; Heterobranchia

1. Introduction

The genus *Nemesis* Furfaro & Mariottini, 2021, was introduced based on evidence from a recent integrative systematic study [1] (Furfaro & Mariottini, 2021) for an independent lineage of aeolid nudibranchs, and *Dondice banyulensis* Portmann & Sandmeier, 1960 [2], was established as the type species. The newly identified lineage is currently monospecific and characterized by (i) the central cusp of the radular tooth that is not marked and a little longer than lateral denticles, (ii) the long distal and proximal deferent ducts of the male portion of the reproductive system and (iii) its inability to autotomise the cerata when stressed by possible predators. Just after the publication of our paper, Luigi Romani (Lucca, Italy) sent to us a letter (e-mail: 18.06.2021) where he noted that in our recent manuscript, we have evidently missed the existence of a senior homonym, *Nemesis* Risso, 1826 [3] (International Commission on Zoological Nomenclature - ICZN, 1999: Article 53.2) [4]. In fact, in 1826, Risso established this genus for a group of Copepoda (Arthropoda, Crustacea) and according to the Principle of Priority (ICZN, 1999: Article 52.3) [4], when two or more names are homonyms, only the senior may be used as a valid name. Therefore, a new replacement name is here proposed under the Article 60.3 of ICZN. Furthermore, Philippe Bouchet (Paris, France) pointed out to us that *Nanuca* Er. Marcus, 1957 [5], has priority over *Dondice* Er. Marcus, 1958 [6]. Consequently, the species in this clade should be classified under *Nanuca*, mostly as new combinations.

2. Results and Discussion

Taxonomy
Familia Myrrhinidae Bergh, 1905 [7]
Genus *Nemesignis* **nom. nov.** pro *Nemesis* Furfaro & Mariottini 2021 (non Risso, 1826)
urn:lsid:zoobank.org:pub:DEED49D6-F89B-4D68-A8A3-E1071197264C
Type species. *Dondice banyulensis* Portmann & Sandmeier, 1960.

Etymology. The genus name *Nemesignis* comes from the union of the Greek word *Nemesis*, that recalls the homonymous Greek goddess and her role of compensatory justice, with the Latin word *Ignis*, that is the fire that burns and blazes, linked to the fiery red colour of the type species of the genus.

Included species. *N. banyulensis* (Portmann & Sandmeier, 1960).

Genus *Nanuca* Er. Marcus, 1957

Type species. *Nanuca sebastiani* Er. Marcus, 1957

= *Dondice* Er. Marcus, 1958 (type species: *Caloria occidentalis* Engel, 1925 [8])

Included species. *Nanuca sebastiani* Er. Marcus, 1957, *Nanuca galaxiana* (Millen & Hermosillo, 2012) **comb. nov.** [9], *N. occidentalis* (Engel, 1925) **comb. nov.**, *N. parguerensis* (Brandon & Cutress, 1985) **comb. nov.** [10], *N. trainitoi* (Furfaro & Mariottini, 2020) **comb. nov.** [11].

3. Conclusions

The presence of a senior homonym, which has priority over the recently stated *Nemesis* Furfaro & Mariottini, 2021, made this latter genus name as invalid and invoked the need for a replacement name according to the rules of the ICZN. Therefore, *Nemesignis* **nom. nov.** is here proposed as the new replacement name, under the Article 60.3 of ICZN and consequently, *Nemesignis banyulensis* (Portmann & Sandmeier, 1960) is its type species. Finally, since *Nanuca* Er. Marcus, 1957 has priority over *Dondice* Er. Marcus, 1958, the species in this clade should be classified under *Nanuca*, as *Nanuca galaxiana* (Millen & Hermosillo, 2012) **comb. nov.**, *N. occidentalis* (Engel, 1925) **comb. nov.**, *N. parguerensis* (Brandon & Cutress, 1985) **comb. nov.**, *N. trainitoi* (Furfaro & Mariottini, 2020) **comb. nov.**

Author Contributions: Conceptualization, G.F. and P.M.; validation, G.F. and P.M.; writing—original draft preparation, G.F.; writing—review and editing, P.M.; supervision, G.F. and P.M. Both authors have read and agreed to the published version of the manuscript.

Funding: This research received no external funding.

Institutional Review Board Statement: Not applicable.

Acknowledgments: We are grateful to Luigi Romani (Lucca, Italy) for alerting us of this homonymy and to Philippe Bouchet (Paris, France) for his suggestions. We thank the two anonymous reviewers that helped to improve the manuscript. Finally, we would like to deeply thank Marco Oliverio (Rome, Italy) for his constant help.

Conflicts of Interest: The authors declare no conflict of interest.

References

1. Furfaro, G.; Mariottini, P. Looking at the nudibranch family Myrrhinidae (Gastropoda, Heterobranchia) from a mitochondrial '2D folding structure' point of view. *Life* **2021**, *11*, 583. [CrossRef] [PubMed]
2. Portmann, A.; Sandmeier, E. *Dondice banyulensis* sp. nov. un Eolidien nouveau de la Méditerranée. *Rev. Suisse Zool.* **1960**, *67*, 159–168. [CrossRef]
3. Risso, A. *Histoire Naturelle des Principales Productions de l'Europe Mèridionale et Particulièrement de Celles des Environs de Nice et des Alpes Maritimes*; F.G. Levrault: Paris, France, 1826; Volume 5.
4. ICZN [International Commission on Zoological Nomenclature]. *International Code of Zoological Nomenclature*, 4th ed.; International Trust for Zoological Nomenclature: London, UK, 1999; p. xxix + 306.
5. Marcus, E. On opisthobranchia from Brazil (2). *Zool. J. Linn. Soc.* **1957**, *43*, 390–486. [CrossRef]
6. Marcus, E. On western Atlantic opisthobranchiate gastropods. *Am. Mus. Novit.* **1958**, *1906*, 1–82.
7. Bergh, L.S.R. Die Opisthobranchiata der Siboga-expedition. *Siboga-Exped.* **1905**, *50*, 1–248.
8. Engel, H. Westindische opisthobranchiate Mollusken. Bijdragen tot de kennis der fauna van Curaçao. Resultaten eener reis van Dr. C. J. van der Horst in 1920. *Bijdr. Tot Dierkd.* **1925**, *24*, 33–80. [CrossRef]
9. Millen, S.V.; Hermosillo, A. Three new species of aeolid nudibranchs (Opisthobranchia) from the Pacific coast of Mexico, Panama and the Indopacific, with a redescription and redesignation of a fourth species. *Veliger* **2012**, *51*, 145–164.
10. Brandon, M.; Cutress, C.E. A new *Dondice* (Opisthobranchia: Favorinidae) predator of *Cassiopea* in southwest Puerto Rico. *Bull. Mar. Sci.* **1985**, *36*, 139–144.
11. Furfaro, G.; Mariottini, P. A new *Dondice* Marcus Er. 1958 (Gastropoda: Nudibranchia) from the Mediterranean Sea reveals interesting insights into the phylogenetic history of a group of Facelinidae taxa. *Zootaxa* **2020**, *4731*, 1–22. [CrossRef] [PubMed]

Article

Mitochondrial DNA Analysis Clarifies Taxonomic Status of the Northernmost Snow Sheep (*Ovis nivicola*) Population

Arsen V. Dotsev [1,*], Elisabeth Kunz [2], Veronika R. Kharzinova [1], Innokentiy M. Okhlopkov [3], Feng-Hua Lv [4], Meng-Hua Li [4], Andrey N. Rodionov [1], Alexey V. Shakhin [1], Taras P. Sipko [5], Dmitry G. Medvedev [6], Elena A. Gladyr [1], Vugar A. Bagirov [1], Gottfried Brem [1,7], Ivica Medugorac [2] and Natalia A. Zinovieva [1,*]

1 L.K. Ernst Federal Science Center for Animal Husbandry, 142132 Moscow, Russia; veronika0784@mail.ru (V.R.K.); rodiand@yandex.ru (A.N.R.); alexshahin@mail.ru (A.V.S.); elenagladyr@mail.ru (E.A.G.); vugarbagirov@mail.ru (V.A.B.); gottfried.brem@agrobiogen.de (G.B.)
2 Population Genomics Group, Department of Veterinary Sciences, LMU Munich, 80539 Munich, Germany; Elisabeth.Kunz@gen.vetmed.uni-muenchen.de (E.K.); Ivica.Medjugorac@gen.vetmed.uni-muenchen.de (I.M.)
3 Institute for Biological Problems of Cryolithozone, 677000 Yakutsk, Russia; imo-ibpc@yandex.ru
4 College of Animal Science and Technology, China Agricultural University, Beijing 100193, China; lvfenghua@cau.edu.cn (F.-H.L.); menghua.li@ioz.ac.cn (M.-H.L.)
5 A.N. Severtsov Institute of Ecology and Evolution of the Russian Academy of Sciences, 119071 Moscow, Russia; sipkotp@mail.ru
6 Department of Game Management and Bioecology, Irkutsk State University of Agriculture, 664038 Irkutsk, Russia; dmimedvedev@yandex.ru
7 Institut für Tierzucht und Genetik, University of Veterinary Medicine (VMU), A-1210 Vienna, Austria
* Correspondence: asnd@mail.ru (A.V.D.); n_zinovieva@mail.ru (N.A.Z.); Tel.: +7-49-6765-1104 (A.V.D.); +7-49-6765-1404 (N.A.Z.)

Citation: Dotsev, A.V.; Kunz, E.; Kharzinova, V.R.; Okhlopkov, I.M.; Lv, F.-H.; Li, M.-H.; Rodionov, A.N.; Shakhin, A.V.; Sipko, T.P.; Medvedev, D.G.; et al. Mitochondrial DNA Analysis Clarifies Taxonomic Status of the Northernmost Snow Sheep (*Ovis nivicola*) Population. *Life* **2021**, *11*, 252. https://doi.org/10.3390/life11030252

Academic Editors: Andrea Luchetti and Federico Plazzi

Received: 19 February 2021
Accepted: 16 March 2021
Published: 18 March 2021

Publisher's Note: MDPI stays neutral with regard to jurisdictional claims in published maps and institutional affiliations.

Copyright: © 2021 by the authors. Licensee MDPI, Basel, Switzerland. This article is an open access article distributed under the terms and conditions of the Creative Commons Attribution (CC BY) license (https://creativecommons.org/licenses/by/4.0/).

Abstract: Currently, the intraspecific taxonomy of snow sheep (*Ovis nivicola*) is controversial and needs to be specified using DNA molecular genetic markers. In our previous work using whole-genome single nucleotide polymorphism (SNP) analysis, we found that the population inhabiting Kharaulakh Ridge was genetically different from the other populations of Yakut subspecies to which it was usually referred. Here, our study was aimed at the clarification of taxonomic status of Kharaulakh snow sheep using mitochondrial cytochrome b gene. A total of 87 specimens from five different geographic locations of Yakut snow sheep as well as 20 specimens of other recognized subspecies were included in this study. We identified 19 haplotypes, two of which belonged to the population from Kharaulakh Ridge. Median-joining network and Bayesian tree analyses revealed that Kharaulakh population clustered separately from all the other Yakut snow sheep. The divergence time between Kharaulakh population and Yakut snow sheep was estimated as 0.48 ± 0.19 MYA. Thus, the study of the mtDNA *cytb* sequences confirmed the results of genome-wide SNP analysis. Taking into account the high degree of divergence of Kharaulakh snow sheep from other groups, identified by both nuclear and mitochondrial DNA markers, we propose to classify the Kharaulakh population as a separate subspecies.

Keywords: wild sheep; bighorn; taxonomy; mtDNA; cytochrome b; Yakut snow sheep; *Ovis nivicola lydekkeri*

1. Introduction

Snow sheep (*Ovis nivicola*) (Video S1) is an endemic species in northeastern Siberia and the Russian Far East. Along with the North American mountain sheep (bighorn) (*Ovis canadensis*) and Dall sheep (*Ovis dalli*), it is referred to the subgenus *Pachyceros* of the genus *Ovis* [1]. The intraspecific taxonomy of snow sheep is highly controversial and currently remains insufficiently studied [2–4]. While some scientists believe that snow sheep should not be divided into subspecies, but rather be considered as a subspecies of North American wild sheep [5], others distinguish up to seven subspecies [6]. According to the Chernyavsky classification [4], which is also officially recognized by the International Council for Game and Wildlife Conservation (CIC), four subspecies are distinguished including Kamchatka *(O. n. nivicola)*, Koryak *(O. n. koriakorum)*, Putorana *(O. n. borealis)* and Yakut *(O. n. lydekkeri)* [7].

To date, the taxonomic divisions of snow sheep *(O. nivicola)* into subspecies have been based on the morphological differences between individuals. However, morphological characteristics are influenced by both the origin of the animals and the environmental conditions (food supply, temperature, etc.). Therefore, the use of this approach can be misleading in the determination of population structure. Thus, differences between individuals of different origins can be leveled while they are in similar environmental conditions. On the contrary, animals of the same or similar genetic origin can develop distinctive characteristics due to their adaptation to various environmental conditions.

The investigation of DNA polymorphisms makes it possible to identify "true" genetic differences between individuals, and it is becoming more widely used in research on issues of biological systematics [8]. The most commonly used methodological approach is to study the polymorphisms of mitochondrial DNA (mtDNA). The advantage of mtDNA is the absence of recombination and the maternal type of inheritance as well as the possibility to extract mtDNA from the small amounts of biological samples since animal cells contain numerous copies of it. For example, using mtDNA, three species of tahr were assigned to separate monotypic genera-*Hemitragus jemlahicus*, *Nilgiritragus hylocrius* and *Arabitragus jayakari*, while by morphological characteristics they were classified as the different subspecies of the same species [9]. The development of DNA chips for simultaneous analysis of several thousand or even hundreds of thousands of single-nucleotide polymorphisms (SNPs) in the genome of farm animals made it possible to use this highly informative tools to study the genomes of related wild species [10,11], including the bighorn [12] and snow sheep [13]. In the latter work, it was revealed that populations of Yakut snow sheep *(O. n lydekkeri)*, traditionally considered as a single subspecies, were represented by individuals with different origins. In particular, a different origin of the snow sheep inhabiting the most northern area of the snow sheep habitats—the Kharaulakh Ridge—from other populations of the Yakut snow sheep was established [13]. The additional research using other types of DNA variability is necessary to confirm the status of Kharaulakh sheep as an independent subspecies. Because of broad applications in phylogenetic studies and recognitions of results by the scientific community, the mtDNA polymorphisms are the most suitable type of sequence variability for this purpose. The use of mitochondrial DNA is of particular interest in the study of *O. nivicola* due to its maternal heritability. It is a well-known fact that snow sheep females are much less migratory than the males. As a rule, ewes remain in the habitat where they were born while rams can migrate for distances up to 100–150 km to search for females [3].

Thus, our present work was aimed at clarifying the taxonomic status of the Kharaulakh population of snow sheep based on the analysis of the polymorphism of the mitochondrial cytochrome B (*cytb*) gene.

2. Materials and Methods

The animals of five different geographic locations of Yakut snow sheep *(O. n. lydekkeri)* including Kharaulakh Ridge-Tiksi bay (TIK, $n = 21$), Orulgan (ORU, $n = 25$), Central Verkhoyansk (VER, $n = 23$), Suntar-Khayata (SKH, $n = 11$) and Momsky (MOM, $n = 7$) ridges as well as the samples of the other three most recognized subspecies, including Kamchatka (KAM, $n = 9$), Koryak (KOR, $n = 8$), Putorana (PUT, $n = 3$), were selected for the study (Figure 1). Muscle tissue samples of snow sheep were collected under permits issued by the Department of Hunting of the Republic Sakha (Yakutia) during scientific expeditions. Some samples were taken from trophy hunters and indigenous peoples' representatives, who are licensed to hunt snow sheep for personal consumption according to the Federal Law of the Russian Federation. We were able to obtain only a few samples, which are suitable for deriving DNA of the appropriate quality, from the Red Book Putorana subspecies from animals which died due to natural causes.

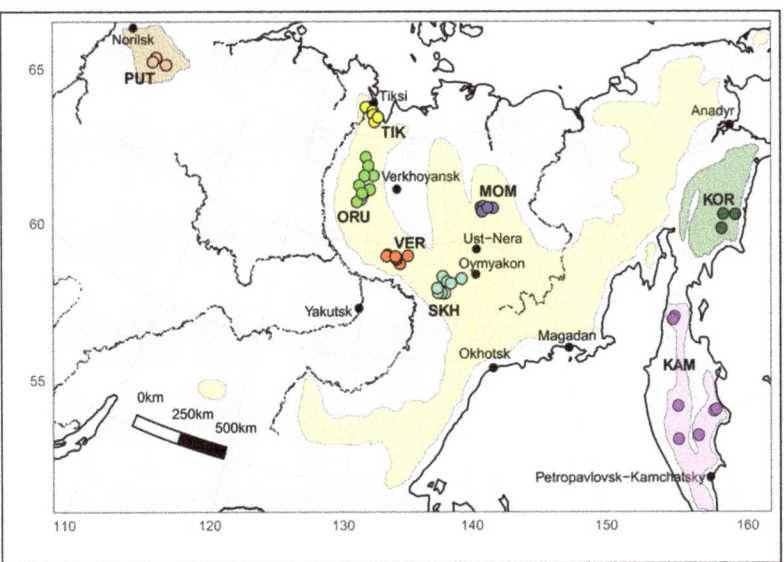

Figure 1. Map with sampling sites of snow sheep used in this study and the area of the species habitat. TIK = Kharaulakh Ridge–Tiksi Bay, ORU = Orulgan, VER = Central Verkhoyansk, SKH = Suntar-Khayata, MOM = Momsky, KAM = Kamchatka snow sheep, KOR = Koryak, PUT = Putorana.

DNA extraction was carried out using Nexttec columns (Nexttec Biotechnology GmbH, Leverkusen, Germany) in accordance with the manufacturer's recommendations. The whole sequences of *cytb* gene of Yakut snow sheep were determined by Sanger sequencing of two overlapping fragments (overlap area of about 50 bp). The *cytb* sequences of Kamchatka, Koryak, and Putorana snow sheep were defined using next generation sequencing (NGS) technology [14]. For this purpose, three overlapping mtDNA fragments were amplified (overlapping region more than 290 bp) with a length of 6.5, 5.7, and 6.7 kb. The obtained polymerase chain reaction (PCR) products were purified and used to prepare the libraries for sequencing, which were then sequenced by 100 bp paired-end procedure on a HiSeq 1500 (Illumina). To verify that the results obtained from Sanger sequencing and NGS technology are comparable, a part of Yakut snow sheep samples ($n = 20$) were sequenced by the two methods. It was shown that all the compared sequences were identical. Mitos WebServer [15] was used to annotate the mitochondrial genome. The *cytb* sequence was recovered from the complete mtDNA sequence after its alignment, performed using the MUSCLE algorithm [16] in the MEGA 7.0.26 software [17].

The *cytb* gene sequences of North American wild sheep were downloaded from the GenBank NCBI database (www.ncbi.nlm.nih.gov (accessed on 15 December 2020)): Dall sheep *(O. dalli)* (ODA, $n = 3$), Rocky Mountain bighorn *(O. canadensis canadensis)* (OCC, $n = 8$) and desert bighorn sheep *(O. canadensis nelsoni)* (OCN, $n = 11$) were added to the final dataset. A complete list of these samples with their GenBank NCBI accession numbers is presented in Table 1.

Table 1. List of North American wild sheep mtDNA *cytb* gene sequences downloaded from the GenBank NCBI database (www.ncbi.nlm.nih.gov (accessed on 15 December 2020)).

#	Species	n	GenBank Accession Number	References
1	Rocky Mountain bighorn sheep (*Ovis canadensis canadensis*)	8	EU365985, EU366063, EU366064, EU366065, EU366066, EU366067, FJ936176, FJ936177	Rezaei H.R. et al. [18]
2	Desert bighorn sheep (*Ovis canadensis nelsoni*)	11	EU366059, EU366060, EU366061, EU366062, FJ936178, FJ936179, FJ936180, FJ936181, FJ936182, FJ936183,	Rezaei H.R. et al. [18]
			HM222706	Naidu A. et al. [19]
3	Dall sheep (*Ovis dalli*)	3	MH779627	Dotsev A. et al. [14]
			EU365992, FJ936184	Rezaei H.R. et al. [18]

To construct a median joining network [20], PopART 1.7 software [21] was used. For Bayesian phylogenetic reconstruction, one sample with the most frequent haplotype from each population was selected. The population from Orulgan Ridge was represented by two samples belonging to the central Verkhoyansk-VER/ORU and to the Kharaulakh Range-TIK/ORU haplotypes. The construction of the Bayesian phylogenetic tree was carried out using the program BEAST 2.5 [22] with subsequent visualization in FigTree 1.4.2 (http://tree.bio.ed.ac.uk/software/figtree (accessed on 15 December 2020)). Determination of the best models of evolution was carried out separately for each nucleotide in the program PartitionFinder 2 [23] using the Akaike information corrected criterion (AICc) [24]. The most optimal were the evolutionary models HKY + I, HKY and HKY + G, respectively, for the first, second and third codons of the *cytb* gene. Calculations of pairwise F_{ST} as well as AMOVA analysis were carried out in the program Arlequin 3.5.2.2 [25]. Genetic distances based on the Kimura-2-parameter model [26] were calculated using the MEGA 7.0.26 software. Further, based on these distances, a Neighbor-Net phylogenetic tree was constructed in the SplitsTree 4.14.6 program [27]. DnaSP 6.12.01 program [28] was used to calculate genetic diversity parameters: number of polymorphic sites (S), average number of nucleotide differences (K), number of haplotypes (H), haplotype diversity (Hd), nucleotide diversity (π). We also tested the hypothesis of population expansion calculating Fu's neutrality statistic Fs [29] and Tajima's D [30] test in DnaSP 6.12.01. Demographic histories of snow sheep populations were inferred by pairwise mismatch distribution analyses [31] and computed under a constant population size model in DnaSP 6.12.01.

Divergence time was estimated using BEAST 2.5 software. To calibrate molecular clock, we added a prior to the model assuming that the two species of the outgroup–*O. canadensis* and *O. dalli* have been split 1.5 million years ago (MYA). This calibration node was retrieved as median time of divergence from the TimeTree web resource (http://www.timetree.org (accessed on 15 December 2020)) [32].

R packages "maps" [33] and "ggplot2" [34] were used to create maps with sampling sites.

3. Results and Discussion

In order to clarify the taxonomic status of the Kharaulakh population of snow sheep, which, according to the results of SNP genome-wide analysis was proposed as an independent subspecies [13], we performed a study of the whole cytochrome B (*cytb*) gene sequence in four mostly recognized subspecies of snow sheep including the Yakut (*O. n. lydekkeri*), Kamchatka (*O. n. nivicola*), Koryak (*O. n. koriakorum*) and Putorana (*O. n. borealis*). The *cytb* sequences of Dall sheep (*O. dalli*) and two subspecies of the bighorn sheep (*O. canadensis canadensis*) and (*O. canadensis nelsoni*) were included in the final dataset as outgroups.

In total, we identified 19 haplotypes in 107 samples of snow sheep. Moreover, *O. n. koriakorum* and *O. n. borealis*, as well as the *O. n. lydekkeri* population from the Momsky Ridge were each represented by a single haplotype. The values of genetic diversity parameters (Table 2) in Yakut snow sheep populations were consistent with data obtained in a study based on SNP markers. Thus, haplotype diversity (Hd) on the Verkhoyansk Range increased southwards, from 0.381 ± 0.101 in Kharaulakh to 0.746 ± 0.098 in the Suntar-Khayata population. The lowest values of nucleotide diversity (π) and the average number of nucleotide differences (K) were also observed in the Kharaulakh population of snow sheep.

Table 2. Genetic diversity and neutrality indices of snow sheep populations calculated from nucleotide sequence of mitochondrial *cytb* gene.

Population	n	S	K (±SD)	H	Hd (±SD)	π (±SD)	Tajima's D	Fu's Fs
TIK	21	1	0.381 ± 0.375	2	0.381 ± 0.101	0.00033 ± 0.00037	0.65593	0.94374
ORU	25	5	1.893 ± 1.117	3	0.507 ± 0.075	0.00166 ± 0.00109	1.23135	3.39988
VER	23	7	1.636 ± 1.002	5	0.640 ± 0.065	0.00144 ± 0.00098	−0.43378	0.34957
SKH	11	8	3.018 ± 1.703	4	0.746 ± 0.098	0.00265 ± 0.00169	0.43451	1.79212
MOM	7	0	0	1	0	0	-	-
KAM	9	7	2.722 ± 1.592	5	0.861 ± 0.087	0.00239 ± 0.00158	0.25402	−0.16693
KOR	8	0	0	1	0	0	-	-
PUT	3	0	0	1	0	0	-	-

Population: TIK = Kharaulakh Ridge—Tiksi Bay, ORU = Orulgan, VER = Central Verkhoyansk, SKH = Suntar-Khayata, MOM = Momsky, KAM = Kamchatka, KOR = Koryak, PUT = Putorana; n—number of samples; S—number of polymorphic sites; K—average number of nucleotide differences; SD—standard deviation; H—number of haplotypes; Hd—haplotype diversity; π—nucleotide diversity; Tajima's D—value of Tajima's neutrality test, Fu's Fs—the value of Fu's neutrality test.

The results of Tajima's D and Fu's Fs tests were statistically insignificant in all the populations, so we could not reject the null hypothesis, which indicates deviations from neutrality and suggests recent population expansion.

The median joining network (Figure 2) showed that the group inhabiting the mountains of Kamchatka–*O. n. nivicola* was the most separated from the others and was the closest to the historical ancestor of the modern snow sheep. All the other samples were derived from a single haplotype currently occurring in the central Verkhoyansk Range—VER and on the Orulgan Ridge—ORU. Among the samples from the Kharaulakh Ridge, we identified two haplotypes, which differed from each other by a single mutation. One of these haplotypes was also found in animals from the Orulgan Ridge. Thus, the presence in ORU population of haplotypes which are differed from each other by four nucleotide substitutions confirms the admixed origin of this population, as it was revealed by the study on multiple SNP markers [13]. A hypothetical haplotype, linking the population from the Kharaulakh Ridge with the Putorana snow sheep, was established.

Genetic distances based on the Kimura-2-parameter model (Table 3) confirmed the most genetic distance of the Kamchatka snow sheep from the other studied groups.

Table 3. Genetic distances between the studied snow sheep populations and North American wild sheep.

Population	TIK	ORU	VER	SKH	MOM	KOR	KAM	PUT	OCC	OCN	ODA
TIK	*	0.002	0.005	0.008	0.005	0.006	0.012	0.006	0.025	0.022	0.025
ORU	0.438	*	0.004	0.006	0.004	0.004	0.010	0.005	0.024	0.021	0.024
VER	0.831	0.570	*	0.004	0.003	0.003	0.009	0.005	0.023	0.021	0.024
SKH	0.850	0.661	0.560	*	0.006	0.005	0.011	0.007	0.025	0.025	0.028
MOM	0.950	0.689	0.651	0.741	*	0.004	0.010	0.006	0.024	0.022	0.025
KOR	0.959	0.718	0.665	0.714	1.000	*	0.009	0.005	0.023	0.021	0.024
KAM	0.919	0.813	0.804	0.771	0.868	0.865	*	0.011	0.020	0.020	0.021
PUT	0.950	0.704	0.741	0.708	1.000	1.000	0.834	*	0.026	0.025	0.028
OCC	0.972	0.924	0.933	0.913	0.967	0.968	0.904	0.958	*	0.012	0.013
OCN	0.972	0.934	0.941	0.926	0.966	0.967	0.912	0.959	0.769	*	0.006
ODA	0.981	0.929	0.935	0.899	0.981	0.982	0.884	0.966	0.876	0.886	*

Below diagonal are shown pairwise F_{ST} and above diagonal genetic distances (based on the Kimura-2-parameter model). The significant F_{ST} values are represented in bold ($p < 0.05$). Population: TIK = Kharaulakh Ridge–Tiksi Bay, ORU = Orulgan, VER = Central Verkhoyansk, SKH = Suntar-Khayata, MOM = Momsky, KAM = Kamchatka, KOR = Koryak, PUT = Putorana, ODA = Dall sheep, OCC = Rocky Mountain bighorn, OCN = desert bighorn sheep. Asterisk: zero.

The Central Verkhoyansk, the Suntar-Khayata, and the Momsky populations had similar genetic distances with the Kharaulakh population (0.005, 0.008, 0.005, respectively) and with the population of Putorana Plateau (0.005, 0.007, 0.006), and lower with Koryak (0.003, 0.005, 0.004). Pairwise F_{ST} values revealed high genetic differentiation between the populations of snow sheep. The lowest F_{ST} values were observed between TIK and ORU–0.438 and the highest between TIK and KOR-0.959. It was shown that F_{ST} values between TIK and populations of *O. n. lydekkeri* (VER, SKH, MOM) ranged from 0.831 (TIK and VER) to 0.950 (TIK and MOM) and were significantly higher than within the other Yakut snow sheep populations—from 0.560 (VER and SKH) to 0.741 (SKH and MOM). These results are consistent with the study of Yakut snow sheep based on whole-genome SNP markers [13]. Based on genetic distances and pairwise F_{ST}, we constructed Neighbor-Net trees (Figure 3).

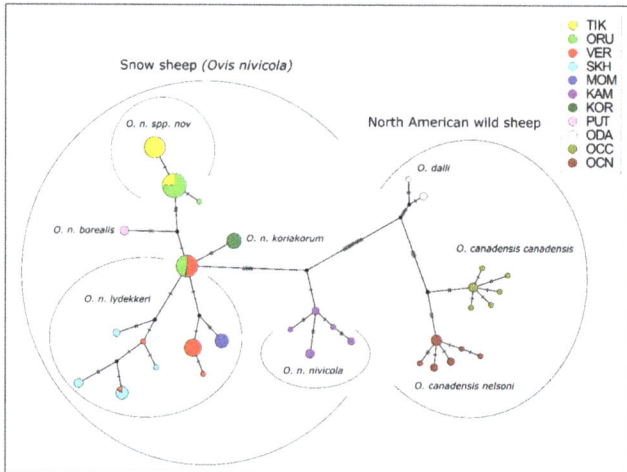

Figure 2. Median joining network of snow sheep (*Ovis nivicola*) and North American wild sheep haplotypes, based on analysis of mtDNA *cytb* gene polymorphism. The diameter of the circle corresponds to the number of individuals belonging to this haplotype. The number of transverse lines indicates the number of nucleotide substitutions. The black circles at the branching points of the network are hypothetical haplotypes. TIK = Kharaulakh Ridge–Tiksi Bay, ORU = Orulgan, VER = Central Verkhoyansk, SKH = Suntar-Khayata, MOM = Momsky, KAM = Kamchatka, KOR = Koryak, PUT = Putorana, ODA = Dall sheep, OCC = Rocky Mountain bighorn, OCN = desert bighorn sheep.

Similar conclusions can be drawn from the analysis of the Bayesian phylogenetic tree (Figure 4): two Kharaulakh haplotypes formed a clade with the Putorana snow sheep.

All the populations of Yakut (VER, SKH, MOM) along with Koryak snow sheep were placed in another clade. The Kamchatka population was the most distant from all the other snow sheep groups. To estimate approximate divergence time between the populations of snow sheep, we calibrated molecular clock, considering pairwise divergence time for *O. canadensis* and *O. dalli*–1.5 million years ago (MYA), calculated as a median from the previous studies [1,18,35,36] and given on the web resource TimeTree [32]. According to our model, we obtained the divergence time between *O. nivicola* and North American wild sheep as being around 2 MYA, which agreed with the median time for these species indicated on the web resource TimeTree–1.94 MYA. The Kharaulakh population (TIK) diverged from the closest group from Putorana Plateau (PUT) around 0.3 ± 0.13 MYA and from Yakut snow sheep populations (VER, SKH, MOM)–0.48 ± 0.19 MYA. The most distant Kamchatka population was split from all the other populations of snow sheep around 0.93 ± 0.35 MYA. All the above mentioned clades were supported with high posterior probability values: TIK-PUT–0.91, TIK-Yakut snow sheep–1, KAM-all the other populations of snow sheep–1 (Figure S1).

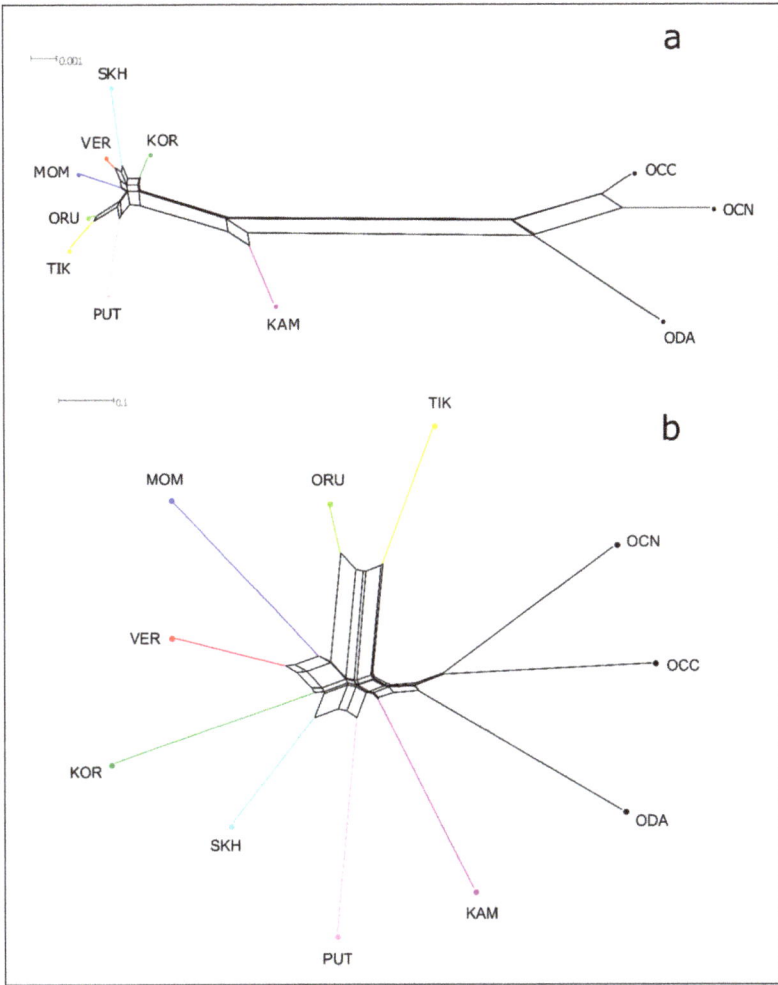

Figure 3. Neighbor-Net tree of snow sheep (*Ovis nivicola*) and North American wild sheep, based on genetic distances based on the Kimura-2-parameter model (**a**) and pairwise F_{ST} (**b**). TIK = Kharaulakh Ridge–Tiksi Bay, ORU = Orulgan, VER = Central Verkhoyansk, SKH = Suntar-Khayata, MOM = Momsky, KAM = Kamchatka, KOR = Koryak, PUT = Putorana, ODA = Dall sheep, OCC = Rocky Mountain bighorn, OCN = desert bighorn sheep.

The results of AMOVA, which was conducted for four populations of *O. n. lydekkeri*: TIK, VER, SKH and MOM, further supported genetic differentiation between the populations of Yakut snow sheep with significant variation of 78.8% (Table 4). Only 21.2% of genetic variation was found within populations. ORU was not included in this test due to its admixed origin.

Table 4. The results of AMOVA for the Yakut populations of *O. nivicola* based on the cytochrome b gene.

Source of Variation	d.f.	SS	VC	V%
Among populations	3	105.035	2.36458	78.8
Within populations	58	36.9	0.63621	21.2
Total	61	141.935	3.00080	

d.f = degrees of freedom; SS = sum of squares; VC = variance components; V% = percent of variation.

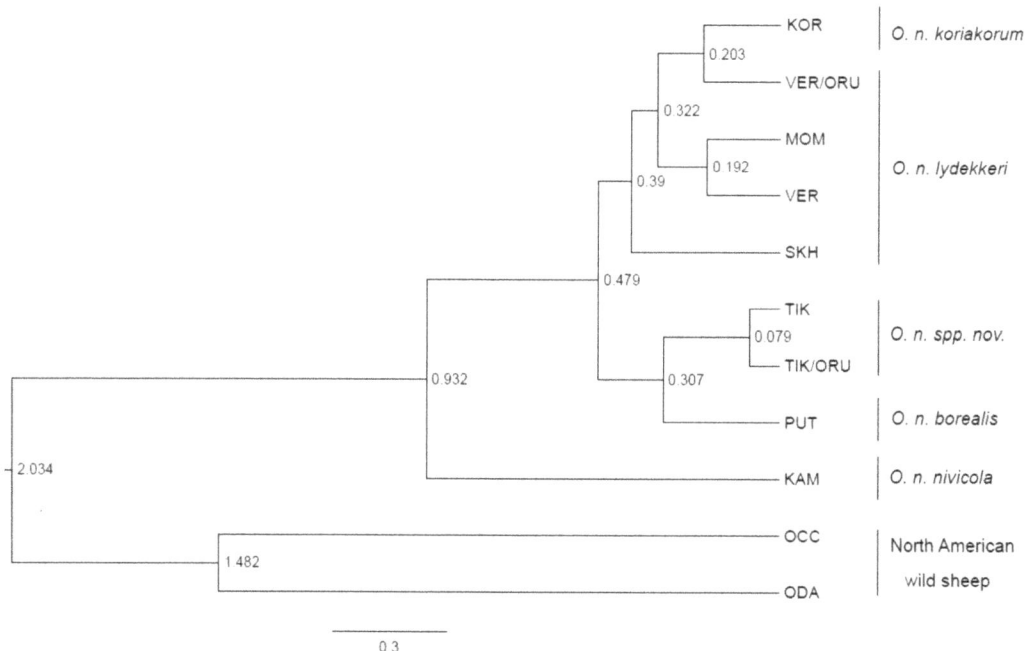

Figure 4. Bayesian phylogenetic tree of snow sheep (*Ovis nivicola*) and North American wild sheep indicating divergence time (in MYA) estimates based on the mtDNA *cytb* gene.

The mismatch distribution (the distribution of the number of pairwise differences between sequences) revealed that in TIK the observed curve agreed with the expected constant population size model (Figure 5). For all the other populations, mismatch distributions were multimodal. These results may reflect the fact that the TIK population, which inhabits the northernmost periphery of the species area, evolved independently, without admixture with other groups. All the other populations could survive during periods of glaciation in refugia and subsequently expand their areas, mixing with other populations. For example, the population from Orulgan Ridge (ORU) was formed by admixture of the Kharauakh population with groups from the Central Verkhoyansk Range, as it was shown by both nuclear [13] and mitochondrial DNA studies.

According to our research results, we can divide populations of snow sheep into three major groups. The first, and the most distant group, is presented by sheep from Kamchatka. The second group joints the Putorana and Kharaulakh populations, and the third one includes Yakut snow sheep along with representatives of the Koryak subspecies. Our results are not entirely consistent with the traditional subdivision of snow sheep, which was based mainly on the morphological characteristics. Thus, the Kharaulakh population should not be considered as *O. n. lydekkeri* and, rather, be classified as a separate subspecies. The status of Koryak subspecies should be further explored using more samples and different types of genetic markers.

Finally, the results of our study should lead to reassessment of snow sheep protection programs in Yakutia. At present, the most numerous Yakut snow sheep subspecies is protected only in the Momsky Natural Park and resource reserves: "Orulgan-Sis" (Orulgan Ridge), "Verkhneindigirsky" (Chersky Ridge), "Kele" (Central part of Verkhoyansk Range). In these territories, environments for the preservation of natural resources (by limiting economic activity) and conditions necessary for the protection of plants and animals are created [37]. The studied population of snow sheep on the Kharaulakh Ridge is protected only on the territory of the federal state natural reserve "Ust-Lensky", where hunting for this species is limited to certain periods and lasts from 1 August to 30 November. Indigenous peoples are allowed to hunt snow sheep for traditional activities also only during these periods. The recognition of the Kharaulakh population as a separate subspecies will make it possible to estimate the census size and organize new conservation areas in breeding and feeding grounds. The establishment of a new resource reserve in the Kharaulakh Ridge is essential for conservation management of the northernmost snow sheep population.

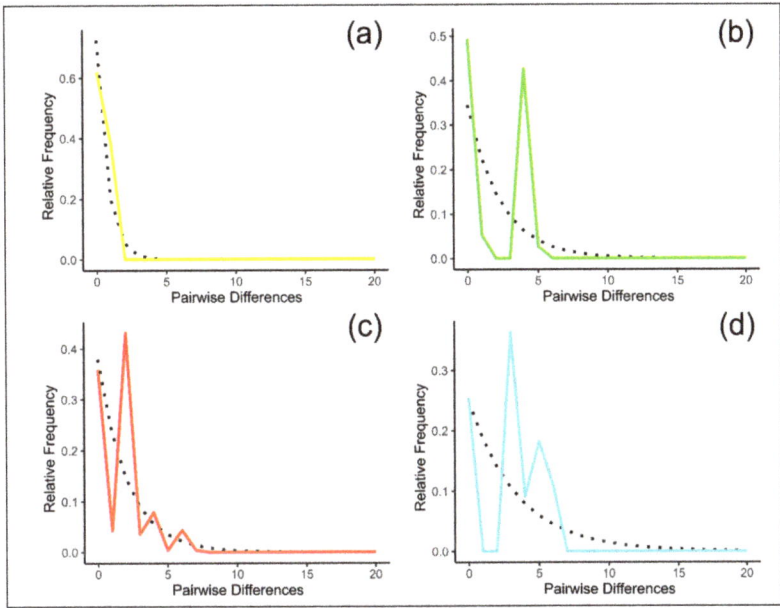

Figure 5. Mismatch distribution for the populations of *Ovis nivicola*. The dash line represents the expected distribution under the constant population size model and the solid line is the observed pairwise difference. (**a**) TIK, (**b**) ORU, (**c**) VER and (**d**) SKH population. TIK = Kharaulakh Ridge–Tiksi, ORU = Orulgan, VER = Central Verkhoyansk, SKH = Suntar-Khayata ridges.

4. Conclusions

Our study of the whole *cytB* sequence in the four most recognized subspecies of Asian snow sheep (*Ovis nivicola*) showed the most genetic distance of the Kamchatka population (*O. n. nivicola*). The haplotypes of all other populations were originated from a single haplotype currently found in the central Verkhoyansk Range and the Orulgan Ridge. It was shown that, in terms of the number of nucleotide substitutions, the Kharaulakh population differs from this "main" Yakut haplotype even more than the officially recognized Putorana and Koryak subspecies. The Orulgan population has the admixture origin and is represented by two major haplotypes, differing from each other by four nucleotide substitutions. Thus, the study of the mtDNA *cytb* sequences confirmed the results of genome-wide SNP research. Taking into account the high degree of divergence of Kharaulakh snow sheep from other groups, identified by both nuclear and mitochondrial DNA markers, we propose to classify the Kharaulakh population as a separate subspecies. The results of our study can be used in biodiversity conservation programs.

Supplementary Materials: The following are available online at https://www.mdpi.com/2075-1 729/11/3/252/s1, Figure S1: Bayesian phylogenetic tree of snow sheep (*Ovis nivicola*) and North American wild sheep indicating posterior probability values, Video S1: A representative of the northernmost population of snow sheep (Kharaulakh Ridge). Video: Bendersky E.V.

Author Contributions: Conceptualization, A.V.D., N.A.Z.; methodology, A.V.D., E.K., F.-H.L., M.-H.L., A.N.R., V.A.B., E.A.G., I.M., and G.B.; software, A.V.D.; E.K., A.V.S., validation, A.V.D., A.V.S., formal analysis, A.V.D. and N.A.Z.; investigation, A.V.D. and N.A.Z.; resources, I.M.O., T.P.S., D.G.M.; data curation, A.V.D.; writing—original draft preparation, A.V.D.; writing—review and editing, V.R.K., N.A.Z.; visualization, A.V.D. and N.A.Z.; supervision, N.A.Z.; project administration, A.V.D.; funding acquisition, A.V.D. and N.A.Z. All authors have read and agreed to the published version of the manuscript.

Funding: This study was supported by the Russian Scientific Foundation within project No. 21-46-00001 (NGS sequencing) and the Russian Ministry of Science and Higher Education within theme 0445-2019-0024 (DNA collection and Sanger sequencing).

Institutional Review Board Statement: All procedures were conducted according to the ethical guidelines of the L.K. Ernst Federal Sci-ence Center for Animal Husbandry. The protocol was approved by the Commission on the Ethics of Animal Experiments of the L.K. Ernst Federal Science Center for Animal Husbandry (Protocol Number: 2021/2). Biomaterials from Bioresource Collection of the L.K. Ernst Federal Science Cen-ter for Animal Husbandry, supported by the Russian Ministry of Science and Higher Education were used in this study. The muscle tissue samples of snow sheep were collected during scientific expeditions after obtaining collection permits granted by the Department of Hunting of the Re-public of Sakha in compliance with the Russian Federation Law No. 209-FZ of 24 July 2009.

Informed Consent Statement: Not applicable.

Data Availability Statement: The *cytb* gene sequences of snow sheep obtained for this study were deposited in GenBank NCBI database (www.ncbi.nlm.nih.gov (accessed on 15 December 2020)). Accession numbers: MW736905 - MW737011.

Acknowledgments: We are grateful to Bendersky E.V. and the Mountain Hunters Club (www.kgo-club.ru (accessed on 15 December 2020)) for providing part of the samples of snow sheep. We thank Gritsyshin V.A. for the laboratory help.

Conflicts of Interest: The authors declare no conflict of interest. The funders had no role in the design of the study; in the collection, analyses, or interpretation of data; in the writing of the manuscript, or in the decision to publish the results.

References

1. Bunch, T.D.; Wu, C.; Zhang, Y.-P.; Wang, S. Phylogenetic Analysis of Snow Sheep (*Ovis nivicola*) and Closely Related Taxa. *J. Hered.* **2006**, *97*, 21–30. [CrossRef] [PubMed]
2. Revin, Y.V.; Sopin, L.V.; Zheleznov, N.K. *Snow Sheep*; Nauka: Novosibirsk, Russia, 1988. (In Russian)
3. Zheleznov-Chukotsky, N.K. *Ecology of Snow Sheep of Northern Asia*; Nauka: Moscow, Russia, 1994. (In Russian)
4. Chernyavsky, F.B. On taxonomy and history of snow sheep (subgenus Pachyceros, Artiodactyla). *J. Zool.* **2004**, *83*, 1059–1070. (In Russian)
5. Danilkin, A.A. *Mammals of Russia and Adjacent Regions: Hollow-Horned Ruminants (Bovidae)*; KMK: Moscow, Russia, 2005. (In Russian)
6. Zheleznov-Chukotsky, N.K. New subspecies of snow sheep in Russian northern Asia. In *6th World Congress on Mountain Un-gulates and 5th International Symposium on Mouflon*; Book of Abstracts; Ministry of the Interior: Nicosia, Cyprus, 2016; p. 95.
7. Damm, G.R.; Franco, N. *CIC Caprinae Atlas of the World*; CIC International Council for Game and Wildlife Conservation: Budakeszi, Hungary; Rowland Ward Publications: Johannesburg, South Africa, 2014.
8. Abramson, N.I. Molecular and conventional phylogenetics. Towards the common ground. *Proc. Zool. Inst. Russ. Acad. Sci.* **2013**, *2*, 219–229. (In Russian)
9. Ropiquet, A.; Hassanin, A. Molecular evidence for the polyphyly of the genus Hemitragus (Mammalia, Bovidae). *Mol. Phylogenetics Evol.* **2005**, *36*, 154–168. [CrossRef] [PubMed]
10. Tokarska, M.; Marshall, T.; Kowalczyk, R.; Wójcik, J.M.; Pertoldi, C.; Kristensen, T.N.; Loeschcke, V.; Gregersen, V.R.; Bendixen, C. Effectiveness of microsatellite and SNP markers for parentage and identity analysis in species with low genetic diversity: The case of European bison. *Heredity* **2009**, *103*, 326–332. [CrossRef] [PubMed]
11. Kharzinova, V.R.; Dotsev, A.V.; Deniskova, T.E.; Solovieva, A.D.; Fedorov, V.I.; Layshev, K.A.; Romanenko, T.M.; Okhlopkov, I.M.; Wimmers, K.; Reyer, H.; et al. Genetic diversity and population structure of domestic and wild reindeer (*Rangifer tarandus* L. 1758): A novel approach using BovineHD BeadChip. *PLoS ONE* **2018**, *13*, e0207944. [CrossRef]
12. Miller, J.; Poissant, J.; Kijas, J.; Coltman, D. The International Sheep Genomics Consortium. A genome-wide set of SNPs detects population substructure and long range linkage disequilibrium in wild sheep. *Mol. Ecol. Resour.* **2010**, *11*, 314–322. [CrossRef] [PubMed]

13. Dotsev, A.V.; Deniskova, T.E.; Okhlopkov, I.M.; Mészáros, G.; Sölkner, J.; Reyer, H.; Wimmers, K.; Brem, G.; Zinovieva, N.A. Genome-wide SNP analysis unveils genetic structure and phylogeographic history of snow sheep (*Ovis nivicola*) populations inhabiting the Verkhoyansk Mountains and Momsky Ridge (*northeastern Siberia*). *Ecol. Evol.* **2018**, *8*, 8000–8010. [CrossRef] [PubMed]
14. Dotsev, A.V.; Kunz, E.; Shakhin, A.V.; Petrov, S.N.; Kostyunina, O.V.; Okhlopkov, I.M.; Deniskova, T.E.; Barbato, M.; Bagirov, V.A.; Medvedev, D.G.; et al. The first complete mitochondrial genomes of snow sheep (*Ovis nivicola*) and thinhorn sheep (*Ovis dalli*) and their phylogenetic implications for the genus Ovis. *Mitochondrial DNA Part B* **2019**, *4*, 1332–1333. [CrossRef]
15. Bernt, M.; Donath, A.; Jühling, F.; Externbrink, F.; Florentz, C.; Fritzsch, G.; Pütz, J.; Middendorf, M.; Stadler, P.F. MITOS: Improved de novo metazoan mitochondrial genome annotation. *Mol. Phylogenetics Evol.* **2013**, *69*, 313–319. [CrossRef]
16. Edgar, R.C. MUSCLE: Multiple sequence alignment with high accuracy and high throughput. *Nucleic Acids Res.* **2004**, *32*, 1792–1797. [CrossRef]
17. Kumar, S.; Stecher, G.; Tamura, K. MEGA7: Molecular Evolutionary Genetics Analysis Version 7.0 for Bigger Datasets. *Mol. Biol. Evol.* **2016**, *33*, 1870–1874. [CrossRef]
18. Rezaei, H.R.; Naderi, S.; Chintauan-Marquier, I.C.; Taberlet, P.; Virk, A.T.; Naghash, H.R.; Rioux, D.; Kaboli, M.; Pompanon, F. Evolution and taxonomy of the wild species of the genus Ovis (Mammalia, Artiodactyla, Bovidae). *Mol. Phylogenet. Evol.* **2010**, *54*, 315–326. [CrossRef] [PubMed]
19. Naidu, A.; Fitak, R.R.; Munguia-Vega, A.; Culver, M. Novel primers for complete mitochondrial cytochrome b gene sequencing in mammals. *Mol. Ecol. Resour.* **2012**, *12*, 191–196. [CrossRef] [PubMed]
20. Bandelt, H.J.; Forster, P.; Rohl, A. Median-joining networks for inferring intraspecific phylogenies. *Mol. Biol. Evol.* **1999**, *16*, 37–48. [CrossRef]
21. Leigh, J.W.; Bryant, D. Popart: Full-feature software for haplotype network construction. *Methods Ecol. Evol.* **2015**, *6*, 1110–1116. [CrossRef]
22. Bouckaert, R.; Vaughan, T.G.; Barido-Sottani, J.; Duchêne, S.; Fourment, M.; Gavryushkina, A.; Heled, J.; Jones, G.; Kühnert, D.; De Maio, N.; et al. BEAST 2.5: An advanced software platform for Bayesian evolutionary analysis. *PLoS Comput. Biol.* **2019**, *15*, e1006650. [CrossRef] [PubMed]
23. Lanfear, R.; Frandsen, P.B.; Wright, A.M.; Senfeld, T.; Calcott, B. PartitionFinder 2: New Methods for Selecting Partitioned Models of Evolution for Molecular and Morphological Phylogenetic Analyses. *Mol. Biol. Evol.* **2017**, *34*, 772–773. [CrossRef] [PubMed]
24. Akaike, H. A new look at the statistical model identification. *IEEE Trans. Autom. Control.* **1974**, *19*, 716–723. [CrossRef]
25. Excoffier, L.; Lischer, H.E.L. Arlequin suite ver 3.5: A new series of programs to perform population genetics analyses under Linux and Windows. *Mol. Ecol. Resour.* **2010**, *10*, 564–567. [CrossRef]
26. Kimura, M. A simple method for estimating evolutionary rates of base substitutions through comparative studies of nucleotide sequences. *J. Mol. Evol.* **1980**, *16*, 111–120. [CrossRef]
27. Huson, D.H.; Bryant, D. Application of Phylogenetic Networks in Evolutionary Studies. *Mol. Biol. Evol.* **2005**, *23*, 254–267. [CrossRef]
28. Rozas, J.; Ferrer-Mata, A.; Sánchez-DelBarrio, J.C.; Guirao-Rico, S.; Librado, P.; Ramos-Onsins, S.E.; Sánchez-Gracia, A. DnaSP 6: DNA Sequence Polymorphism Analysis of Large Data Sets. *Mol. Biol. Evol.* **2017**, *34*, 3299–3302. [CrossRef]
29. Fu, Y.-X. Statistical Tests of Neutrality of Mutations against Population Growth, Hitchhiking and Background Selection. *Genetics* **1997**, *147*, 915–925. [CrossRef]
30. Tajima, F. Statistical method for testing the neutral mutation hypothesis by DNA polymorphism. *Genetics* **1989**, *123*, 585–595. [CrossRef] [PubMed]
31. Rogers, A.R.; Harpending, H. Population growth makes waves in the distribution of pairwise genetic differences. *Mol. Biol. Evol.* **1992**, *9*, 552–569. [CrossRef] [PubMed]
32. Kumar, S.; Stecher, G.; Suleski, M.; Hedges, S.B. TimeTree: A Resource for Timelines, Timetrees, and Divergence Times. *Mol. Biol. Evol.* **2017**, *34*, 1812–1819. [CrossRef] [PubMed]
33. Becker, R.A.; Wilks, A.R.; Brownrigg, R.; Minka, T.P.; Deckmyn, A. Maps: Draw Geographical Maps; R Package Version 3.3.0. 2018. Available online: https://CRAN.R-project.org/package=maps (accessed on 22 December 2020).
34. Wickham, H. *ggplot2: Elegant Graphics for Data Analysis*; Springer: Berlin/Heidelberg, Germany, 2009; p. 213. [CrossRef]
35. Fritz, S.A.; Bininda-Emonds, O.R.P.; Purvis, A. Geographical variation in predictors of mammalian extinction risk: Big is bad, but only in the tropics. *Ecol. Lett.* **2009**, *12*, 538–549. [CrossRef] [PubMed]
36. Humphreys, A.M.; Barraclough, T.G. The evolutionary reality of higher taxa in mammals. *Proc. R. Soc. B Boil. Sci.* **2014**, *281*, 20132750. [CrossRef]
37. Bagirov, V.A.; Okhlopkov, I.M.; Zinovieva, N.A. *Yakut Snow Sheep: Genetic Diversity and Ways of Conservation of the Gene Pool*; Publishing House of L.K. Ernst Federal Research Center for Animal Husbandry: Dubrovitsy, Russia, 2016. (In Russian)

Article

Unique Mitochondrial Single Nucleotide Polymorphisms Demonstrate Resolution Potential to Discriminate *Theileria parva* Vaccine and Buffalo-Derived Strains

Micky M. Mwamuye [1,*], Isaiah Obara [1], Khawla Elati [1], David Odongo [2], Mohammed A. Bakheit [3], Frans Jongejan [4] and Ard M. Nijhof [1,*]

1. Institute for Parasitology and Tropical Veterinary Medicine, Freie Universität Berlin, Robert-von-Ostertag-Str. 7-13, 14163 Berlin, Germany; iobara@zedat.fu-berlin.de (I.O.); khawla.elati@fu-berlin.de (K.E.)
2. School of Biological Sciences, University of Nairobi, P.O. Box 30197-00100 Nairobi, Kenya; david.odongo@uonbi.ac.ke
3. Department of Parasitology, Faculty of Veterinary Medicine, University of Khartoum, P.O. Box 321-11115 Khartoum, Sudan; mabakheit@uofk.edu
4. Vectors and Vector-Borne Diseases Research Programme, Department of Veterinary Tropical Diseases, Faculty of Veterinary Science, University of Pretoria, Private Bag X04, 0110 Onderstepoort, South Africa; frans.jongejan@up.ac.za
* Correspondence: micky.mwamuye@fu-berlin.de (M.M.M.); ard.nijhof@fu-berlin.de (A.M.N.); Tel.: +49-30-838-62326 (A.M.N.)

Received: 16 November 2020; Accepted: 8 December 2020; Published: 8 December 2020

Abstract: Distinct pathogenic and epidemiological features underlie different *Theileria parva* strains resulting in different clinical manifestations of East Coast Fever and Corridor Disease in susceptible cattle. Unclear delineation of these strains limits the control of these diseases in endemic areas. Hence, an accurate characterization of strains can improve the treatment and prevention approaches as well as investigate their origin. Here, we describe a set of single nucleotide polymorphisms (SNPs) based on 13 near-complete mitogenomes of *T. parva* strains originating from East and Southern Africa, including the live vaccine stock strains. We identified 11 SNPs that are non-preferentially distributed within the coding and non-coding regions, all of which are synonymous except for two within the *cytochrome b* gene of buffalo-derived strains. Our analysis ascertains haplotype-specific mutations that segregate the different vaccine and the buffalo-derived strains except *T. parva*-Muguga and Serengeti-transformed strains suggesting a shared lineage between the latter two vaccine strains. Phylogenetic analyses including the mitogenomes of other *Theileria* species: *T. annulata*, *T. taurotragi*, and *T. lestoquardi*, with the latter two sequenced in this study for the first time, were congruent with nuclear-encoded genes. Importantly, we describe seven *T. parva* haplotypes characterized by synonymous SNPs and parsimony-informative characters with the other three transforming species mitogenomes. We anticipate that tracking *T. parva* mitochondrial haplotypes from this study will provide insight into the parasite's epidemiological dynamics and underpin current control efforts.

Keywords: *Theileria parva*; mitogenomes; haplotypes; SNPs; live vaccine

1. Introduction

The protozoan parasite *Theileria parva* that causes East Coast fever (ECF) and Corridor Disease (CD) is considered among the most debilitating tick-borne pathogens in cattle over its endemic range in East, Central, and Southern Africa [1]. In typical ECF symptoms, the disease severity is mainly due to the parasites' ability to transform host lymphocytes [2]. Parasitized lymphocytes proliferate uncontrollably and disseminate the dividing parasite into multiple host tissues. Their accumulation in the lungs triggers severe vasculitis, eventually resulting in respiratory failure with death occurring within three to four weeks of infection [3,4]. With mortalities of up to 100% in susceptible animals, an estimated one million die per year from an estimated risk population of 28 million cattle mainly belonging to livestock farmers with economically constrained livelihoods [5].

Thus, control of the parasite is urgent to livelihood improvement efforts among resource-poor farmers in sub-Saharan Africa, as highlighted by the World Organization for Animal Health (OIE) [6]. Current control methods include strict tick control measures to curtail pathogen transmission. However, this approach relies heavily on acaricide use, which is unsustainable in the long-run due to acaricide resistance challenges, and toxicity concerns in food and the environment [7]. Anti-theilerial chemotherapy is effective but only with early detection of the disease, which is impractical under field conditions [8,9].

Early observations that cattle acquire long-term immunity when challenged with infected ticks under a long-acting antibiotic treatment opened avenues for the development of an alternative control method based on live parasite stocks, which is called the infection and treatment method (ITM) [10,11]. ITM consists of inoculating cattle with cryopreserved *T. parva* sporozoites combined with simultaneous treatment with long-acting oxytetracyclines [12]. Early experiments revealed that there were varying cross-reactivities between geographical strains [13,14]. Due to this limitation, a cocktail of three immunizing parasite stocks known as the 'Muguga cocktail', comprising Serengeti-transformed, Kiambu 5, and Muguga strains, were combined to achieve broad protection against diverse field isolates [11]. Several other strains have been immunologically profiled to identify an isolate that cross-reacts to diverse field strains in ECF endemic areas. Among the identified strains was a Marikebuni stock isolated from the Kenyan Coast that showed cross-protection against several eastern African strains and, a Boleni strain from Zimbabwe which, apart from a cross-reactivity against Eastern and Central African strains, induced mild infections, hence eliminating the need for antibiotic use in ITM protocol [12].

Historically, ECF is traced to have originated from East Africa and spread southwards, first being reported in present-day Zimbabwe and eventually into South Africa [1]. Yet, it is notable that *T. parva* strains from different geographic regions have varying immunological profiles and epidemiological features. For example, an ability to induce a carrier state in which recovered animals remain infective to ticks has been demonstrated in some strains, enabling transmission between cattle by the vector tick, *Rhipicephalus appendiculatus* [15–17]. This persistence of vaccine strains raised initial concerns about spreading foreign parasite genotypes into endemic countries free of the vaccine parasite stocks, thereby possibly disrupting endemic stability [18].

By contrast, it is also known that some parasite strains, particularly of African buffalo (*Syncerus caffer*) origin, induce limited parasitosis and parasitemia, are non-persistent and not efficiently transmissible between cattle hosts [19]. These strains are known to cause a more acute clinical syndrome called Corridor Disease in areas where susceptible cattle are exposed to vector ticks infected on buffalo, which are the primary mammalian carrier hosts [20,21]. Based on its unique clinical presentation, which differs from classical ECF, these particular strains were initially recognized as *Theileria parva lawrencei* in earlier literature; however, this nomenclature was subsequently abolished with increasing molecular and antigenic data confirming similarities between the two strain populations [22–24]. Further, these data have revealed that cattle transmissible strains are a separately maintained subset population of those found in buffalo, and to differentiate between the two populations, *T. parva* strains are arbitrarily considered to be either of buffalo or cattle-derived for epidemiological reasons [25].

However, the genetic underpinnings of these strain differences are yet to be fully unraveled, and a precise delineation of the various genotypes is lacking [23]. This is partly because of the parasite's biology, which renders it technically unamenable for genomic studies, especially in obtaining pure parasite DNA free from host-DNA contamination [26]. An accurate determination of the origin (buffalo or cattle derived) and geographic spread of strains will help intervention and control efforts. Additionally, accurate characterization of *T. parva* strains will help to track their frequency and distribution in specific populations, and to characterize breakthroughs in areas of live vaccine field deployments. Further, since *T. parva* has sexual reproductive phases that are associated with genetic recombination [27], unraveling the parasite genotypes could enhance the understanding of the long-term effects of live vaccine components in the field.

Owing to limited or no recombination, uniparental inheritance patterns and a high substitution rate relative to nuclear genomes, mitochondrial genome studies on related apicomplexan parasites have provided clues of the geographical origin and variants of parasites [28,29]. However, the utility of mitochondrial genomes in *T. parva* in delimiting the strains and their geographical origin remains unexplored. In this study, we sequenced the mitochondrial genomes of ten *T. parva* strains, found within the parasite's currently known endemic range, as well as some characterized isolates used as vaccine strains. We also included the mitogenomes of nine other *T. parva* isolates assembled from their whole-genome data that are publicly available [30]. Further, this study assessed the divergence of *T. parva* from the closely related host-leukocyte transforming species *T. annulata*, *T. taurotragi* and *T. lestoquardi* with an aim to identify phylogenetically informative mitochondrial characters.

2. Materials and Methods

2.1. Source of Isolates

The parasite material for the different strains consisted of infected frozen ground-up tick supernatants (GUTS), salivary glands (SG), cattle whole blood, or infected lymphocyte cell lines (Table 1). The GUTS and SG were collected from archived *T. parva*-infected *R. appendiculatus* specimens from early live vaccination projects. Parasite DNA from GUTS, SG, and cell culture sample sources was extracted using the NucleoSpin Tissue kit (Macherey-Nagel, Düren, Germany), whereas the NucleoSpin Blood Mini (Macherey-Nagel) was used to extract DNA from blood samples.

Table 1. Parasite material and origin of isolates used in the current study.

Strain/Isolate	Origin	Material Used	Year Created *	Reference
T. parva (Serengeti-transformed)	Tanzania	GUTS	1981	[31]
T. parva (Boleni)	Zimbabwe	GUTS	1980	[32]
T. parva (Pugu I)	Tanzania	Cell culture	1977	n.a.
T. parva lawrencei (Manyara)	Tanzania	GUTS	1980	[33]
T. parva (Satinsyi)	Rwanda	GUTS	1981	n.a.
T. parva (Kiambu)	Kenya	Salivary glands	1980	[34]
T. parva (Marikebuni)	Kenya	GUTS	1985	[35]
T. parva (Marula)	Kenya	Blood	2000	[20]
T. parva (Muguga)	Kenya	Salivary glands	1991	[36]
T. parva (Onderstepoort)	South Africa	GUTS	1988	[37]
T. taurotragi	Tanzania	Blood	2003	[38]
T. lestoquardi (Atbara)	Sudan	Cell culture	2001	[39]

* Indicates the year the material used for this study was created, which may differ from the time when the parasite was first isolated for some strains. n.a.: not available.

2.2. Next-Generation Sequencing (NGS) T. parva Datasets

We additionally obtained publicly available whole-genome datasets of nine *T. parva* strains (DRR002439-46), downloaded in FASTQ from the NCBI (SRA accession number: DRA000613) for assembly of their mitogenomes sequences (Supplementary Table S2). The details of the parasite strains are described in a previous study [30]. All NGS datasets comprised 36 nucleotide, single-end sequence runs performed on the Illumina GAII Analyzer [30].

2.3. Mitogenome Amplification and Sequencing

Primers were designed based on an alignment of *T. parva* and *T. annulata* mitogenomes available in the GenBank (Accession nos. AB499089 and NW_001091933, respectively). A 5808 bp fragment was amplified from all isolated DNA extracts (0.5–5 ng) in 25 µL reaction volumes comprising; 0.5 U of S7 Fusion polymerase (Biozym Scientific, Hessisch Oldendorf, Germany), 5× GC Phusion buffer (ThermoFisher Scientific GmbH, Darmstadt, Germany), 200 mM of dNTPs mix, and 0.5 µM of each primer (Supplementary Table S1). The cycling conditions were as follows: 98 °C for 30 s, followed by 44 cycles of 98 °C for 15 s, 60 °C for 25 s, and 72 °C for 4 min. The final extension step was maintained at 72 °C for 10 min. Amplification of expected ~5.8 kb fragments was confirmed on 1.5% agarose gels stained with GRGreen DNA stain (Excellgene, Monthey, Switzerland) under UV trans-illumination. The amplicons were purified using GeneJET PCR Purification Kit (ThermoFisher Scientific GmbH, Darmstadt, Germany) before cloning using the Strataclone blunt vector (Agilent Technologies, USA) under the manufacturer's instructions. The plasmid was purified using the GenUP™ Plasmid Kit (biotechrabbit GmbH, Berlin, Germany) and evaluated for targeted inserts based on the *Eco*RI digestion (ThermoFisher Scientific GmbH, Darmstadt, Germany). The plasmids were sequenced by Sanger technology using standard vector primers (LGC Genomics GmbH, Berlin, Germany) and 10 primers designed in this study to amplify overlapping regions of the mitogenome (Supplementary Table S1).

2.4. Assembly, Mapping, and Annotation

The Sanger generated sequences were assembled in Geneious prime 2020.2.3 (www.geneious.com) by creating consensus sequences from the approximately 1000 bp overlapping reads aligned to a reference mitogenome (GenBank Accession: AB499089), which is based on *T. parva* Muguga vaccine strain. Similarly, the Illumina NGS reads were mapped with reference to (AB499089) using the Geneious mapper under medium-low sensitivity with fine-tuning of at least five iterations. Consensus sequences were generated from contigs based on a threshold of at least 60% of the adjusted chromatogram quality of contributing bases, while ignoring reads mapped to multiple locations on the reference. The same GenBank reference was used to map and annotate protein-coding genes (PCGs) and the known rRNA genes based on nucleotide similarities.

2.5. Phylogenetic Analysis and Identification of Informative Single Nucleotide Polymorphisms (SNPs)

Mitogenome sequence alignments generated using MAFFT [40] as well as concatenated alignments of *cox1* and *cob* sequences with additional GenBank retrieved sequences of non-transforming *Theileria* spp. and *Babesia* spp. were used to infer maximum-likelihood phylogenies. We selected best-fit models for nucleotide substitution based on the lowest Bayesian information (BIC) scores calculated using the jModel Test 2 program and tested nodal support with 100 bootstrap replicates [41]. Phylogenetic trees were generated using PhyML implemented as a plugin within the Geneious software platform [42].

To avoid the challenges of missing data due to incomplete read coverage of the Illumina assemblies, we used only the Sanger data to generate the multiple alignments used for the SNPs detection. We aligned the ten *T. parva* Sanger-generated mitogenome sequences together with three other host-transforming species; *T. taurotragi*, *T. lestoquardi* and *T. annulata* (retrieved from GenBank: NW_001091933). SNPs were identified in Geneious prime with reference to the *T. parva* Muguga GenBank AB499089 sequence under default settings, with analysis of the variants on protein translations based on Mold-Protozoan Mitochondrial genetic code. We determined informative SNPs for the 13 mitogenomes under the parsimony optimality criterion with equal weights for all characters using PAUP*4.0 software [43].

2.6. T. parva Mitogenomes Haplotypes Definition and Network Analysis

Using a modified approach from [30], a second set of SNPs with consideration to the ten *T. parva* Sanger mitogenomes was extracted from the initial parsimony-informative SNPs. We used DnaSP v.6.12.03 on the second SNP data set to generate *T. parva* haplotype data [44] and a median-joining (M-J) network was constructed using Network V. 10 software (https://www.fluxus-technology.com/) under default settings to examine relationships among the *T. parva* mitogenomes [45]. Of the Illumina assembled mitogenomes, strains that had missing data with respect to the second SNP data set were excluded from the haplotype analysis.

3. Results

At least ten bidirectional overlapping Sanger reads were obtained for each strain, which were assembled into mitogenomes sequences ranging in size from 5800 to 5811 bp. The sequences are archived in NCBI's GenBank under accession numbers MW172707-MW172717; MW218514. The content and gene order for all ten *T. parva*, one *T. taurotragi*, and one *T. lestoquardi* mitogenomes were consistent with previous data, comprising three PCGs, fragmented rRNAs, and no tRNA [46] (Figure 1).

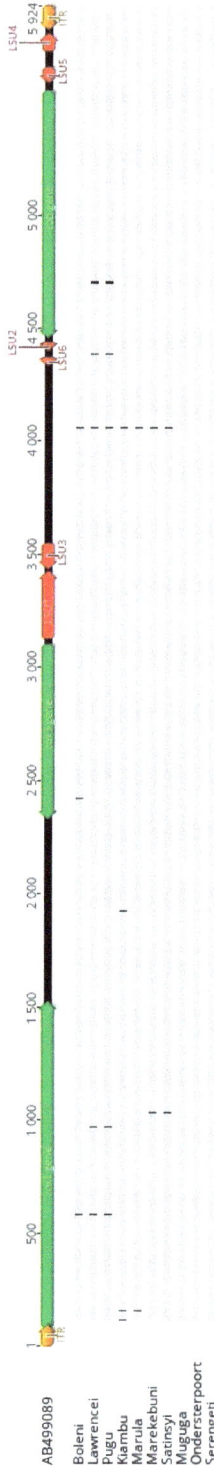

Figure 1. Linear map of *T. parva* mitochondrial genome and alignment showing the distribution of variants (SNPs) across the mitogenome sequences obtained by Sanger sequencing. cox1 and cox3: cytochrome oxidase subunits; cob: Cytochrome b; LSU: large subunit; ITR: Inverted terminal repeat region; SNPs: single nucleotide polymorphisms; vertical markings indicate polymorphisms in respective nucleotide sequence relative to the reference sequence AB499089 above.

3.1. Divergence of T. parva from Other Host-Lymphocyte Transforming Theileria sp.

Due to length variations, we considered 5793 positions in the multiple alignment of the sanger-sequenced *T. parva* mitogenomes and the three additional host-lymphocyte transforming *Theileria* spp. Of the positions considered, there were 42 indels and 1036 SNPs. However, only 662 of the SNPs were parsimony informative across all mitogenome sequences. In terms of percentage identities of the PCG, cob was the most diverse gene, having 73.2–79.6% identity between *T. parva* Muguga strain and the three host-lymphocyte transforming *Theileria* (Figure 2). Phylogenetic analyses were congruent both using the whole mitogenomes and the concatenated gene sequences. In all instances, *T. annulata* and *T. lestoquardi* consistently formed an outgroup clade to *T. parva* and *T. taurotragi* (Figure 3; Appendix A, Figure A1).

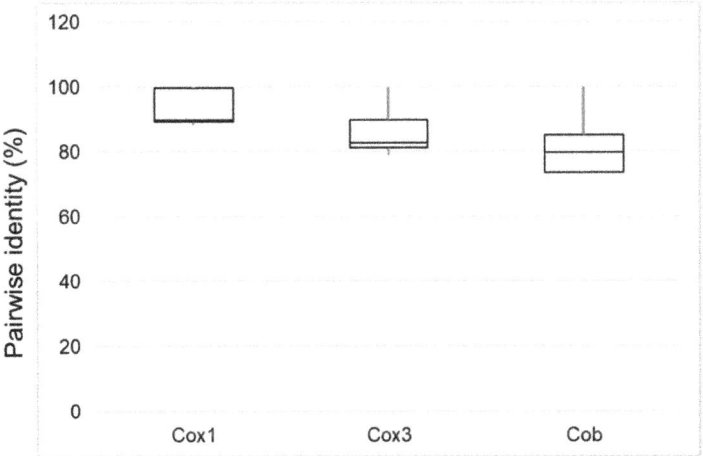

Figure 2. Percentage pairwise identity of the three protein-coding genes across the 14 mitogenomes analyzed in this study. The 25th and 75th percentiles are represented by the box limits; lines across the boxes indicate the median; whiskers extend to the maximum and minimum (%) identity values.

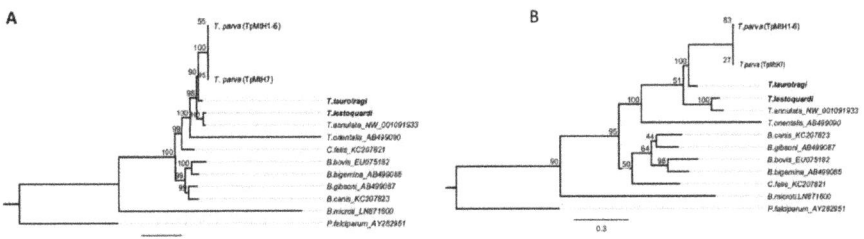

Figure 3. Maximum likelihood phylogeny based on (**A**) near-complete whole mitogenome sequences (~5.8 kb) and (**B**) cob sequences (~1.1 kb). The nucleotide substitution models for the tree constructions as determined by the lowest Bayesian information (BIC) values were TVM + G and GTR + G, respectively. Bootstrap values are based on 100 replicates. Sequences from this study are in bold.

3.2. T. parva Haplotype Analysis

We used the extracted second set of SNPs, which comprised nine informative SNPs for *T. parva* haplotype analysis. As previously noted, we included three Illumina assembled mitogenomes (Nyakizu from Rwanda, MandaliZ22H10, and Buffalo Z5E5 from Zambia) that had data on all the determined nine informative SNPs, irrespective of the other missing regions lacking reads coverage.

In total, 13 *T. parva* strains were considered for the haplotype analysis. The SNPs segregated the *T. parva* strains used into seven haplotypes, which we have identified in this study by assigning the TpmtH prefix numerically beginning with Muguga as a reference sequence (Figure 4). The Muguga strain isolate was assigned into one haplotype identical to Onderstepoort and Serengeti isolates (TpMtH1). Similarly, *T. parva lawrencei* (Manyara) isolated from an African buffalo, and Pugu I, both from Tanzania, together with a Zambian buffalo isolate (Buffalo Z5E5), formed one haplotype (TpMtH7) that was characterized by two SNPs within both the *cox* 1 and *cob* genes. We presumed Pugu I to have originated from buffalo *T. parva* based on analysis of its sporozoite surface (p67) antigen gene, which showed that it lacked the typical 129 bp deletion that is present in cattle transmissible *T. parva* [47,48]. The p67 sequence generated (See Appendix A for primer information and cycling conditions) for the Pugu I isolate is archived under accession no MW183674 in the GenBank.

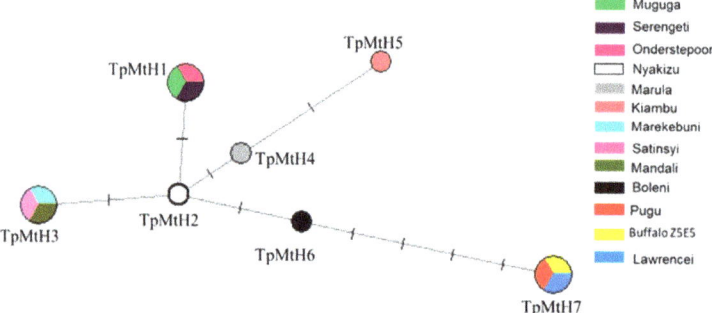

Figure 4. A median-joining (MJ) parsimony network for the 13 *T. parva* haplotype mitogenome sequences. Node labels TpMtH1-7 represents the unique haplotypes. Lines between nodes indicate mutation points. Larger and fractionated nodes indicate shared haplotypes with multiple strains, each marked with a different color key, as shown.

Further, the Marikebuni strain originating from the Kenyan coast, Mandali from Zambia, and Satinsyi strain from Rwanda formed one haplotype (TpMtH3) characterized by two SNP transitions relative to *T. parva* Muguga (Table 2; Figure 4). The Marula and Kiambu-V isolates formed independent haplotypes (TpMtH4 and 5), but differed with one SNP position (119) between them (Figure 5). The Boleni isolate (TpMtH6) possessed a transversion mutation within the *cox1* gene (SNP 584). This transversion mutation was also notable within the buffalo haplotype (TpMtH7). Interestingly, all haplotypes deviate from TpMtH1 by a transition (A→G) at SNP position 4060, which lies in a currently functionally unknown region, but appears to be ancestral in the other transforming *Theileria* (Table 2; Figure 5). This transition is the single defining SNP of the Nyakizu (Rwanda) strain from Muguga, and makes TmMtH2 a central node from which all other haplotypes deviate. However, there was no apparent differentiation by geographic origin as the M-J network nodes associated with multiple haplotypes clustered isolates of diverse origin (Figure 4).

Table 2. SNPs among the seven haplotypes based on the GenBank reference AB499089. The reference sequence matches haplotype TpMtH1 in the present study.

Haplotype	Gene	Variant Type	Change	Codon Change	Codon Position	AA Change	Protein Effect
	Cox1						
TpMtH3		Transition	C→T	GCC→GCT	951		
TpMtH4		Transition	C→T	CTG→TTG	76		
TpMtH5		Transition	G→A	GTG→GTA	36		
		Transition	C→T	CTG→TTG	76		
TpMtH6		Transversion	A→C	GTA→GTC	501		
TpMtH7		Transversion	A→C	GTA→GTC	501		
		Transition	C→T	TAC→TAT	891		
	Cox3						
TpMtH6		Transition	T→C	CAA→CAG	555		
	Cob						
TpMtH7		Transition	A→G	GTT→GCT	848	V→A	Substitution
		Transition	A→G	GTA→GCA	851	V→A	Substitution
	Intergenic				SNP position		
TpMtH2		Transition	A→G		4060		
TpMtH3		Transition	A→G		4060		
TpMtH4		Transition	A→G		4060		
TpMtH5		Transition	T→C		1924		
		Transition	A→G		4060		
TpMtH6		Transition	A→G		4060		
TpMtH7		Transition	A→G		4060		
		Transversion	T→A		4382		

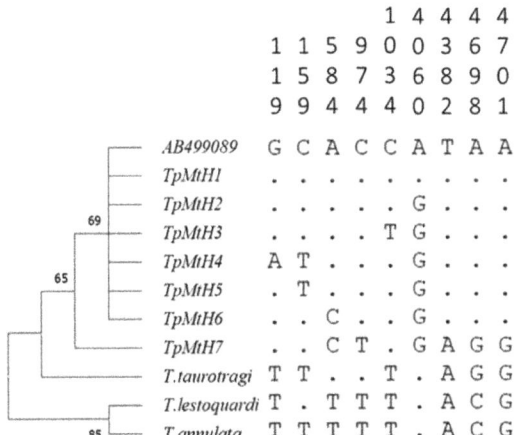

Figure 5. Phylogenetic grouping of the seven *T. parva* haplotypes identified in this study. The neighbor-joining tree is constructed based on the Jukes–Cantor Model using 9 *T. parva* only SNPs out of 662 informative SNPs extracted from 13-mitogenome sequences alignment that included; *T. taurotragi*, *T. annulata*, and *T. lestoquardi*. Numbers behind the nodes indicate bootstrap values based on 1000 replicates. The positions are relative to the AB499089 *T. parva* Muguga mitogenome sequence. Orthologous positions in the three other transforming *Theileria* are shown for comparison.

3.3. Intraspecific Divergence among T. parva Strains

Relative to the GenBank reference AB499089, an alignment of 10 Sanger-sequenced *T. parva* mitogenomes showed variation at 11 sites, all of which were SNPs with no indels observed (Figure 1). Of the 11 SNPs, only three were found within the intergenic region and involved one transversion within the haplotype associated with buffalo *T. parva*. In total, there were three transversion SNPs positions, two of which were observed within the *cob* gene of the buffalo-associated haplotype. Among the genes, *cox3* was most conserved with only a single SNPs position within the Boleni mitogenome, while the *cox1* gene had five mutated positions, all of which were synonymous. The remaining two

SNPs that were found within the *cob* gene sequences were non-synonymous and resulted in two contiguous amino acid substitutions in their predicted proteins (Table 2). Both substitutions involved the valine codon, which was replaced by an alanine amino acid codon. Notably, these substitutions were only in the haplotype associated with the buffalo *T. parva* isolates.

4. Discussion

In this study, we describe promising mitogenome-based SNPs that demonstrate precision and convenience in characterizing *T. parva* strains. Previously identified nuclear-based markers mainly based on a panel of mini- and micro-satellites are sometimes biased due to selective amplification of predominant strain clonotypes during passages through cattle and ticks [27,49]. In addition, since the design of the initial markers relied on the genome of *T. parva* Muguga stock, some markers are possibly biased in detecting diversity within this stock [50]. Although whole-genome SNPs analysis has been demonstrated to have high discriminatory power in typing vaccine strains, it is yet to find field applications [50,51]. Additionally, obtaining pure parasite DNA for whole-genome sequencing, especially for buffalo-derived *T. parva* is a hurdle due to its biology and may be complicated in the field where the parasite exists as a mixed diverse population [23,27].

Mitochondrial genomes and their individual genes have been extensively used to study phylogeny and applied in species identification and delimitation across broad taxonomic levels [52]. The majority of apicomplexan mitochondrial genomes that have been sequenced to date exhibit an extreme size reduction, containing at most three protein genes (*cox1*, *cox3*, and *cob*) and fragmented rRNA genes [53]. The extreme mitogenomes size reduction and a faster coalescence make mtDNA attractive to study differentiated *T. parva* population strains. Indeed, our analysis revealed haplotype-defining SNPs within the *T. parva* mitogenomes, which are parsimonious with other host-leukocytes transforming *Theileria* species. Based on median joining (MJ) parsimony analysis, the *T. parva* mtDNA sequences generated were segregated into an unambiguous network, congruent with the existence of multiple linked lineages. With respect to the *T. parva* Muguga isolate haplotype, each haplogroup was defined by synonymous nucleotide changes, except for non-synonymous changes leading to amino acid substitutions within the *cob* of the buffalo-associated strains; *T. parva lawrencei*, *T. parva* Buffalo Z5E5, and one field isolate, *T. parva* Pugu I.

Our analysis identifies *T. parva* Muguga and Serengeti-transformed as belonging to the same haplotype characterized by nine defining SNPs positions from the strains used in this study. This is not surprising as previous studies have demonstrated a similar monoclonal antibody profile and conservation on their known *T. parva* antigen coding genes [50,54]. Further, the two strains are strikingly similar at the whole genome level, with only 420 non-synonymous substitutions in Serengeti-transformed relative to the Muguga reference genome reported [50]. These substitutions occur in a paltry 53 genes (out of over 4000 *T. parva* genes) mainly within polymorphic multicopy gene families and ATP-binding cassette transporter genes located in subtelomeric ends [50]. With the almost similar identity of the two strains, our results, in addition to previous studies, question the necessity of both Muguga and Serengeti-transformed in the trivalent cocktail instead of a divalent cocktail containing either of the two and Kiambu-V. Interestingly, both Muguga and Serengeti-transformed *T. parva* strains also shared the same haplotype with a historical isolate *T. parva* Onderstepoort, a laboratory maintained stock isolated in 1937 on the farm Schoonspruit in the Transvaal, South Africa prior to ECF eradication in this country [1,37,55]. Earlier analyses on three *T. parva* antigen proteins; the Polymorphic immunodominant molecule (PIM), sporozoite surface protein (p67), and p104, have shown that these nuclear-encoded antigen genes are, in fact, identical to those of the Muguga parasite [50,56]. It is thus conceivable that the ECF-causing strains derive from a common lineage that can be inferred at the mitochondrial genome.

An important finding of this study is the clustering of buffalo-derived *T. parva* strains under one haplotype (TpMtH7) with the same nine SNPs. It is noteworthy that the buffalo strains used in this study originate from two different countries (Zambia and Tanzania), while the field isolate (Pugu I) was isolated during vaccine field trials in Tanzania. And although a Kenyan Buffalo (*T. parva* LAWR) from the NGS assembly was not included in the haplotype analysis due to missing data on SNP position (159), all its other SNPs positions also matched haplotype (TpMtH7) (data not shown). The buffalo has long been recognized as the natural reservoir of *T. parva*. The *T. parva* strains maintained in cattle are considered a subset population from that maintained in buffalo [23,25]. However, there has not been a definitive genetic basis to differentiate what constitutes a buffalo-derived *T. parva* and a cattle-derived *T. parva* or whether their designation as a single species is justified [23,56]. The available approach of their differentiation based on the p67 alleles only provides a preliminary indication of presumptive exposure of cattle to buffalo *T. parva* based on alleles-2, 3, and 4, which are considered highly probable to be of buffalo origin in contrast to allele-1 that is found in cattle transmitted *T. parva*, but does not necessarily preclude its presence in buffalo [48,49,57]. Our analysis suggests strain defining mitochondrial SNPs that are potential markers for buffalo-derived *T. parva* lineages.

Noticeably, the Boleni strain formed a separate haplotype (TpMt6) that shared a transversion mutation within the *coxI* gene with the buffalo haplotype. This strain was isolated from Zimbabwe from a farm that had experienced a severe theileriosis outbreak in January 1978 [31]. Under the now obsolete trinomial nomenclature of *T. parva*, it was named *Theileria parva bovis*, which was associated to what was referred to as January disease [58]. The delineation of this strain from our data is thus a significant find as it agrees with the epidemiological distinctions that have been apparent from earlier investigations on theilerioses caused by *T. parva*. Further, our analysis identifies the Kenyan-Marekebuni, Zambian-Mandali, and Rwandese Satinsyi strains as one haplotype (TpMtH3). Although the shared haplotypes from widely separated regions may suggest a lack of geographical differentiation of the haplotypes, our observations could also be because of a limited sample size as well as through spread by carrier animals.

A high level of interspecies divergence among the transforming *Theileria* is observed that is characterized by up to 42 indels with respect to *T. parva* Muguga. However, a limited polymorphism is observed amongst the 13 *T. parva* mitogenomes analyzed, which is also observed in other apicomplexan species such as *Plasmodium falciparum* [59]. Of the eleven *T. parva* SNPs observed, only nine were informative. We modestly suppose this may be convenient compared to whole-genome-based SNPs in which up to >120,000 SNPs have been observed in buffalo strains alone [30]. Additionally, we think our approach to defining SNPs that are foremost parsimonious with other leukocyte transforming *Theileria* provides initial indications on the potential of the identified SNPs to be informative for typing of recently diverged field *T. parva* strains from common leukocyte transforming ancestor. Nonetheless, further investigation to test the utility of the SNPs is necessary with a larger field population across the *T. parva* endemic range, especially in wildlife-livestock areas where 'breakthrough' infections against the trivalent live vaccine are known to occur.

The phylogenetic analysis of both the full-length near-complete mitochondrial genomes and the concatenated *cox1* and *cob* genes place *T. parva* and *T. taurotragi* in one clade, consistent with previous analyses using nuclear genes such as the 18S RNA gene. The same phylogenetic tree topology is maintained with the sporozoite surface protein gene and its orthologues in respective leukocyte host-transforming species [60]. Thus, the mitogenomes data's observed congruency with nuclear-based data rules out possibilities of inheritance patterns specific to mitochondria in our analysis.

Our data indicate *T. annulata* and *T. lestoquardi* form an outgroup clade among the transforming parasites, reflecting an allopatric speciation separation from *T. parva* and *T. taurotragi*, and conforms to their currently known demography. Noticeably, *T. taurotragi* was initially described as a parasite of the eland (*Taurotragus oryx*) [61], but is also reported to cause infections in cattle in the known endemic range (Eastern, Central, and Southern Africa) of *T. parva* and its tick vector *R. appendiculatus*, alongside other tick vectors [38]. As such, co-infections of *T. taurotragi* and *T. parva* are, in fact, frequently common [62]. While the pathogenicity of *T. taurotragi* in cattle is not clearly understood, it has been shown to transform a wide range of host cells in in vitro studies [63], and has been associated with cases of cerebral theileriosis (BCT) [38,64].

Similarly sympatric, *T. annulata* and *T. lestoquardi*, occur within the same currently known endemic range (N. Africa, S. Asia, and S. Europe) and are transmitted by ticks belonging to *Hyalomma* genus. Both are important parasites responsible for heavy economic losses and have an intertwined epidemiology that poses interpretation challenges in their overlap in affected countries [65]. *Theileria lestoquardi* is a parasite of small ungulates and causes malignant ovine theileriosis, while *T. annulata* causes bovine tropical theileriosis but also co-infects with the former in sheep [65–67].

In conclusion, this study catalogs SNPs based on mitogenomes of characterized *T. parva* strains and vaccine stocks that can facilitate their tracking in the field. We identify haplotypes defined by SNPs that are initially parsimonious among transforming *Theileria*; *T. parva*, *T. annulata*, *T. taurotragi*, and *T. lestoquardi* mitogenomes, the latter two reported herein for the first time. We anticipate that the knowledge of the circulating haplotypes with reference to the live vaccine strains haplotypes will be insightful in characterizing *T. parva* epidemiology with important implications for control, and have a predictive value on the success of live vaccine deployments besides characterization of breakthrough infections.

Supplementary Materials: The following are available online at http://www.mdpi.com/2075-1729/10/10/334/s1. Table S1: Amplification and sequencing primer sequences, Table S2: Summary of NGS reads (SRA accession number: DRA000613) mapped to *T. parva* Muguga-mitochondrial sequence (AB499089).

Author Contributions: Conceptualization, M.M.M. and A.M.N.; methodology, M.M.M.; validation, A.M.N., M.M.M., and I.O.; formal analysis, M.M.M.; investigation, M.M.M. and K.E.; resources, A.M.N., M.A.B., D.O., and F.J.; data curation, M.M.M.; writing—original draft preparation, M.M.M.; writing—review and editing, M.M.M., I.O., K.E., D.O., M.A.B., F.J., and A.M.N.; visualization, M.M.M.; supervision, A.M.N. and I.O.; funding acquisition, A.M.N. All authors have read and agreed to the published version of the manuscript.

Funding: The work was supported by the Deutsche Forschungsgemeinschaft (DFG) (Project No. CL166/4-2). MMM was supported by a doctoral scholarship of the Kenyan government through the National Research Fund (NRF) and the Germany Academic Exchange Service (DAAD). AMN received financial support from the Federal Ministry of Education and Research (BMBF) under project number 01KI1720 as part of the 'Research Network Zoonotic Infectious Diseases'.

Acknowledgments: Open Access funding provided by the Freie Universität Berlin.

Conflicts of Interest: The authors declare no conflict of interest.

Appendix A

The pugu *p67* gene was amplified by PCR in a 25 µL reaction using published primers [48]. Briefly, the amplification reaction contained; 5× High-Fidelity (HF) Phusion buffer (ThermoFisher Scientific), 200 µM dNTPs (Biozym Scientific GmbH), 0.5 µM forward primer (IL 6133: 5'-ACAAACACAATCCCAAGTTC-3'), 0.5 µM reverse primers; IL 7922: 5'-CCTTTACTACGTTGGCG-3'), 0.02 U/µL HF Phusion DNA polymerase, sterile PCR-grade water (Carl Roth GmbH, Karlsruhe), and 2 µL of template DNA. The amplification cycle was set as follows; 98 °C for 30 s, followed by 35 cycles of 98 °C for 30 s, an annealing step at 58 °C for 40 s, and elongation at 72 °C for 1 min. The final extension step was 10 min at 72 °C. The p67 amplicon was purified using GeneJET PCR Purification Kit (ThermoFisher Scientific), and sequenced using Sanger technology (LGC Genomics, Berlin). The obtained sequence was translated and evaluated for the presence/absence of the 43 amino-acid deletions used to categorize cattle and buffalo derived p67 sequences.

Appendix B

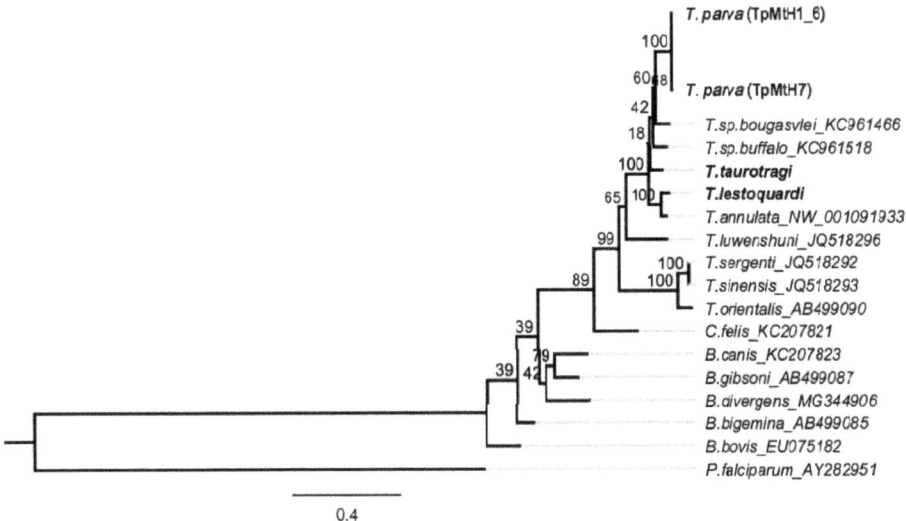

Figure A1. Maximum likelihood phylogeny based on *cox1* gene sequences (~1.4 kb). The nucleotide substitution model for the tree constructions as determined by the lowest BIC values was GTR + I + G. Bootstrap values are based on 100 replicates. Sequences from this study are in bold.

References

1. Norval, R.A.I.; Perry, B.D. *The Epidemiology of Theileriosis in Africa*; Academic Press: London, UK, 1992; pp. 1–41.
2. Tretina, K.; Gotia, H.T. *Theileria*-transformed bovine leukocytes have cancer hallmarks. *Trends Parasitol.* **2015**, *31*, 306–314. [CrossRef]
3. Fry, L.M.; Schneider, D.A. East Coast Fever Caused by *Theileria parva* is characterized by macrophage activation associated with vasculitis and respiratory failure. *PLoS ONE* **2016**, *11*, e0156004. [CrossRef]
4. Irvin, A.D.; Mwamachi, D.M. Clinical and diagnostic features of East Coast Fever (*Theileria parva*) infection of Cattle. *Vet. Rec.* **1983**, *113*, 192–198. [CrossRef]
5. Gachohi, J.; Skilton, R. Epidemiology of East Coast Fever (*Theileria parva* Infection) in Kenya: Past, present and the future. *Parasit. Vectors* **2012**, *5*, 194. [CrossRef]
6. Grace, D.; Songe, M. Impact of neglected diseases on animal productivity and public health in Africa. In Proceedings of the 21st Conference of the OIE Regional Commission for Africa, Rabat, Morocco, 16–20 February 2015.
7. Abbas, R.Z.; Zaman, M.A. Acaricide resistance in cattle ticks and approaches to its Management: The state of play. *Vet. Parasitol.* **2014**, *203*, 6–20. [CrossRef]
8. Mbwambo, H.A.; Sudi, F.F. Comparative studies of the efficacy of parvaquone and parvaquone-plus-frusemide in the treatment of *Theileria parva* infection (East Coast Fever) in cattle. *Vet. Parasitol.* **2002**, *108*, 195–205. [CrossRef]
9. D'haese, L.; Penne, K. Economics of theileriosis control in Zambia. *Trop. Med. Int. Health* **1999**, *4*, A49–A57. [CrossRef]
10. Neitz, W.O. Aureomycin in *Thieileria parva* Infection. *Nature* **1953**, *171*, 34–35. [CrossRef] [PubMed]
11. Radley, D.E.; Brown, C.G.D. East Coast Fever: 3. Chemoprophylactic immunization of cattle using oxytetracycline and a combination of theilerial strains. *Vet. Parasitol.* **1975**, *1*, 51–60. [CrossRef]
12. Morzaria, S.P.; Nene, V. Vaccines against *Theileria parva*. *Ann. N. Y. Acad. Sci.* **2006**, *916*, 464–473. [CrossRef] [PubMed]

13. Young, A.S.; Brown, C.G.D. Observations on the cross-immunity between *Theileria lawrencei* (Serengeti) and *Theileria parva* (Muguga) in Cattle. *Int. J. Parasitol.* **1973**, *3*, 723–728. [CrossRef]
14. Cunningham, M.P.; Brown, C.G.D. Theileriosis: The exposure of immunised cattle in a *Theileria lawrencei* enzootic Area. *Trop. Anim. Health Prod.* **1974**, *6*, 39–43. [CrossRef] [PubMed]
15. Young, A.S.; Leitch, B.L. The occurrence of a *Theileria parva* carrier state in cattle from an East Coast Fever endemic area of Kenya. In *Advances in the Control of Theileriosis*; Irvin, A.D., Cunningham, M.P., Eds.; Martinus Nijhoff Publishers: The Hague, The Netherlands, 1981; pp. 60–62.
16. Skilton, R.A.; Bishop, R.P. The persistence of *Theileria parva* infection in cattle immunized using two Stocks which differ in their ability to induce a carrier State: Analysis using a novel blood spot PCR Assay. *Parasitology* **2002**, *124*, 265–276. [CrossRef] [PubMed]
17. Olds, C.L.; Mason, K.L. *Rhipicephalus appendiculatus* ticks transmit *Theileria parva* from persistently infected cattle in the absence of detectable parasitemia: Implications for East Coast Fever epidemiology. *Parasit. Vectors* **2018**, *11*, 126. [CrossRef] [PubMed]
18. Nene, V.; Kiara, H. The Biology of *Theileria parva* and control of East Coast Fever—Current status and future trends. *Ticks Tick. Borne Dis.* **2016**, *7*, 549–564. [CrossRef]
19. Jura, W.G.Z.; Losos, G.J. A Comparative study of the diseases in cattle caused by *Theileria lawrencei* and *Theileria parva*. 1. Clinical signs and parasitological observations. *Vet. Parasitol.* **1980**, *7*, 275–286. [CrossRef]
20. Bishop, R.P.; Hemmink, J.D. The african buffalo parasite *Theileria*. sp. (Buffalo) can infect and immortalize cattle leukocytes and encodes divergent orthologues of *Theileria parva* antigen genes. *Int. J. Parasitol. Parasites Wildl.* **2015**, *4*, 333–342. [CrossRef]
21. Young, A.S.; Brown, C.G. Establishment of an experimental field population of *Theileria lawrencei*-infected ticks maintained by african buffalo (*Syncerus caffer*). *J. Parasitol.* **1977**, *63*, 903–907. [CrossRef]
22. Allsopp, B.A.; Baylish, H.A. Discrimination between six species of *Theileria* using oligonucleotide probes which detect small subunit ribosomal RNA sequences. *Parasitology* **1993**, *107*, 157–165. [CrossRef]
23. Morrison, W.I.; Hemmink, J.D. *Theileria parva*: A parasite of african buffalo, which has adapted to infect and undergo transmission in cattle. *Int. J. Parasitol.* **2020**, *50*, 403–412. [CrossRef]
24. Maritim, A.C.; Young, A.S. Transformation of *Theileria parva* derived from african buffalo (*Syncerus caffer*) by tick passage in cattle and its use in infection and treatment immunization. *Vet. Parasitol.* **1992**, *43*, 1–14. [CrossRef]
25. Pelle, R.; Graham, S.P. Two *Theileria parva* CD8 T cell antigen genes are more variable in buffalo than cattle parasites, but differ in pattern of sequence diversity. *PLoS ONE* **2011**, *6*, e19015. [CrossRef] [PubMed]
26. Palmateer, N.C.; Tretina, K. Capture-based enrichment of *Theileria parva* DNA enables full genome assembly of first buffalo-derived strain and reveals exceptional intra-specific genetic diversity. *PLOS Negl. Trop. Dis* **2020**, *14*, e0008781. [CrossRef] [PubMed]
27. Katzer, F.; Ngugi, D. Extensive genotypic diversity in a recombining population of the apicomplexan parasite *Theileria parva*. *Infect. Immun.* **2006**, *74*, 5456–5464. [CrossRef]
28. Schmedes, S.E.; Patel, D. Using the *Plasmodium* mitochondrial genome for classifying mixed-species infections and inferring the geographical origin of *P. falciparum* parasites imported to the U.S. *PLoS ONE* **2019**, *14*, e0215754. [CrossRef]
29. Joy, D.A.; Feng, X. Early origin and recent expansion of *Plasmodium falciparum*. *Science* **2003**, *300*, 318–321. [CrossRef]
30. Hayashida, K.; Abe, T. Whole-Genome Sequencing of *Theileria parva* strains provides insight into parasite migration and diversification in the African Continent. *DNA Res.* **2013**, *20*, 209–220. [CrossRef]
31. Young, A.S.; Purnell, R.E. Transmission of *Theileria lawrencei* (Serengeti) by the ixodid tick, *Rhipicephalus appendiculatus*. *Trop. Anim. Health Prod.* **1973**, *5*, 146–152. [CrossRef]
32. Lawrence, J.; Mackenzie, P.K. Isolation of a non-pathogenic *Theileria* of cattle transmitted by *Rhipicephalus appendiculatus*. *Zimbabwe Vet. J.* **1980**, *11*, 27–35.
33. Schreuder, B.E.; Uilenberg, G. studies on Theileriidae (Sporozoa) in Tanzania. VIII. Experiments with african buffalo (*Syncerus caffer*). *Trop. Parasitol.* **1977**, *28*, 367–371.
34. Irvin, A.D.; Purnell, R.E. The application of an indirect method of infecting ticks with piroplasms for use in the isolation of field infections. *Br. Vet. J.* **1974**, *130*, 280–287. [CrossRef]

35. Irvin, A.D.; Dobbelaere, D.A. Immunisation against East Coast Fever: Correlation between monoclonal antibody profiles of *Theileria parva* stocks and cross immunity in Vivo. *Res. Vet. Sci.* **1983**, *35*, 341–346. [CrossRef]
36. Brocklesby, D.W.; Barnett, S.F.; Scott, G.R. Morbidity and mortality rates in East Coast Fever (*Theileria parva* infection) and their application to drug screening procedures. *Br. Vet. J.* **1961**, *117*, 529–531. [CrossRef]
37. Neitz, W.O. Studies on East Coast Fever. *S. Afr. Sci.* **1948**, *1*, 133–135.
38. Catalano, D.; Biasibetti, E. "Ormilo Disease" a disorder of zebu cattle in Tanzania: Bovine Cerebral Theileriosis or new protozoan disease? *Trop. Anim. Health Prod.* **2015**, *47*, 895–901. [CrossRef]
39. Bakheit, M.A.; Endl, E. Purification of macroschizonts of a sudanese isolate of *Theileria lestoquardi* (*T. lestoquardi* [Atbara]). *Ann. N. Y. Acad. Sci.* **2006**, *1081*, 453–462. [CrossRef]
40. Katoh, K.; Standley, D.M. MAFFT Multiple sequence alignment software version 7: Improvements in performance and usability. *Mol. Biol. Evol.* **2013**, *30*, 772–780. [CrossRef]
41. Darriba, D.; Taboada, G.L. JModelTest 2: More Models, New Heuristics and Parallel Computing. *Nat. Methods* **2012**, *30*, 772. [CrossRef]
42. Guindon, S.; Dufayard, J.F. New algorithms and methods to estimate maximum-likelihood phylogenies: Assessing the performance of PhyML 3.0. *Syst. Biol.* **2010**, *59*, 307–321. [CrossRef]
43. Swofford, D. *PAUP* Phylogenetic Analysis Using Parsimony (*and Other Methods). Version 4*; Sinauer Associates: Sunderland, MA, USA, 2003.
44. Rozas, J.; Ferrer-Mata, A.; Sanchez-DelBarrio, J.C.; Guirao-Rico, S.; Librado, P.; Ramos-Onsins, S.E.; Sanchez-Gracia, A. DnaSP 6: DNA sequence polymorphism analysis of large data sets. *Mol. Biol. Evol.* **2017**, *34*, 3299–3302. [CrossRef]
45. Bandelt, H.J.; Forster, P. Median-Joining networks for inferring intraspecific Phylogenies. *Mol. Biol. Evol.* **1999**, *16*, 37–48. [CrossRef] [PubMed]
46. Hikosaka, K.; Watanabe, Y. Divergence of the mitochondrial genome structure in the apicomplexan parasites, *Babesia* and *Theileria*. *Mol. Biol. Evol.* **2009**, *27*, 1107–1116. [CrossRef] [PubMed]
47. Obara, I.; Ulrike, S.; Musoke, T.; Spooner, P.R.; Jabbar, A.; Odongo, D.; Kemp, S.; Silva, J.C.; Bishop, R.P. Molecular evolution of a central region containing B Cell epitopes in the gene encoding the P67 sporozoite antigen within a field population of *Theileria parva*. *Parasitol. Res.* **2015**, *114*, 1729–1737. [CrossRef] [PubMed]
48. Nene, V.; Musoke, A. Conservation of the sporozoite p67 vaccine atigen in cattle-derived *Theileria parva* stocks with different cross-immunity profiles. *Infect. Immun.* **1996**, *64*, 2056–2061. [CrossRef] [PubMed]
49. Oura, C.A.L.; Odongo, D.O. A Panel of microsatellite and minisatellite markers for the characterisation of field isolates of *Theileria parva*. *Int. J. Parasitol.* **2003**, *33*, 1641–1653. [CrossRef]
50. Norling, M.; Bishop, R.P. The genomes of three stocks comprising the most widely utilized live sporozoite *Theileria parva* vaccine exhibit very different degrees and patterns of sequence divergence. *BMC Genom.* **2015**, *16*, 729. [CrossRef]
51. Bishop, R.P.; Odongo, D. A review of recent research on *Theileria parva*: Implications for the Infection and Ireatment vaccination method for control of East Coast Fever. *Transbound. Emerg. Dis.* **2020**, *67*, 56–67. [CrossRef]
52. Bernt, M.; Braband, A. Genetic aspects of mitochondrial genome evolution. *Mol. Phylogenet Evol.* **2013**, *69*, 328–338. [CrossRef]
53. Hikosaka, K.; Kita, K.; Tanabe, K. Diversity of Mitochondrial Genome Structure in the Phylum Apicomplexa. *Mol. Biochem. Parasitol.* **2013**, *188*, 26–33. [CrossRef]
54. Bishop, R.; Geysen, D. Molecular and immunological characterisation of *Theileria parva* stocks which are components of the 'muguga cocktail' used for vaccination against east coast fever in cattle. *Vet Parasitol.* **2001**, *94*, 227–237. [CrossRef]
55. Neitz, W.O. Theilerioses, Gonderioses and Cytauxzoonoses: A Review. *Onderstepoort J. Vet. Res.* **1957**, *27*, 275–326.
56. Sibeko, K.P.; Collins, N.E. Analyses of genes encoding *Theileria parva* p104 and Polymorphic Immunodominant Molecule (PIM) reveal evidence of the presence of cattle-type alleles in the South African *T. parva* Population. *Vet. Parasitol.* **2011**, *181*, 120–130. [CrossRef] [PubMed]
57. Sitt, T.; Henson, S. Similar levels of diversity in the gene encoding the p67 sporozoite surface protein of *Theileria parva* are observed in blood samples from buffalo and cattle naturally infected from buffalo. *Vet. Parasitol.* **2019**, *269*, 21–27. [CrossRef] [PubMed]

58. Uilenberg, G.; Perié, N.M. Causal agents of bovine theileriosis in southern Africa. *Trop. Anim. Health Prod.* **1982**, *14*, 127–140. [CrossRef] [PubMed]
59. Preston, M.D.; Campino, S. A barcode of organellar genome polymorphisms identifies the Geographic Origin of *Plasmodium falciparum* strains. *Nat. Commun.* **2014**, *5*, 4052. [CrossRef]
60. Sivakumar, T.; Hayashida, K. Evolution and genetic diversity of *Theileria*. *Infect. Genet. Evol.* **2014**, *27*, 250–256. [CrossRef]
61. Brocklesby, D.W.; Martin, H. A new parasite of the eland. *Vet. Rec.* **1960**, *72*, 331–332.
62. Njiiri, N.E.; de Bronsvoort, B.M.C. The epidemiology of tick-borne haemoparasites as determined by the Reverse Line Blot Hybridization assay in an intensively studied cohort of calves in western Kenya. *Vet. Parasitol.* **2015**, *210*, 69–76. [CrossRef]
63. Stagg, D.A.; Young, A.S. Infection of mammalian cells with *Theileria* Species. *Parasitology* **1983**, *86*, 243–254. [CrossRef]
64. Biasibetti, E.; Sferra, C.; Lynen, G.; Di Giulio, G.; De Meneghi, D.; Tomassone, L.; Valenza, F.; Capucchio, M.T. Severe Meningeal Fibrinoid Vasculitis Associated with *Theileria taurotragi* Infection in Two Short-Horned Zebu Cattle. *Trop. Anim. Health Prod.* **2016**, *48*, 1297–1299. [CrossRef]
65. Brown, C.G.D.; Ilhan, T. *Theileria lestoquardi* and *T. annulata* in cattle, sheep, and goats: In vitro and in vivo Studies [a]. *Ann. N. Y. Acad. Sci.* **1998**, *849*, 44–51. [PubMed]
66. Al-Hamidhi, S.; Weir, W. *Theileria lestoquardi* displays reduced genetic diversity relative to sympatric *Theileria annulata* in Oman. *Infect. Genet. Evol.* **2016**, *43*, 297–306. [CrossRef] [PubMed]
67. Bishop, R.; Musoke, A. *Theileria*: Intracellular Protozoan Parasites of Wild and Domestic Ruminants Transmitted by Ixodid Ticks. *Parasitology* **2004**, *129*, S271–S283. [CrossRef] [PubMed]

Publisher's Note: MDPI stays neutral with regard to jurisdictional claims in published maps and institutional affiliations.

© 2020 by the authors. Licensee MDPI, Basel, Switzerland. This article is an open access article distributed under the terms and conditions of the Creative Commons Attribution (CC BY) license (http://creativecommons.org/licenses/by/4.0/).

Article

The Patterns and Puzzles of Genetic Diversity of Endangered Freshwater Mussel *Unio crassus* Philipsson, 1788 Populations from Vistula and Neman Drainages (Eastern Central Europe)

Adrianna Kilikowska [1], Monika Mioduchowska [1,2], Anna Wysocka [1], Agnieszka Kaczmarczyk-Ziemba [1], Joanna Rychlińska [1], Katarzyna Zając [3,*], Tadeusz Zając [3], Povilas Ivinskis [4] and Jerzy Sell [1]

1. Department of Genetics and Biosystematics, Faculty of Biology, University of Gdańsk, Wita Stwosza 59, 80-308 Gdańsk, Poland; adrianna.kilikowska@biol.ug.edu.pl (A.K.); monika.mioduchowska@ug.edu.pl (M.M.); anna.wysocka@ug.edu.pl (A.W.); agnieszka.kaczmarczyk-ziemba@ug.edu.pl (A.K.-Z.); joannry@gmail.com (J.R.); jerzy.sell@biol.ug.edu.pl (J.S.)
2. Department of Marine Plankton Research, University of Gdansk, Piłsudskiego 46, 81-378 Gdynia, Poland
3. Institute of Nature Conservation, Polish Academy of Sciences, 31-120 Kraków, Poland; tzajac@iop.krakow.pl
4. Nature Research Centre, Akademijos 2, LT-08412 Vilnius, Lithuania; ivinskis@ekoi.lt
* Correspondence: kzajac@iop.krakow.pl

Received: 16 June 2020; Accepted: 16 July 2020; Published: 21 July 2020

Abstract: Mussels of the family Unionidae are important components of freshwater ecosystems. Alarmingly, the International Union for Conservation of Nature and Natural Resources Red List of Threatened Species identifies almost 200 unionid species as extinct, endangered, or threatened. Their decline is the result of human impact on freshwater habitats, and the decrease of host fish populations. The Thick Shelled River Mussel *Unio crassus* Philipsson, 1788 is one of the examples that has been reported to show a dramatic decline of populations. Hierarchical organization of riverine systems is supposed to reflect the genetic structure of populations inhabiting them. The main goal of this study was an assessment of the *U. crassus* genetic diversity in river ecosystems using hierarchical analysis. Different molecular markers, the nuclear ribosomal internal transcribed spacer ITS region, and mitochondrial DNA genes (*cox1* and *ndh1*), were used to examine the distribution of *U. crassus* among-population genetic variation at multiple spatial scales (within rivers, among rivers within drainages, and between drainages of the Neman and Vistula rivers). We found high genetic structure between both drainages suggesting that in the case of the analyzed *U. crassus* populations we were dealing with at least two different genetic units. Only about 4% of the mtDNA variation was due to differences among populations within drainages. However, comparison of population differentiation within drainages for mtDNA also showed some genetic structure among populations within the Vistula drainage. Only one haplotype was shared among all Polish populations whereas the remainder were unique for each population despite the hydrological connection. Interestingly, some haplotypes were present in both drainages. In the case of *U. crassus* populations under study, the Mantel test revealed a relatively strong relationship between genetic and geographical distances. However, in detail, the pattern of genetic diversity seems to be much more complicated. Therefore, we suggest that the observed pattern of *U. crassus* genetic diversity distribution is shaped by both historical and current factors i.e. different routes of post glacial colonization and history of drainage systems, historical gene flow, and more recent habitat fragmentation due to anthropogenic factors.

Keywords: *Unio crassus*; freshwater mussels; population genetics; genetic diversity; mtDNA; ITS

1. Introduction

Mussels of the family Unionidae with 680 described species [1] are important components of freshwater ecosystems. Alarmingly, the International Union for Conservation of Nature and Natural Resources Red List identifies almost 200 species of this family as extinct, endangered, or threatened [2] which makes unionids one of the most endangered groups of invertebrates in the world [3,4]. Their decline is the result of ever-increasing human impact on freshwater habitats, such as the regulation and impoundment of rivers or water pollution. The decline of host fish populations also has an effect on extirpation of freshwater mussels as their larvae (glochidia) are parasites on fish, which serve to complete their life cycle and disperse progeny [3,5–9].

In Europe, the remarkable decline suffered by freshwater mussel populations has attracted the attention of conservation organizations [10]. Over recent years some comprehensive studies have been published concerning biology, ecology, phylogeny, and conservation status of European freshwater mussels (e.g., [11–23]). This is all the more important in that the situation of freshwater bivalves in Europe is even more alarming than in North America. Among European freshwater mussel species, 75% of the species are categorized as Threatened or near Threatened [21]. In comparison with other taxa in Europe (e.g., fish, amphibians, birds, mammals), the real situation is poorly understood and especially alarming due to much lower species biodiversity within the whole superfamily Unionoidea.

Much of the global awareness of freshwater mussel decline stems from North American Unionoidea, which constitutes the continent's most imperiled fauna, but much more numerous in terms of species than the rest of the continents [24,25]. Over 70% of North American species are considered imperiled at some level [24] and more than 25 species are presumed extinct (see [4,26] for details). Consequently, many more studies have been focused on Unionoidea species from North America and many of these have reported upon molecular based phylogenies of this group of mollusks (e.g., [27–36]).

The Thick Shelled River Mussel, *Unio crassus* Philipsson, 1788 is one of the examples that has been reported to show a dramatic decline of populations within the western and central European part of its range and thus has become a major target species for conservation [37–40]. In the eastern part of the mussel range the situation is better [41]. Up to the 20th century *U. crassus* occupied and colonized a wide range of habitats and was considered the most abundant unionid species in central and northern Europe ([42] and references therein). In Poland, up to the last century it had been also a dominant species in many rivers, reaching extremely high densities [43]. However, today its numbers have fallen dramatically, in line with the deterioration of water quality. Although potentially harmful effects of anthropogenic activities, including water pollution and flow modification, have been reduced in central Europe over the last decades, *U. crassus* populations are not recovering accordingly ([44] and references therein). Consequently, it has been listed as endangered (category EN) in the IUCN Red List of threatened species [2] as well as protected in Europe e.g., in the European Union by the enclosed Appendixes II and IV of the Habitats Directive.

Preservation of biodiversity requires not only the protection of individual taxa but also the preservation of genetic diversity. Although molecular data seem to be critical for the conservation management of imperiled freshwater mussels, the knowledge about genetic diversity of *U. crassus* populations is still very scarce. It seems that first Nagel and Badino (2001) [45] reported on the genetic variability within and between *U. crassus* populations. However, the basic aim of that study was to solve the taxonomic and phylogenetic problems within Unionidae. Moreover, the study was based on allozyme markers, which are widely known to underestimate genetic variation, especially on the recent time scale and to be influenced by natural selection.

Since then, other molecular investigations including a limited number of *U. crassus* individuals have been reported. However, these studies analyzed taxonomic uncertainties at the genus or family level, establish phylogenetic relationships, biogeographic patterns or evolutionary history of European unionids rather than focusing on population genetics of this species (e.g., [21,46–48]). More precisely, Prié and Puillandre (2014) [49] used mitochondrial DNA (mtDNA) data of *Unio* species from France (including limited samples of *U. crassus*) to clarify the unionid taxonomy in this country.

Similar taxonomic studies have been performed in the area covering Russia and Ukraine [50,51]. Except for analyses based on mtDNA, there have also been investigations of the phylogenetic relationships within unionids using sequences of nuclear DNA (nDNA) and the transcribed spacers ITS1 and ITS2 [52]. Some *U. crassus* sequences have been used to evaluate an application of the ITS region as a phylogenetic marker [12]. Using the distribution data and multi-locus phylogeny (COI, 16S rRNA, 28S rRNA) Bolotov et al. (2020) [41] described the actual taxonomic richness of Unionidae in Russia and Kazakhstan with distribution patterns for each genus and species including *U. crassus*. Feind et al. (2018) [53] reported results of the *U. crassus* populations structure analyses in six major drainage systems in Germany and Sweden using a set of nine microsatellite markers. Additionally, mitochondrial sequences of *U. crassus* have been used to establish the utility of paternally-inherited markers in phylogeographical studies [54].

Thus, given the extremely limited genetic data on populations of *U. crassus*, studies of the genetic diversity and population structure of this species are needed. Especially that, recent analyses of another endangered freshwater mussel—*Margaritifera margaritifera* (Linnaeus, 1758)—have demonstrated that knowledge of the genetic structure of populations can be extremely useful for their conservation [3,5,6,55–57].

In general, genetic variability of a species is partitioned into variation within and among populations and groups of populations. The opposing forces of genetic drift and gene flow determine the relative proportion of neutral genetic variation within each of these two components [58]. Genetic drift results in loss of genetic diversity within a population and at the same time promotes differentiation of populations [59]. Conversely, gene flow among populations tends to increase variation within populations, while minimizing differences among these populations. In aquatic invertebrates, dispersal is a major component of gene flow among populations [60]. The effects of genetic drift occur more rapidly in small populations, and isolation promotes population differentiation [61]. Reduction of a within-population variation due to genetic drift increases the probability of extinction [62–64]. In addition, genetic changes due to isolation will increase the genetic structure among populations. Thus, preservation of the genetic diversity within a conservation framework requires understanding of both within-population genetic variation and patterns of variation among populations across the species geographical range landscape. Such knowledge seems to be extremely important in the case of endangered freshwater mussels.

Because of the hierarchical organization of riverine systems (first order streams giving rise to second order streams, etc.), populations of organisms within these systems are likely to have a genetic structure that reflects such a type of organization. Thus, hierarchical analysis of genetic variation is a particularly useful approach for genetic diversity assessment of riverine organisms ([32] and references therein).

In the present study, partial DNA sequence data of two mitochondrial genes—NADH dehydrogenase subunit 1 (*ndh1*) and cytochrome oxidase subunit I (*cox1*) as well as the entire ITS region of the nuclear ribosomal DNA (herein called nrDNA)—were used to:

(i) test the correlation between the genetic differentiation and geographic isolation of *U. crassus* populations using a hierarchical approach,
(ii) examine the distribution of genetic variation among populations of *U. crassus* at multiple spatial scales (within rivers, among rivers within drainages, and across drainages),
(iii) test the role of current vs historical gene flow into the distribution of *U. crassus* genetic diversity on the basis of populations representing currently isolated Neman and Vistula drainages (Central Europe) with the reported ancient connection,
(iv) asses the phylogenetic affinities among *U. crassus* populations.

2. Materials and Methods

2.1. Sample Collection and Identification

The conservation status of *U. crassus* categorized as Endangered species prevented us from dealing with large numbers of specimens, which were sometimes below the recommended figures for population structure analyses. Nevertheless, this is currently an insurmountable problem that should not dissuade us from obtaining as much data as possible on these endangered species. In total, 99 specimens representing *U. crassus* species were collected from 6 localities in Poland and 6 in Lithuania (Figure 1, Tables S1 and S2). Moreover, individuals of *Unio tumidus* Philipsson, 1788 were collected from the Pilica river and used as an outgroup in the analyses.

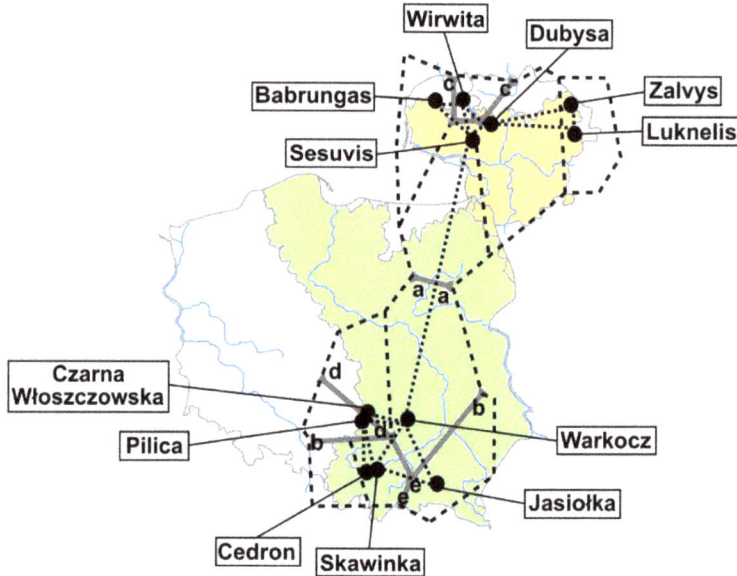

Figure 1. Sampling sites and spatial analysis based on Monmonier's maximum difference algorithm [65] for detecting genetic barriers (genetic breaks) among populations of *U. crassus*. Vistula and Neman drainages are marked with different colors, identified barriers are indicated as bold grey lines (the first five barriers are shown, from "a" to "e"), the Delaunay triangulation is visualized as black dotted lines. Locality codes: CED–Cedron, CZW–Czarna Włoszczowska, JAS–Jasiołka, PIL–Pilica, SKA–Skawinka, WAR–Warkocz, BAB–Babrungas, DUB–Dubysa, LUK–Luknelis, SES–Sesuvis, VIR–Virvita, ZAL–Zalvys.

Samples of *U. crassus* were collected from habitats localized in central and southern Poland, from EU protected areas of the Natura 2000. Additional samples were also collected from selected rivers in Lithuania, also within the Natura 2000 areas. Details on sampling localities are presented in Table S2. The available reports on the monitoring results indicate that overall conservation status of the *U. crassus* is unsatisfactory—assessments of conservation status of the species under Article 17 of the Habitats Directive in period 2013–2018 for Poland, Lithuania and EU are available at "Article 17 web tool" [66].

In the sampling design (Figure 1, Table S2) we considered the distribution of among-population genetic variation at multiple spatial scales: i) within rivers and its tributaries (the Skawinka river with its tributary the Cedron; the Pilica river with its tributary the Czarna Włoszczowska); ii) among rivers within drainages: The Vistula drainage (Skawinka, Cedron, Pilica, Czarna Włoszczowska, Jasiołka, Warkocz); also having regard to division for the Upper (Cedron, Jasiołka, Skawinka, Warkocz) and the

Middle Vistula (Pilica, Czarna Włoszczowska,), and the Neman drainage (Babrungas, Luknelis, Dubysa, Zalvys); Virvicia (tributary of Venta) was also included in the following analyses of Neman drainage due to the Windawski canal of only 15 km in length connecting Dubysa and Venta); iii) across drainages.

Specimens were collected and transported to the laboratory alive in water from the individual localities. Identification down to the species level was based upon morphological diagnostic characters provided by Piechocki and Dyduch-Falniowska (1993) [67]. A small fragment of somatic tissue (gills) was taken from each individual and then immediately frozen at −80 °C. Individuals from Lithuania were preserved in 96% ethanol.

2.2. Extraction, PCR Amplification and Sequencing

DNA was extracted from a small piece of gill tissue of each specimen using a modified phenol/chloroform method [68]. Since only somatic tissue was used, we assumed that we had extracted only F-type mitochondria.

The entire ITS region of nrDNA was amplified using the primers ITS4 (5'– TCCTCCGCTT ATTGATATGC–3') and ITS5 (5'–GGAAGTAAAAGTCGTAACAAGG–3') of [69], annealing to the 5' end of 28S and 3' end of 18S rRNA genes, respectively. Parts of the NADH dehydrogenase, subunit 1 (*ndh1*) and cytochrome c oxidase subunit 1 (*cox1*) genes from mtDNA were amplified with the primer combinations: for *ndh1* 5'–TGGCAGAAAAGTGCATCAGATTTAAGC–3' and 5'–GCTATTAGTAGGTCGT ATCG–3' [70,71], for *cox1* 5'–GTTCCACAAATCATAAGGATATTGG–3' and 5'–TACACCTCAGGGTGACCAAA AAACCA–3' [72], respectively.

All PCR reactions were performed in 25 mL volumes with 0.4 µL 5 µM of each primer and about 10 ng of template using a cycling profile of 95 °C for 5 min, followed by 30 cycles of 95 °C for 30 s, 48 °C, 50 °C, 51 °C for 60 s, and 72 °C for 60 s. The mentioned annealing temperature was used for the *cox1*, *ndh1*, and nrDNA, respectively. For some specimens it was impossible to amplify all three fragments of DNA (Table S1). PCR products were cleaned up by exonuclease I (20 U/µL, Thermo Scientific) and alkaline phosphatase FastAP (1 U/µL, Thermo Scientific) treatment according to the manufacturer's guidelines, and sequenced directly in both directions using the same primers as at the amplification stage.

2.3. DNA Sequence Analysis

To verify the identity of the amplified region, BLAST [73] searches at the National Center for Biotechnology Information NCBI were performed. Sequences were quality checked and trimmed to the same length in BioEdit version 7.2.5 [74] and a consensus sequence was created for each individual. In the case of *cox1* and *ndh1* genes, the sequences could be unambiguously aligned without inserting gaps. The sequences of nrDNA were aligned using Clustal Omega [75] with default settings. Although the nrDNA region alignment had several indels, all specimens yielded sequences that were readily readable without cloning and were therefore included in the analyses. In total, 242 new sequences of *U. crassus* were obtained and deposited in GenBank (Table S1): 83 sequences of *cox1* (614 bp-long), 94 sequences of *ndh1* (859 bp-long), and 65 sequences of the nrDNA region (879 bp-long).

Data from each of the three regions were analyzed separately. Moreover, when the sequences of both mtDNA genes (*cox1* and *ndh1*) were known for particular individuals (Table S1), the data were concatenated and analyzed as a single mtDNA locus. To analyze the mitochondrial gene pairs as a single locus, congruence among tree topologies of *cox1* and *ndh1* regions was assessed by the partition homogeneity test in PAUP*.

Two data sets were analyzed for each DNA region; one consisted of only the *U. crassus* sequences, while the second set comprised also *U. tumidus* as an outgroup (newly obtained *U. tumidus* sequences were also deposited in GenBank–Acc. Nos: KJ525923–KJ525927; KJ525965, KJ525966).

Alignment statistics and DNA polymorphism, quantified as the number of haplotypes (n), haplotypic diversity (h) and nucleotide diversity (π) were calculated in DnaSP v5.10.01 [76].

We used the Arlequin v.3.5. software [77] for analysis of molecular variation. This software takes into account both haplotype frequency and molecular sequence divergence. Genetic differentiation was analyzed within-drainage for the Vistula (VIS) and Neman (NEM) drainages. We calculated genetic variation partitioned into three hierarchical levels using analysis of molecular variance (AMOVA) [77]: within populations (Φ_{ST}), between populations within rivers and its tributaries (Φ_{SR}), and among rivers (Φ_{RT}). We then calculated among-drainage variation using three hierarchical levels: within populations (Φ_{ST}), among populations within drainages (Φ_{SD}), and between drainages (Φ_{DT}). The number of migrants and pairwise differentiation between populations were calculated using Φ_{ST}, a direct analogue of Wright's F_{ST} for nucleotide sequence divergence. Also, these estimates were calculated for data grouped according to drainage (VIS, NEM) and hydrological division (rivers of southern Poland representing the Upper Vistula, rivers of central Poland representing the Middle Vistula). The significance of this estimate was calculated using 10,000 permutations of the data and was considered significant at $P = 0.05$.

To test for isolation by distance, the shortest geographic distance between populations was calculated using the Earth great circle distances calculator [78] and F_{ST} values for population pairs were tested for correlation with distance using Mantel's test [79] as implemented in Arlequin v.3.5.

For mtDNA, we calculated a network of all individual gene sequences using the median-joining (MJ) algorithm implemented in Network v. 4.5.1.6 software [80] The method groups related haplotypes through median vectors into a tree or network. Different settings for the homoplasy level parameter (ε) were tested, and ε = 20 was eventually used. To account for differences in substitution rates the weight of 1 for transitions and 2 for transversions was applied. Ambiguous relationships were resolved with a Maximum Parsimony (MP) heuristic algorithm. With this analysis, we could determine the relationship among haplotypes and the frequencies of these haplotypes in our sampling.

In addition, we estimated the phylogenetic relationship among *U. crassus* haplotypes with *U. tumidus* sequences as an outgroup. First, the most appropriate model of sequences evolution was determined, and the nucleotide substitution parameters were estimated by jModelTest 2 [81] using Bayesian Information Criterion. For the concatenated mtDNA data, the preferred model of nucleotide substitution was the Hasegawa, Kishino, Yano 85 (HKY) [82] model. The likelihood-estimated transition/transversion ratio was 2.211. In the case of the nrDNA region, the selected model was the Kimura's two-parameter model (K2P) [83] with gamma-distributed rate heterogeneity ($\Gamma = 0.512$). The likelihood-estimated transition/transversion ratio was 0.882. Maximum likelihood trees were calculated using Mega v.7.0 [84] under the general settings of the selected models with 1000 bootstraps.

Identification of barriers to gene-flow between populations of *U. crassus* was conducted using Barrier v2.2 [65]. The geographical map was created with a dual structure of the Voronoi diagram [85], the Delaunay triangulation method [86], which allowed populations with a set of triangles to be connected. This analysis was based on coordinates of the sampling sites. The Fst values calculated by Arlequin v.3.5. were used as a matrix data to link computational geometry of the Delaunay network. Finally, in the geometric network, genetic boundaries (in hierarchical order, from "a" to "e") were identified and calculated according to the Monmonier's maximum difference algorithm [87].

For estimating ancestral population dynamics through time on the basis of mtDNA sequences, BEAST 1.7. was used [88]. BEAST uses standard Markov chain Monte Carlo (MCMC) sampling procedures to estimate a posterior distribution of effective population size through time directly from a sample of gene sequences, given any specified nucleotide substitution model. The data set for this analysis consisted of the alignment of all obtained sequences, not just unique haplotypes. Three demographic models: constant size, exponential population growth, and Bayesian Skyline were tested. Each of these was run with the eight possible site models: HKY or GTR (General Time Reversible) with either equal rates, proportion of invariable sites or gamma distributed variable rates. To ensure convergence and achieve a high Effective Sample Size (ESS) value (at least 300) for every estimated parameter, each analysis was run at least in quadruplicate. After examination of the log files in Tracer [88], the results from all the runs were combined in LogCombiner 1.7.5 [88], for each model respectively, removing the non-stationary burning data. The best demographic and site models were a posteriori selected, through Bayes Factor comparisons, using the method of Newton and Raftery (1994) with the modification proposed by Suchard, Weiss, and Sinsheimer (2001) as implemented in Tracer [89,90]. The models implementing exponential population growth were slightly better than constant size or the bulk synchronous parallel (BSP) models, but without significant differences.

3. Results

Bearing in mind that all three genes could not be analyzed in all specimens, 243 sequences (83 *cox1*, 95 *ndh1*, and 65 ITS region) were obtained for the 99 specimens examined (Table S1). Only somatic tissue was sampled, and there was no evidence of heteroplasmy or doubly uniparental inheritance DUI [91].

The partition homogeneity test (as implemented in PAUP) showed no significant differences between the phylogenies reconstructed from *cox1* and *ndh1* (P = 0.35), such that the data sets could be combined.

3.1. Mitochondrial Gene Regions

The alignment of the two combined mitochondrial regions of *U. crassus* produced 1473 characters (614 for *cox1* and 859 for *ndh1*), 46 of which were variable and 41 parsimony informative.

The results of mtDNA polymorphism analyses and relative frequencies of haplotypes for separate and combined mtDNA data are presented in Table 1. Seventeen mitochondrial haplotypes (h = 0.688, π = 0.012) were found among 79 individuals of *U. crassus* from 12 localities in Poland and Lithuania where sequences for both genes (*cox1* and *ndh1*) were available.

Table 1. Polymorphism estimates and relative frequencies of haplotypes for separate and combined mtDNA data of *U. crassus* from the studied localities. Locality code: CED–Cedron, CZW–Czarna Włoszczowska, JAS–Jasiołka, PIL–Pilica, SKA–Skawinka, WAR–Warkocz, BAB–Babrungas, DUB–Dubysa, LUK–Luknelis, SES–Sesuvis, VIR–Virvita, ZAL–Zalvys; N–number of individuals; n–number of haplotypes; π–nucleotide diversity; HD (SE)–standard error.

Haplotype	Vistula Drainage						Neman Drainage					
cox1	PIL	CZW	WAR	CED	SKA	JAS	BAB	DUB	LUK	SES	VIR	ZAL
C1	0.500	0.875	0.727	1.000	1.000	0.833						
C2		0.125										
C3						0.167						
C4	0.400						0.250		1.000			
C5	0.100											
C6			0.273									
C7							0.125			0.333	1.000	0.400
C8							0.500			0.667		0.600
C9							0.125					
C10								0.500				
C11								0.500				
N	10	8	11	12	8	12	8	2	1	3	3	5
n	3	2	2	1	1	2	4	2	1	2	1	2
HD	0.644	0.250	0.436	0.000	0.000	0.303	0.750	-	-	0.667	0.000	0.600
(SE)	(0.101)	(0.180)	(0.133)			(0.147)	(0.139)			(0.314)		(0.175)
π	0.016	0.008	0.012	0.000	0.000	0.001	0.003	-	-	0.001	0.000	0.001
(SE)	(0.009)	(0.005)	(0.007)			(0.001)	(0.002)			(0.001)		(0.001)
ndh1	PIL	CZW	WAR	CED	SKA	JAS	BAB	DUB	LUK	SES	VIR	ZAL
N1	0.500	0.750	0.727	0.833	1.000	1.000				0.200		
N2				0.083								
N3		0.125		0.083								
N4	0.200	0.125	0.272				0.500	0.667	0.571	0.200		0.250
N5	0.100											
N6	0.100											
N7							0.125			0.200	1.000	0.250

Table 1. Cont.

Haplotype	Vistula Drainage						Neman Drainage					
N8							0.375	0.333	0.429	0.400		0.500
N	9	8	11	12	9	10	8	3	7	5	5	8
n	4	3	2	3	1	1	3	2	2	4	1	3
HD	0.694	0.4643	0.436	0.318	0.000	0.0000	0.679	0.667	0.571	0.900	0.000	0.714
(SE)	(0.147)	(0.200)	(0.133)	(0.164)			(0.122)	(0.314)	(0.119)	(0.161)		(0.123)
π	0.012	0.006	0.009	0.001	0.000	0.000	0.001	0.001	0.001	0.010	0.000	0.001
(SE)	(0.007)	(0.003)	(0.005)	(0.001)			(0.001)	(0.001)	(0.001)	(0.006)		(0.001)
cox1+ ndh1	PIL	CZW	WAR	CED	SKA	JAS	BAB	DUB	LUK	SES	VIR	ZAL
CN1	0.375	0.750	0.727	0.833	1.000	0.800						
CN2				0.083								
CN3		0.125		0.083								
CN4		0.125										
CN5						0.200						
CN6	0.250						0.250		1.000			
CN7	0.125											
CN8	0.125											
CN9	0.125											
CN10			0.273									
CN11							0.125			0.333	1.000	0.200
CN12							0.375			0.667		0.600
CN13							0.125					
CN14							0.125					
CN15								0.500				
CN16								0.500				
CN17												0.200
N	8	8	11	12	8	10	8	2	1	3	3	5
n	5	3	2	3	1	2	5	2	1	2	1	2
HD	0.857	0.464	0.436	0.318	0.000	0.356	0.857	-	-	0.667	0.000	0.700
(SE)	(0.108)	(0.200)	(0.133)	(0.164)		(0.159)	(0.108)			(0.314)		(0.218)
π	0.014	0.007	0.011	0.000	0.000	0.000	0.002	-	-	0.001	0.000	0.001
(SE)	(0.008)	(0.004)	(0.006)	(0.000)	(0.000)	(0.000)	(0.001)			(0.001)		(0.001)

Below, we present the results of analyses for combined mtDNA data and selected results for both mitochondrial regions (cox1 and ndh1) analyzed separately.

3.2. Within Population Variability

Analyses of within population structure revealed a high number of mtDNA haplotypes unique to a particular population (Table 1). In the case of mtDNA haplotype richness, Polish and Lithuanian populations averaged 2.7 and 2.3 haplotypes respectively (Table 1). The sequence divergence within-population was rather low with a maximum value (π = 0.014) for Pilica (Table 1).

3.3. Among Population Structure

For mtDNA data, only one haplotype (CN6) was observed in populations from both drainages—Vistula and Neman. Only one haplotype (CN1) was shared among all Polish populations but we did not find any haplotype common to all Lithuanian populations. However, in the case of ndh1 polymorphism, N4 and N8 haplotypes were observed in almost all Lithuanian populations (with the exception of Virvicia). The relative frequencies of shared mtDNA haplotypes varied from 0.083 to 1.000 (Table 1).

A hierarchical AMOVA conducted over the 12 populations of the two drainages indicated that approximately 77% of the variation arose from genetic variation between populations from the Neman and Vistula drainages (Φ_{DT} = 0,771; P = 0.003), whereas only about 4% of the variation was due to differences among populations within drainages (Φ_{SD} = 0.189; P < 0.001). Correspondingly, there was significant mtDNA structure among all *U. crassus* populations (Φ_{ST} = 0.813; P < 0.001).

Comparison of population differentiation within drainages for mtDNA showed also some genetic structure among populations within the Vistula drainage (Φ_{ST} = 0.221; P = 0.007). However, the estimated level of diversity was much lower than in the case of between drainages comparison.

The population structure occurred also on the finest spatial scale: between populations from river Pilica and its tributary Czarna Włoszczowska (Φ_{ST} = 0.181 but P = 0.100). Interestingly, strong genetic differentiation also occurred between populations from rivers located in Central and Southern Poland belonging to the Middle and Upper Vistula drainages, respectively (Φ_{ST} = 0.302, P = 0.000).

In the case of the Neman drainage, the Φ_{ST} estimate value (Φ_{ST} = 0.146) indicated a moderate level of genetic structure among analyzed populations, however the value was not significant (P = 0.107).

Pairwise Φ_{ST} sample comparisons and gene flow estimations (Nm) between populations are summarized in Table 2. The Mantel's test results (Figure 2) indicated, in general, a significant relationship between pairwise Φ_{ST} estimates and geographical distances for comparisons among all locations (r^2 = 0.622, P < 0.001). However, in some cases it was observed that pairwise Φ_{ST} estimates did not increase with geographic distance between sampling sites, and distant populations did not show higher genetic structure (i.e., populations from Pilica and Luknelis representing different drainages; Table 2). On the contrary, as already mentioned adjacent populations from Pilica and its tributary Czarna Włoszczowska showed significant differentiation (Φ_{ST} = 0.181; Table 2).

Table 2. The estimates of Φ_{ST} (below diagonal) and number of migrants (Nm, above diagonal) between pairs of populations of *U. crassus* from different localities. Locality code: CED–Cedron, CZW–Czarna Włoszczowska, JAS–Jasiołka, PIL–Pilica, SKA–Skawinka, WAR–Warkocz, BAB–Babrungas, DUB–Dubysa, LUK–Luknelis, SES–Sesuvis, VIR–VIRVITA, ZAL–Zalvys. Fairy gray background indicates Vistula drainage; estimates between Skawinka with its tributary Cedron and Pilica with its tributary Czarna Włoszczowska were marked in bold. Dark grey background indicates Neman drainage. * $p < 0.05$.

Locality Code	CED	CZW	JAS	PIL	SKA	WAR	BAB	DUB	LUK	SES	VIR	ZAL
CED		10.726	7.635	10.726	inf	1.889	0.018	0.010	0.005	0.008	0.003	0.009
CZW	0.044		15.139	**2.263**	inf	inf	0.118	0.184	0.213	0.155	0.146	0.128
JAS	0.061	0.032		0.567	5.714	2.203	0.019	0.012	0.005	0.009	0.004	0.010
PIL	0.502 *	**0.181**	0.469 *		0.658	31.104	0.776	2.185	inf	1.304	1.252	0.898
SKA	−0.038	0.000	0.080	0.432 *		2.750	0.019	0.008	0.000	0.006	0.000	0.008
WAR	0.209 *	−0.047	0.185	0.016	0.154		0.291	0.486	0.682	0.400	0.389	0.330
BAB	0.966 *	0.809 *	0.962 *	0.392 *	0.963 *	0.632 *		9.900	inf	inf	1.066	inf
DUB	0.980 *	0.730	0.977 *	0.186	0.982 *	0.507 *	0.048		inf	inf	0.536	114.931
LUK	0.991	0.701	0.990	−0.137	1.000	0.423	−0.128	−0.112		0.603	0.000	0.445
SES	0.984 *	0.764 *	0.982 *	0.277 *	0.988 *	0.555 *	−0.053	−0.126	0.453		0.500	inf
VIR	0.992 *	0.774 *	0.992 *	0.285 *	1.000 *	0.562 *	0.319	0.482	1.000	0.500		0.495
ZAL	0.981 *	0.796 *	0.979 *	0.358 *	0.984 *	0.602 *	−0.004	0.004	0.529	−0.336	0.502	

Mismatch analysis of mtDNA haplotypes showed an average of 8.16 bp difference between sequences (1.81% sequence difference), within a range of 1–32 bp difference. These haplotype differences are mirrored in network reconstructions (Figure 3) where we found two evolutionary distinct groups of haplotypes separated by a high number of mutational steps (32 for combined mtDNA data). On the contrary, haplotypes within each of the groups were separated by only one or two mutational steps. Most individuals from Vistula drainage represented one haplotype. The observed distribution of mtDNA haplotypes among *U. crassus* populations from different localities in most cases is congruent with geographical subdivision into Vistula and Neman drainage. The only exceptions are mtDNA haplotypes from several individuals found in central Poland rivers (PIL, CZW, WAR) that clustered with haplotypes from Lithuanian populations.

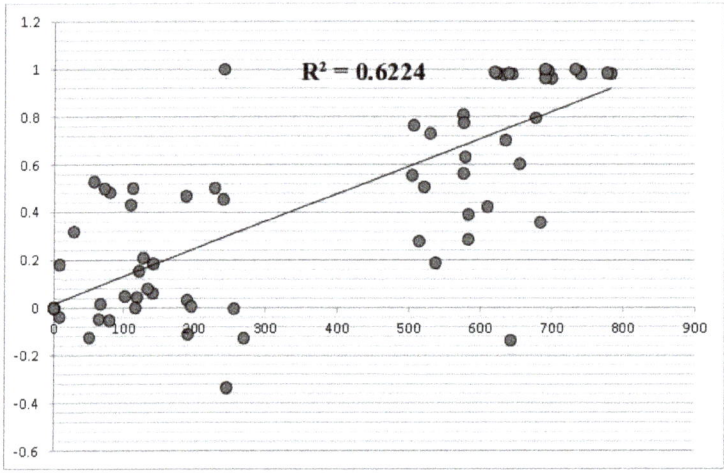

Figure 2. Results of Mantel tests between pairwise geographic distance among sampling sites in km (x axis), and pairwise Fst estimates (y axis) using mtDNA haplotypes.

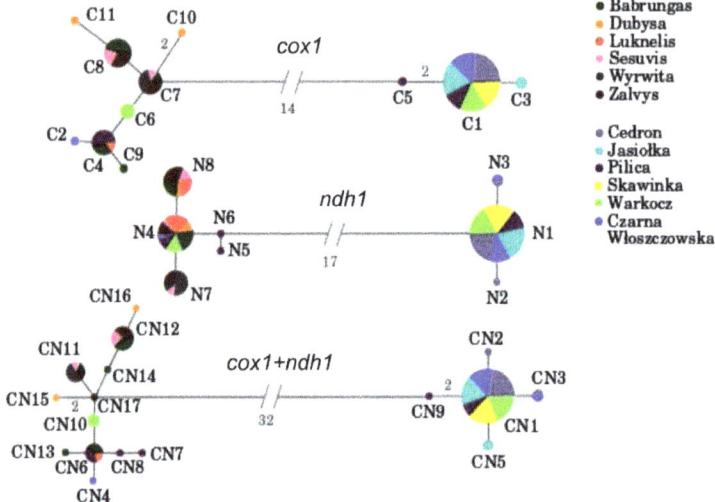

Figure 3. Median-joining network based on mtDNA sequences of *U. crassus*. The size of the circles is proportional to the number of sequences. The mutational steps values are indicated along the lines.

The topology of the ML phylogenetic tree based on concatenated mtDNA data revealed evident phylogeographic structure with strong bootstrap support (Figure 4a). First clade (clade 1) consisted only of haplotypes from all Polish localities (Vistula drainage) whereas in the second (clade 2) haplotypes from both drainages were intermingled. However only haplotypes from the central Poland region: Pilica, its tributary—Czarna Włoszczowska and Warkocz (PIL, CZW, WAR)—were found in the second clade together with Lithuanian haplotypes. The southern Poland haplotypes from Cedron, Jasiołka, and Skawinka (CED, JAS, SKA) were grouped in the first clade. So, neither Polish nor Lithuanian populations relating to the Vistula and Neman rivers, respectively formed monophyletic clade.

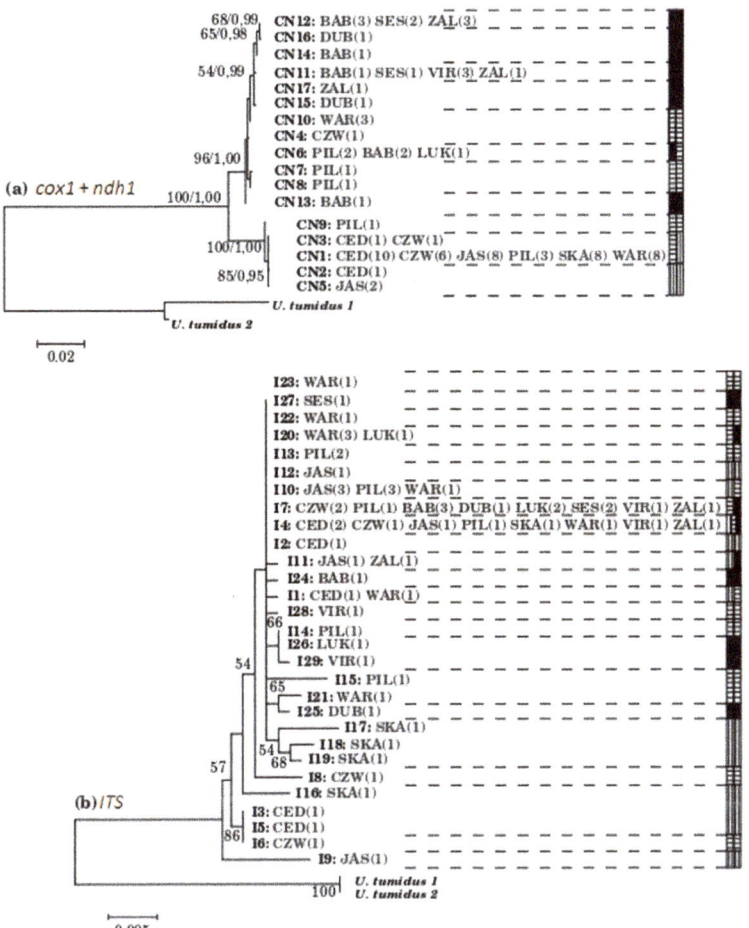

Figure 4. Phylogenetic ML trees of *U. crassus* haplotypes. (**a**) based on concatenated mitochondrial data (*cox1* + *ndh1* regions); (**b**) based on the entire ITS region of the nuclear ribosomal DNA variants of *U. crassus*; gaps were included. The trees were rooted with *Unio tumidus* (Acc. Nos: KJ525923–KJ525927; KJ525965, KJ525966). Bootstrap values higher than 50 are given next to the respective node. The scale bar indicates the number of substitutions per site. Locality codes: CED–Cedron, CZW–Czarna Włoszczowska, JAS–Jasiołka, PIL–Pilica, SKA–Skawinka, WAR–Warkocz, BAB–Babrungas, DUB–Dubysa, LUK–Luknelis, SES–Sesuvis, VIR–Virvita, ZAL–Zalvys. Unique haplotypes were given Arabic numbers. Numbers in brackets indicate the number of individuals from a particular locality. Colors depict geographical regions: Lithuania (black), Central Poland (grey), and Southern Poland (white).

Genetic discontinuities between populations of *U. crassus* were identified as barriers to gene-flow (Figure 1). Using Monmonier's maximum difference algorithm we indicated two significant barriers, described as "a" and "b". Furthermore, three other barriers were included, indicated as "c", "d", and "e", showing where gene-flow was also limited (Figure 1). The first main predicted barrier ("a") separated all populations from the Vistula drainage (Poland) and the Neman drainage (Lithuania). The second main barrier ("b") separated populations from Vistula drainage according to their geographical localization: Middle and Upper Vistula drainages.

The BSP method revealed a flat plot without fluctuation with a relatively recent decrease and rapid increase in the effective population size of *U. crassus* (Figure 5). The apparent decline started at the time needed to accumulate 0.0015 substitutions per site in the analyzed combined mtDNA fragments. The rapid rise started near to 0.0001 substitutions per site. The calculation of the approximate dates, when these two demographic events took place was done according to the molecular rate for the order Unionida (0.265 ± 0.06% per million years) estimated by Froufe et al. (2016) [92]. According to these estimates the decline could have started in the Middle Pleistocene (ca. 566 ka BP), while the rapid incline in the Late Pleistocene, during the Weichselian glaciation (ca. 38 ka BP), before the Last Glacial Maximum.

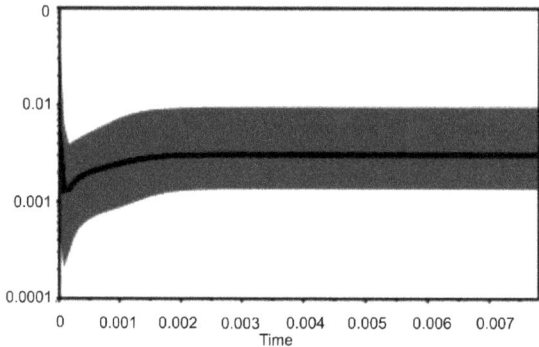

Figure 5. Bayesian skyline plot derived from a set of mtDNA sequences of *U. crassus*. The x axis is in units of time (mutation per site), and the y axis is equal to Neτ (the product of the effective population size and the generation time in mutational units). The median estimates are shown as thick black line, the 95% highest posterior density (HPD) limits are shown by the grey areas.

3.4. The Entire ITS Region of the Nuclear Ribosomal DNA

The alignment of the nrDNA region produced 879 characters and possessed one 11-bp, one 6-bp, and several 1- to 2-bp separate indels. The number of variable and parsimony informative characters was 42 and 13 respectively when gaps were treated as missing information.

The results of nrDNA polymorphism analyses (gaps considered) are presented in Table 1. In total, 29 nuclear sequence variants (h = 0.919, π = 0.006) were found among 62 individuals of *U. crassus* from 12 localities in Poland and Lithuania (Table 1). When sites with gaps were excluded, the number of nuclear sequence variants was reduced to 19 (h = 0.602, π = 0.054, data not shown).

Similar to mtDNA data, we observed nrDNA variants shared by Lithuanian populations and populations from Central Poland (PIL, CZW, WAR): I7 and I20 (Table 1). Interestingly, there were also cases of nrDNA variants found in Lithuanian as well as southern Polish populations: I4 and I11 (Table 1). I4 was shared among all Polish populations and two Lithuanian ones and the other way round—I7 was shared among all Lithuanian and two Polish ones. There was one more nrDNA variant not unique for a single population (Table 1): I10–shared among JAS, PIL, and WAR. The relative frequencies of shared variants varied from 0.111 to 0.750 (Table 1) in particular populations. The haplotype diversity (h) within populations (Table 1) ranges from 0.500 (BAB) to 1.000 (SKA, VIR).

The same topology of the ML phylogenetic tree (Figure 4b) was found regardless of how gaps were treated, but the support values slightly increased when the information from gaps was included. The phylogenetic tree obtained by the ML method did not reveal any evident phylogeographic structure and bootstrap supports for branches were rather low (Figure 4b). The nrDNA sequences variants from different localities within Vistula and Neman drainages were intermingled. Only I9 which possessed 11-bp insertion, occupied an isolated position in the tree.

4. Discussion

The relationship between dispersal and differentiation of European freshwater mussels in the drainage scale has been barely studied so far. Nagel (2000) [93] investigated genetic relationships within *Unio pictorum* (Linnaeus, 1758) from central Europe on the basis of geographical distribution of allele frequencies at 17 enzyme loci. His results in general show that the genetic affinities of the populations are the closest within the same drainage basin. Similarly, North American unionid species show the same pattern of relationships between genetic diversity and the history of drainage system ([93] and references therein). In the present study, a hierarchical AMOVA also indicated that only 4% of variation was due to differences among *U. crassus* populations within drainages while genetic differences between the Neman and Vistula drainages were strongly indicated (77% of variation). Genetic variation was different in the results reported by Feind et al. (2018) [53]. They characterized the genetic constitution of 18 *U. crassus* populations in Germany and Sweden originating from six major drainage systems (Elbe, Rhine, Danube, Schlei-Trave, Eider and Kävlingeåns). Only 9.1% of the variation was due to differences among drainage systems; 6.9% of the variation was explained by differences among populations within drainage systems. Analyses performed on populations of other endangered freshwater mussel, *M. margaritifera*, show different values of genetic differentiation within drainages [5], the lowest in the case of the Elbe and Danube drainages (Fst = 0.121 and Fst = 0.240, respectively) and the highest for the Meuse river (Fst = 0.773). For comparison, in the Vistula and Neman drainages we obtained $\Phi_{ST} = 0.221$ and $\Phi_{ST} = 0.146$, respectively. However, the global fixation index for all the Polish and Lithuanian populations of *U. crassus* studied here was more than twice as high as that obtained by Geist and Kuehn (2005) [5] for *M. margaritifera* ($\Phi_{ST} = 0.813$ versus Fst = 0.374, respectively). It suggests that in the case of *U. crassus* we dealt with at least two different genetic units.

The boundary between genetic diversity patterns of *U. crassus* populations from the Vistula drainage (Poland) and those from the Neman drainage (Lithuania) was revealed in our study by the level of mtDNA markers. Although some minor inconsistencies were observed between genetic diversity patterns revealed by mitochondrial and nuclear markers, the occurrence of two well-defined groups associated with the two considered drainages was fairly well supported (Figures 3 and 4). Independent sources of characters, such as mitochondrial and nuclear DNA, can reflect different, but equally accurate, phylogenetic patterns (e.g., [94–99]). Nevertheless, such incongruence between these two data sources does not provide a criterion by which to select one phylogeny over another. In fact, these disagreements often lead to an enhanced understanding of the evolutionary history of the species in question, including elucidation of patterns of introgression, complex population structure or sex- biased gene flow [100]. Therefore, we suggest that minor differences in relationships among mitochondrial and nuclear haplotypes described here are not the limitation for the proposed inference about relationships of the tested *U. crassus* populations. However, despite already postulated utility of ITS sequences to investigate phylogenetic relationship within Unionidae [12], we did not fully support this idea—the ITS analyses should be backed by investigation of the other genetic markers.

In the study described here, with the few exceptions we identified two distinct groups of *U. crassus* populations inhabiting rivers belonging to two different drainage basins. The results suggest that different evolutionary paths of the analyzed populations may be one of the components that differentiate the genetic units identified here. Another one could be the geographic isolation of populations from various river basins that lead to reduced gene flow and, at the same time, greater influence of genetic drift on genetic structure. The reduction of gene flow promoting genetic differentiation of populations may also be caused by habitat fragmentation due to anthropogenic factors [91]. Here, such an explanation may be proposed in the case of *U. crassus* populations from rivers representing central Poland: Pilica (flowing into the middle Vistula) and its tributary, the Czarna Włoszczowska river. High Φ_{ST} values suggest reduction of gene flow between particular populations, although we should expect them to share the same haplotypes. On the contrary, these populations share only one mtDNA haplotype that is also present in the rest of the Polish populations whereas the remaining haplotypes are unique for each population (Table 1). Studies on other mussel species also indicate the isolation of populations in neighboring river basins and even the lack of gene flow between populations from different tributaries of the same river (e.g., [27]). Moreover, Geist and Kuehn (2005) [5] found a high fragmentation level of the structure of freshwater pearl mussel (*M. margaritifera*) populations inhabiting five main European river basins, the effect of the limited gene flow between these populations. Zanatta, Fraley, and Murphy (2007) [101] in turn observed a high gene flow between populations of a unionid wavy-rayed lampmussel (*Lampsilis fasciola* Rafinesque, 1820) inhabiting the same river basin, indicating the presence of panmixia in this species. Elderkin et al. (2007) [32] also used a hierarchical approach to study population genetic structure of other unionid mussel *Amblema plicata* (Say, 1817) among rivers and drainages. Interestingly, they found low genetic structure among rivers and drainages separated by large geographic distances, what may indicate high effective population size and/or highly agile fish hosts for this species. In the case of *U. crassus* populations in this study, the Mantel test revealed a relatively strong relationship between Φ_{ST} estimates and geographical distances (Figure 2). However, in detail, the pattern of genetic diversity seems to be much more complicated. Therefore, we suggest that the observed pattern of *U. crassus* genetic diversity distribution has been shaped by different factors, i.e. historical and current gene flow.

The distribution of the genetic diversity should reflect the present or past connections within and between drainage systems. Results obtained for *U. crassus* populations in this work partly fulfill this hypothesis as most of the sample populations cluster according to their distribution within the two analyzed drainages. However, there are some mtDNA haplotypes found in central Poland populations (PIL, CZW, WAR) that clustered with haplotypes from Lithuanian populations. Nagel (2000) [93] also found exceptions from general congruence between spatial and genetic distances in the case of European populations of a painter's mussel *U. pictorum*. Similar results were obtained also by Geist and Kuehn (2005) [5] for populations of *M. margaritifera* separated only by 20 km in geographical distance within the same drainage, which were located in different clades on the cladogram. Also, populations from the Danube did not cluster together. Two possibilities were proposed by the authors to explain these irregularities—first, the canals connecting rives and alternatively, tectonic movements affecting the river systems. In our case, the role of canals in promoting gene flow between populations can be seen in the case of the *U. crassus* population from the Babrungas river (tributary of Venta) connected with the Dubysa river by the Windawski Canal of only 15 km in length. Both, Babrungas and Dubysa populations share the same mtDNA haplotypes. On the contrary, the Augustowski Canal connecting Vistula and Neman drainages through the Biebrza river—A tributary of the Narew river, and the Neman river through its tributary, the Czarna Hańcza river—does not seem to influence the observed pattern of genetic diversity of *U. crassus*. However, further investigations including populations of *U. crassus* from northeastern Poland should be conducted to provide full support for this claim.

Humans have profoundly influenced most aquatic ecosystems. For example, formerly separate drainage systems were connected by canals promoting faunal exchange across long established boundaries. At present, anthropogenic factors strongly influence fauna, causing species extinction, changing current distribution ranges, enhancing invasions of alien species that force out indigenous species, or reducing genetic diversity. The anthropogenic pressure, which has caused fragmentation of *U. crassus* habitat and population declines, might well shape consequently genetic architecture and distribution of mtDNA haplotypes of this mussel. As stated by Zając (2004) [43], populations of this mollusk are in Poland quite often isolated and scattered, as well as the overall degree of *U. crassus* preservation is improper mainly due to small population size and unfavorable status of the mussel habitats. Small populations, in turn, are more susceptible to the effect of inbreeding and genetic drift. In our study we included populations reported as declined due to human activity and/or isolated by polluted fragments of rivers at different time scales. Such a population is Jasiołka (JAS), characterized by the presence of one common mtDNA haplotype (CN1) found also in the remaining Polish populations and one private mtDNA haplotype (CN5) (vide Table 1). Besides, human activities such as commercial trade with live fish are believed to influence their genetic composition [93]. In turn, fragmentation of the environment is a barrier to the movement of organisms and is therefore a major threat to species existence both for demographic and genetic reasons [102]. Thus, it can be inferred that the pattern of genetic variability of freshwater mussels and of *U. crassus* is constantly changing. In turn, if current water connections and ongoing gene flow are the main forces driving patterns of genetic diversity of *U. crassus*, populations located nearby (e.g., Skawinka with its tributary Cedron; Pilica with its tributary Czarna Włoszczowska) should be similar genetically and share the same haplotypes of mtDNA whereas the Lithuanian populations should be different from the Polish ones. However, we found some mtDNA haplotypes found in central Poland populations (PIL, CZW, WAR) that clustered with haplotypes from Lithuanian populations (Figure 4). Therefore, we suggest that historical phenomena could most strongly shape genetic diversity of *U. crassus* in Europe.

Pleistocene ice ages have exerted considerable influence on the contemporary genetic diversity patterns of freshwater species. Multiple transgressions and regressions of ice caused the situation that many species had to displace or escape to the ice age refugia. In glacial periods, the northern part of Poland and the area of Lithuania were covered to varying degrees by the glacier (for example, during the Sanian 2 glaciation, this area was almost entirely covered with ice sheets that reached the Carpathians), causing the withdrawal of organisms from this region [103]. Glacial water outflows in different directions also occurred in the Pleistocene, e.g., from the area of the current San and Pilica river basins to the east, to the Dniester river valley [103]. The merging of these river systems has only subsided as a result of the final formation of the Vistula River. In addition, water from the Neman basin drained southwest to the Torun-Eberswalder Urstromtal during the Pomeranian phase of the Vistulian Glaciation. Then, after the glacier retreat, rivers of northern Poland and Lithuania started to flow towards the Baltic Sea (e.g., [104]). In the case of Polish and Lithuanian populations of *U. crassus* the observed intermingling of haplotypes may be connected to the ancient connection between Neman and Vistula drainages as indicated by the geological data. During the Vistulian Glaciation the ice sheet blocked the pre-existing drainage system and caused the development of vast ice-dammed lakes including waters of the ancient Neman River as well as the Polish rivers Biebrza, Narew, and Vistula [103]. Pre-existing connection between the Neman and Vistula waters seems to be also confirmed by morphological similarities between ichtiofauna of their drainages [105]. In turn, if current water connections and ongoing gene flow are the main forces driving patterns of genetic diversity of *Unio* populations, populations located nearby (e.g., Skawinka with its tributary Cedron; Pilica with its tributary Czarna Włoszczowska) should be similar genetically and share haplotypes of mtDNA whereas Lithuanian populations should be different from Polish ones. The genetic diversity and differentiation of *U. crassus* populations revealed during this study can also be explained by colonization from different glacial refugia or postglacial recolonization. Moreover, the observed pattern of genetic diversity may result from specificity between glochidia and host fish vectors.

As larval stages of unionids (glochidia) are obligate parasites of freshwater fish, they were able to migrate and recolonize the European areas during warmer periods only along inland waters together with their host fish species. Thus, the gene flow among populations of freshwater mussels was mostly affected by migration patterns of fish host species. Potential hosts for *U. crassus* include fish species like bullhead (*Cottus gobio* Linnaeus, 1758), Eurasian Minnow (*Phoxinus phoxinus* Linnaeus, 1758), chub (*Squalius cephalus* Linnaeus, 1758), rudd (*Scardinius erythrophthalmus* Linnaeus, 1758) Three-spinned Stickleback (*Gasterosteus aculeatus* Linnaeus, 1758), and perch (*Perca fluviatilis* Linnaeus, 1758) (e.g., [106–108]).

The Ponto-Caspian region (the Black Sea, Sea of Azov and the Caspian Sea) was the main glacial refugium of the recent fish fauna of both Poland and Lithuania [109] which colonized northern Europe through two main river networks: Dniester-San-Vistula and Dnieper-Neman-Vistula [110]. The pattern of drainage systems in Central Europe changed many times in the Pleistocene, but the major route of recolonization of this part of Europe became the Dniester and Dnieper (northern trail) because several species were not able to use the Danube (southern trail) due to the barrier of the Carpathians Mountain Range. The colonization of Europe by fish-hosts of *U. crassus* has been the subject of many studies, for example: *Squalius cephalus* [111], *Perca fluviatilis* [112], *Cottus gobio* [113,114], *Gasterosteus aculeatus* [115]. Nevertheless, it is difficult to point out the simple relationship between the hypothetical postglacial migration pathways of any fish species and the distribution of *U. crassus* evolutionary lines. However, the *U. crassus* expansion pattern has some characteristics of the observed distribution of other species of freshwater mussels. For example, the analysis of DNA microsatellite loci of *M. margaritifera* from central Europe revealed that the diversity of individual populations is not consistent with the modern hydrological network [5]. The genetic variability pattern of *M. margaritifera* was also not compatible with the distribution of genetic variability of *C. gobio*, one of the fish-host species of glochidia in the same area [116]. These differences, however, are not surprising given the possibility of dispersal of the freshwater pearl mussel through other fish hosts, whose migration paths are different (e.g., [117]). A study of Berg et al. (1998) [118] indicates that a varied level of population migration can be expected even within the same river basin. The observed complex pattern of genetic variability distribution was thus explained by the influence of glacial phenomena and links between the presently separated river basins existing in the past [119]. Interestingly, Nagel (2000) [93] presented the dendrogram, based on Nei's genetic distance, in which individuals of *U. pictorum* from the Vistula are separated from those inhabiting other rivers of the Baltic Sea catchment and cluster with populations of the Danube. Similarly, geographic isolation was found in the present study between populations belonging to the same river basin, which also indicated the existence of hydrological barriers in migrations of glochidia hosts. Therefore, we suggest that the genetic relationships within *U. crassus* from Poland and Lithuania reflect paleogeographical relationships between river systems during Pliocene and Pleistocene rather that current gene flow.

Interestingly, in the case of our study the present-day population differentiation of *U. crassus* did not match the present-day drainage systems in the case of the central Poland rivers. Individuals from the southern Poland rivers were localized in "Polish cluster" and did not intermingle with Lithuanian populations. The observed pattern of genetic diversity and differentiation of *U. crassus* populations revealed during this study can be explained either by historical gene flow or different routes of post glacial colonization. In such a case, the region of central Poland could be a contact zone between two haplotype lineages.

We believe that our results suggest the existence of a secondary contact area in central Poland resulting from the recolonization of a given area by populations from separate glacial refugia [120]. The identified suture zone is at the same time a barrier to further expansion of genealogy lines from different glacial refugia and maintains their integrity. Different populations from different evolutionary lines can be distinguished in populations present in the suture zone due to the presence of various haplotypes [121]. In Europe, there are five identified suture zones: eastern and western Europe, central Europe, central Scandinavia, the Alps, and the Pyrenees [120,122,123]. Poland is also a specific suture zone, characterized by the presence of multiple secondary contact zones, distinguished for different refugial lines (compare e.g., data on weasel *Mustela nivalis* Linnaeus, 1758 by McDevitt et al., 2012 [124]).

It is worth mentioning that for *U. crassus*, a number of subspecies with local forms of uncertain taxonomic rank has been widely accepted [125–128]. So, such a high level of genetic diversity between specimens from the Vistula and Neman drainages may support the above-mentioned idea. On the other hand, the results of this study revealed mtDNA haplotype shared by specimens form both drainages and some Polish haplotypes cluster together with the Lithuanian ones. Therefore, the observed genetic structure does not match the present drainage systems, so there is no straight correlation between genetic diversity and geographic region.

In conclusion, the results of our study indicated that in the case of eastern Central European populations of *U. crassus* we dealt with at least two different genetic units. However, the observed genetic structure does not match the present drainage systems. Therefore, we suggest that the observed genetic relationships within *U. crassus* from Poland and Lithuania reflect rather paleogeographical relationships between river systems during the Pliocene and Pleistocene than being the result of the current gene flow. The present-day genetic pattern of *U. crassus* diversity may also be shaped by different routes of post glacial colonization and specificity between glochidia and host fish vectors. More recent habitat fragmentation due to anthropogenic factors may also have contributed to the observed populations structure.

Supplementary Materials: The following are available online at http://www.mdpi.com/2075-1729/10/7/119/s1, Table S1: Information on specimens of *Unio crassus* under study and Table S2. Data on sampling localities from Poland and Lithuania.

Author Contributions: Conceptualization, A.K., K.Z., T.Z., J.S.; Methodology, A.K., M.M., A.W.; Validation, A.K.; Formal Analysis, A.K., M.M., J.R.; Investigation, A.K.; Resources, A.K., K.Z., T.Z., P.I.; Data Curation, A.K., M.M.; Writing—Original Draft Preparation, A.K., A.K.-Z., M.M., A.W.; Writing—Review & Editing, A.K., A.K.-Z., M.M., A.W., Z.A.; Visualization, A.K., A.K.-Z., M.M., J.R.; Supervision, A.K., J.S.; Project Administration, A.K.; Funding Acquisition, A.K., K.Z., T.Z., J.S. All authors have read and agreed to the published version of the manuscript.

Funding: This research was funded by Ministerstwo Nauki i Szkolnictwa Wyższego grant number N N304 363638 and Unwersytet Gdański grant number 1410-5-0376-8.

Acknowledgments: We wish to thank Natalia Gasperowicz, Joanna Kaminiecka, Anna Knuth, and Rafał Ziółkowski for their laboratory work and Tadeusz Namiotko for the final correction. This study was founded by a research grant of University of Gdańsk (No. 1410-5-0376-8) and a research grant of the Polish Ministry of Science and Higher Education (No. N N304 363638). The samples were taken with the permission of Chief Nature Conservator (Permission no DOPozgiz-4200/1-29/244/10/ed).

Conflicts of Interest: The authors declare no conflict of interest.

References

1. Graf, D.L.; Cummings, K.S. Actual and Alleged Freshwater Mussels (Mollusca: Bivalvia: Unionoida) from Madagascar and the Mascarenes, with Description of a New Genus, *Germainaia*. *Proc. Acad. Nat. Sci. Philadelphia* **2009**, *158*, 221–238. [CrossRef]
2. IUCN 2020. The IUCN Red List of Threatened Species. Version 2020-2. Available online: https://www.iucnredlist.org (accessed on 21 July 2020).
3. Machordom, A.; Araujo, R.; Erpenbeck, D.; Ramos, M.-Á. Phylogeography and conservation genetics of endangered European Margaritiferidae (Bivalvia: Unionoidea). *Biol. J. Linn. Soc.* **2003**, *78*, 235–252. [CrossRef]
4. Lydeard, C.; Cowie, R.H.; Ponder, W.F.; Bogan, A.E.; Bouchet, P.; Clark, S.A.; Cummings, K.S.; Frest, T.J.; Gargominy, O.; Herbert, D.G.; et al. The Global Decline of Nonmarine Mollusks. *Bioscience* **2004**, *54*, 321. [CrossRef]
5. Geist, J.; Kuehn, R. Genetic diversity and differentiation of central European freshwater pearl mussel (*Margaritifera margaritifera* L.) populations: Implications for conservation and management. *Mol. Ecol.* **2005**, *14*, 425–439. [CrossRef] [PubMed]
6. Geist, J.; Kuehn, R. Host-parasite interactions in oligotrophic stream ecosystems: The roles of life-history strategy and ecological niche. *Mol. Ecol.* **2008**, *17*, 997–1008. [CrossRef] [PubMed]
7. Douda, K.; Horký, P.; Bílý, M. Host limitation of the thick-shelled river mussel: Identifying the threats to declining affiliate species. *Anim. Conserv.* **2012**, *15*, 536–544. [CrossRef]
8. Taeubert, J.E.; Martinez, A.M.P.; Gum, B.; Geist, J. The relationship between endangered thick-shelled river mussel (*Unio crassus*) and its host fishes. *Biol. Conserv.* **2012**, *155*, 94–103. [CrossRef]
9. Ćmiel, A.M.; Zając, K.; Lipińska, A.M.; Zając, T. Glochidial infestation of fish by the endangered thick-shelled river mussel *Unio crassus*. *Aquat. Conserv. Mar. Freshw. Ecosyst.* **2018**, *28*, 535–544. [CrossRef]
10. Killeen, I.J.; Seddon, M.B.; Holmes, A.M. Molluscan Conservation: A Strategy for the 21st Century. *J. Conchol.* **1998**, *Spec. Publ*, 320.
11. Araujo, R.; Gómez, I.; Machordom, A. The identity and biology of *Unio mancus* Lamarck, 1819 (= *U. elongatulus*) (Bivalvia: Unionidae) in the Iberian Peninsula. *J. Molluscan Stud.* **2005**, *71*, 25–31. [CrossRef]
12. Kallersjo, M.; von Proschwitz, T.; Lundberg, S.; Eldenas, P.; Erseus, C. Evaluation of ITS rDNA as a complement to mitochondrial gene sequences for phylogenetic studies in freshwater mussels: An example using Unionidae from north-western Europe. *Zool. Scr.* **2005**, *34*, 415–424. [CrossRef]

13. Zając, K.; Zając, T.A.; Adamski, P.; Bielański, W.; Ćmiel, A.M.; Lipińska, A.M. Dispersal and mortality of translocated thick-shelled river mussel *Unio crassus* Philipsson, 1788 adults revealed by radio tracking. *Aquat. Conserv. Mar. Freshw. Ecosyst.* **2019**, *29*, 331–340. [CrossRef]
14. Zając, K.; Zając, T.A. Seasonal patterns in the developmental rate of glochidia in the endangered thick-shelled river mussel, *Unio crassus* Philipsson, 1788. *Hydrobiologia* **2020**. [CrossRef]
15. Araujo, R.; Toledo, C.; Machordom, A. Redescription of *Unio gibbus*, A West Palaearctic Freshwater Mussel with Hookless Glochidia. *Malacologia* **2009**, *51*, 131–141. [CrossRef]
16. Reis, J.; Araujo, R. Redescription of *Unio tumidiformis* Castro, 1885 (Bivalvia, Unionidae), an endemism from the south-western Iberian Peninsula. *J. Nat. Hist.* **2009**, *43*, 1929–1945. [CrossRef]
17. Bolotov, I.N.; Aksenova, O.V.; Bakken, T.; Glasby, C.J.; Gofarov, M.Y.; Kondakov, A.V.; Konopleva, E.S.; Lopes-Lima, M.; Lyubas, A.A.; Wang, Y.; et al. Discovery of a silicate rock-boring organism and macrobioerosion in fresh water. *Nat. Commun.* **2018**, *9*, 2882. [CrossRef]
18. Soroka, M. Characteristics of mitochondrial DNA of unionid bivalves (Mollusca: Bivalvia: Unionidae). I. Detection and characteristics of doubly uniparental inheritance (DUI) of unionid mitochondrial DNA. *Folia Malacol.* **2010**, *18*, 147–188. [CrossRef]
19. Froufe, E.; Sobral, C.; Teixeira, A.; Sousa, R.; Varandas, S.C.; Aldridge, D.; Lopes-Lima, M. Genetic diversity of the pan-European freshwater mussel *Anodonta anatina* (Bivalvia: Unionoida) based on CO1: New phylogenetic insights and implications for conservation. *Aquat. Conserv. Mar. Freshw. Ecosyst.* **2014**, *24*, 561–574. [CrossRef]
20. Lopes-Lima, M.; Teixeira, A.; Froufe, E.; Lopes, A.; Varandas, S.; Sousa, R. Biology and conservation of freshwater bivalves: Past, present and future perspectives. *Hydrobiologia* **2014**, *735*, 1–13. [CrossRef]
21. Lopes-Lima, M.; Sousa, R.; Geist, J.; Aldridge, D.C.; Araujo, R.; Bergengren, J.; Bespalaya, Y.; Bódis, E.; Burlakova, L.; Van Damme, D.; et al. Conservation status of freshwater mussels in Europe: State of the art and future challenges. *Biol. Rev.* **2016**, *92*, 572–607. [CrossRef]
22. Zając, K.; Florek, J.; Zając, T.; Adamski, P.; Bielański, W.; Ćmiel, A.M.; Klich, M.; Lipińska, A.M. On the Reintroduction of the Endangered Thick-Shelled River Mussel *Unio crassus*: The Importance of the River's Longitudinal Profile. *Sci. Total Environ.* **2018**, *624*, 273–282. [CrossRef] [PubMed]
23. Prié, V.; Soler, J.; Araujo, R.; Cucherat, X.; Philippe, L.; Patry, N.; Adam, B.; Legrand, N.; Jugé, P.; Richard, N.; et al. Challenging exploration of troubled waters: A decade of surveys of the giant freshwater pearl mussel *Margaritifera auricularia* in Europe. *Hydrobiologia* **2018**, *810*, 157–175. [CrossRef]
24. Williams, J.D.; Warren, M.L.J.; Cummings, K.S.; Harris, J.L.; Neves, R.J. Conservation Status of Freshwater Mussels of the United States and Canada. *Fisheries* **1993**, *18*, 6–22. [CrossRef]
25. Strayer, D.L.; Downing, J.A.; Haag, W.R.; King, T.L.; Layzer, J.B.; Newton, T. Changing perspectives on Pearly mussels, North American's most imperiled animals. *Bioscience* **2004**, *54*, 429–439. [CrossRef]
26. Tedesco, P.A.; Bigorne, R.; Bogan, A.E.; Giam, X.; Jézéquel, C.; Hugueny, B. Estimating how many undescribed species have gone extinct. *Conserv. Biol.* **2014**, *28*, 1360–1370. [CrossRef]
27. King, T.L.; Eackles, M.S.; Gjetvaj, B.; Hoeh, W.R. Intraspecific phylogeography of *Lasmigona subviridis* (Bivalvia: Unionidae): Conservation implications of range discontinuity. *Mol. Ecol.* **1999**, *8*, S65–S78. [CrossRef]
28. Roe, K.J.; Hartfield, P.D.; Lydeard, C. Phylogeographic analysis of the threatened and endangered superconglutinate-producing mussels of the genus *Lampsilis* (Bivalvia: Unionidae). *Mol. Ecol.* **2001**, *10*, 2225–2234. [CrossRef]
29. Krebs, R.A. Combining paternally and maternally inherited mitochondrial DNA for analysis of population structure in mussels. *Mol. Ecol.* **2004**, *13*, 1701–1705. [CrossRef]
30. Mock, K.E.; Brim-Box, J.C.; Miller, M.P.; Downing, M.E.; Hoeh, W.R. Genetic diversity and divergence among freshwater mussel (*Anodonta*) populations in the Bonneville Basin of Utah. *Mol. Ecol.* **2004**, *13*, 1085–1098. [CrossRef]
31. Kelly, M.W.; Rhymer, J.M. Population genetic structure of a rare unionid (*Lampsilis cariosa*) in a recently glaciated landscape. *Conserv. Genet.* **2005**, *6*, 789–802. [CrossRef]
32. Elderkin, C.L.; Christian, A.D.; Vaughn, C.C.; Metcalfe-Smith, J.L.; Berg, D.J. Population genetics of the freshwater mussel, *Amblema plicata* (Say 1817) (Bivalvia: Unionidae): Evidence of high dispersal and post-glacial colonization. *Conserv. Genet.* **2007**, *8*, 355–372. [CrossRef]

33. Burlakova, L.E.; Campbell, D.; Karatayev, A.Y.; Barclay, D. Distribution, genetic analysis and conservation priorities for rare Texas freshwater molluscs in the genera *Fusconaia* and *Pleurobema* (Bivalvia: Unionidae). *Aquat. Biosyst.* **2012**, *8*, 12. [CrossRef] [PubMed]
34. Perkins, M.A.; Johnson, N.A.; Gangloff, M.M. Molecular systematics of the critically-endangered North American spinymussels (Unionidae: *Elliptio* and *Pleurobema*) and description of *Parvaspina* gen. nov. *Conserv. Genet.* **2017**, *18*, 745–757. [CrossRef]
35. Pfeiffer, J.M.; Sharpe, A.E.; Johnson, N.A.; Emery, K.F.; Page, L.M. Molecular phylogeny of the Nearctic and Mesoamerican freshwater mussel genus *Megalonaias*. *Hydrobiologia* **2018**, *811*, 139–151. [CrossRef]
36. Zanatta, D.T.; Stoeckle, B.C.; Inoue, K.; Paquet, A.; Martel, A.L.; Kuehn, R.; Geist, J. High genetic diversity and low differentiation in North American *Margaritifera margaritifera* (Bivalvia: Unionida: Margaritiferidae). *Biol. J. Linn. Soc.* **2018**, *123*, 850–863. [CrossRef]
37. Zettler, M.L.; Jueg, U. The situation of the freshwater mussel *Unio crassus* (Philipsson, 1788) in north-east Germany and its monitoring in terms of the EC Habitats Directive. *Mollusca* **2007**, *25*, 165–174.
38. Geist, J. Strategies for the conservation of endangered freshwater pearl mussels (*Margaritifera margaritifera* L.): A synthesis of Conservation Genetics and Ecology. *Hydrobiologia* **2010**, *644*, 69–88. [CrossRef]
39. Geist, J. Integrative freshwater ecology and biodiversity conservation. *Ecol. Indic.* **2011**, *11*, 1507–1516. [CrossRef]
40. Cuttelod, A.; Seddon, M.; Neubert, E. *European Red List of Non-marine Molluscs*; Publications Office of the European Union: Luxembourg, 2011; ISBN 9789279201981.
41. Bolotov, I.N.; Kondakov, A.V.; Konopleva, E.S.; Vikhrev, I.V.; Aksenova, O.V.; Aksenov, A.S.; Bespalaya, Y.V.; Borovskoy, A.V.; Danilov, P.P.; Dvoryankin, G.A.; et al. Integrative taxonomy, biogeography and conservation of freshwater mussels (Unionidae) in Russia. *Sci. Rep.* **2020**, *10*, 3072. [CrossRef]
42. Taeubert, J.-E.; Gum, B.; Geist, J. Host-specificity of the endangered thick-shelled river mussel (*Unio crassus*, Philipsson 1788) and implications for conservation. *Aquat. Conserv. Mar. Freshw. Ecosyst.* **2012**, *22*, 36–46. [CrossRef]
43. Zając, K. Unio crassus. In *Głowaciński Z. Nowicki J. Polska czerwona Księga Zwierząt. Bezkręgowce*.; Instytut Ochrony Przyrody PAN w Krakowie, Akademia Rolnicza im. A. Cieszkowskiego w Poznaniu: Kraków, Poland, 2004; p. 448.
44. Denic, M.; Stoeckl, K.; Gum, B.; Geist, J. Physicochemical assessment of *Unio crassus* habitat quality in a small upland stream and implications for conservation. *Hydrobiologia* **2014**, *735*, 111–122. [CrossRef]
45. Nagel, K.-O.; Badino, G. Population Genetics and Systematics of European Unionoidea. In *Ecology and Evolution of the Freshwater Mussels Unionoida*; Springer: Berlin/Heidelberg, Germany, 2001; pp. 51–80.
46. Prie, V.; Puillandre, N.; Bouchet, P. Bad taxonomy can kill: Molecular reevaluation of *Unio mancus* Lamarck, 1819 (Bivalvia: Unionidae) and its accepted subspecies. *Knowl. Manag. Aquat. Ecosyst.* **2012**, *405*, 1–18. [CrossRef]
47. Reis, J.; Machordom, A.; Araujo, R. Diversidad morfológica y molecular de los Unionidae (Mollusca, Bivalvia) de Portugal. *Graellsia* **2013**, *69*, 17–36. [CrossRef]
48. Araujo, R.; Buckley, D.; Nagel, K.-O.; García-Jiménez, R.; Machordom, A. Species boundaries, geographic distribution and evolutionary history of the Western Palaearctic freshwater mussels *Unio* (Bivalvia: Unionidae). *Zool. J. Linn. Soc.* **2017**, 1–25. [CrossRef]
49. Prié, V.; Puillandre, N. Molecular phylogeny, taxonomy, and distribution of French *Unio* species (Bivalvia, Unionidae). *Hydrobiologia* **2014**, *735*, 95–110. [CrossRef]
50. Mezhzherin, S.V.; Yanovych, L.M.; Zhalay, Y.I.; Pampura, M.M.; Vasilieva, L.A. Reproductive isolation of two *Unio crassus* Philipsson, 1788 (Bivalvia, Unionidae) vicarious forms with low genetic differentiation level. *Reports Natl. Acad. Sci. Ukr.* **2013**, *2*, 138–143.
51. Klishko, O.; Lopes-Lima, M.; Froufe, E.; Bogan, A.; Vasiliev, L.; Yanovich, L. Taxonomic reassessment of the freshwater mussel genus *Unio* (Bivalvia: Unionidae) in Russia and Ukraine based on morphological and molecular data. *Zootaxa* **2017**, *4286*, 93–112. [CrossRef]
52. Gerke, N.; Tiedemann, R. A PCR-based molecular identification key to the glochidia of European freshwater mussels (Unionidae). *Conserv. Genet.* **2001**, *2*, 287–289. [CrossRef]
53. Feind, S.; Geist, J.; Kuehn, R. Glacial perturbations shaped the genetic population structure of the endangered thick-shelled river mussel (*Unio crassus*, Philipsson 1788) in Central and Northern Europe. *Hydrobiologia* **2018**, *810*, 177–189. [CrossRef]

54. Mioduchowska, M.; Kaczmarczyk, A.; Zając, K.; Zając, T.; Sell, J. Gender-Associated Mitochondrial DNA Heteroplasmy in Somatic Tissues of the Endangered Freshwater Mussel *Unio crassus* (Bivalvia: Unionidae): Implications for Sex Identification and Phylogeographical Studies. *J. Exp. Zool. Part A Ecol. Genet. Physiol.* **2016**, *325*, 610–625. [CrossRef]
55. Geist, J.; Rottmann, O.; Schröder, W.; Kühn, R. Development of microsatellite markers for the endangered freshwater pearl mussel *Margaritifera margaritifera* L. (Bivalvia: Unionoidea). *Mol. Ecol. Notes* **2003**, *3*, 444–446. [CrossRef]
56. Bouza, C.; Castro, J.; Martínez, P.; Amaro, R.; Fernández, C.; Ondina, P.; Outeiro, A.; San Miguel, E. Threatened freshwater pearl mussel *Margaritifera margaritifera* L. in NW Spain: Low and very structured genetic variation in southern peripheral populations assessed using microsatellite markers. *Conserv. Genet.* **2007**, *8*, 937–948. [CrossRef]
57. Geist, J.; Wunderlich, H.; Kuehn, R. Use of mollusc shells for DNA-based molecular analyses. *J. Molluscan Stud.* **2008**, *74*, 337–343. [CrossRef]
58. Hutchison, D.W.; Templeton, A.R. Correlation of Pairwise Genetic and Geographic Distance Measures: Inferring the Relative Influences of Gene Flow and Drift on the Distribution of Genetic Variability. *Evolution* **1999**, *53*, 1898–1914. [CrossRef] [PubMed]
59. Hartl, D.L.; Clark, A.G. *Principles of population genetics*; Sinauer Associates: Sunderland, MA, USA, 1997; ISBN 0878933069.
60. Bilton, D.T.; Freeland, J.R.; Okamura, B. Dispersal in Freshwater Invertebrates. *Annu. Rev. Ecol. Syst.* **2001**, *32*, 159–181. [CrossRef]
61. Newman, D.; Pilson, D. Increased probability of extinction due to decreased genetic effective population size: Experimantal populations of *Clarcia pulchella*. *Evolution* **1997**, *51*, 354–362. [CrossRef]
62. Saccheri, I.; Kuussaari, M.; Kankare, M.; Vikman, P.; Hanski, I. Inbreeding and extinction in a butterfly metapopulation. *Nature* **1998**, *392*, 491–494. [CrossRef]
63. Frankham, R. Genetics and conservation biology. *C. R. Biol.* **2003**, *326* (Suppl. 1), S22–S29. [CrossRef]
64. Reed, D.H.; Frankham, R. Correlation between Fitness and Genetic Diversity. *Conserv. Biol.* **2003**, *17*, 230–237. [CrossRef]
65. Manni, F.; Guérard, E.; Heyer, E. Geographic patterns of (genetic, morphologic, linguistic) variation: How barriers can be detected by using Monmonier's algorithm. *Hum. Biol.* **2004**, *76*, 173–190. [CrossRef]
66. Article 17 web tool. Available online: https://nature-art17.eionet.europa.eu/article17/reports2012 (accessed on 21 July 2020).
67. Piechocki, A.; Dyduch-Falniowska, A. *Mięczaki (Mollusca). Małże (Bivalvia)*; Fauna Słod. Polskie Towarzystwo Hydrobiologiczne. Wydawnictwo Naukowe PWN: Warszawa, Poland, 1993.
68. Sambrook, J.; Russell, D.W. *Molecular Cloning: A Laboratory Manual*; Cold Spring Harbor Laboratory Press: New York, NY, USA, 2001; ISBN 0879695773.
69. White, T.J.; Bruns, S.; Lee, S.; Taylor, J. Amplification and direct sequencing of fungal ribosomal RNA genes for phylogenetics. In *PCR Protocols: A Guide to Methods and Applications*; Innis, M.A., Gelfand, D.H., Sninsky, J.J., White, T.J., Eds.; Academic Press: New York, NY, USA, 1990; pp. 315–322. ISBN 008088671X.
70. Buhay, J.E.; Serb, J.M.; Renee, D.; Parham, Q.; Lydeard, C. Conservation genetics of two endangered unionid bivalve species, *Epioblasma florentina* walkeri and *E. capsaeformis* (Unionidae: Lampsilini). *J. Molluscan Stud.* **2002**, *68*, 385–391. [CrossRef]
71. Serb, J.M.; Lydeard, C. Complete mtDNA Sequence of the North American Freshwater Mussel, *Lampsilis ornata* (Unionidae): An Examination of the Evolution and Phylogenetic Utility of Mitochondrial Genome Organization in Bivalvia (Mollusca). *Mol. Biol. Evol.* **2003**, *20*, 1854–1866. [CrossRef] [PubMed]
72. Folmer, O.; Black, M.; Hoeh, W.; Lutz, R.; Vrijenhoek, R. DNA primers for amplification of mitochondrial cytochrome c oxidase subunit I from diverse metazoan invertebrates. *Mol. Mar. Biol. Biotechnol.* **1994**, *3*, 294–299. [PubMed]
73. Altschul, S.F.; Gish, W.; Miller, W.; Myers, E.W.; Lipman, D.J. Basic Local Alignment Search Tool. *J. Mol. Biol.* **1990**, *215*, 403–410. [CrossRef]
74. Hall, T. BioEdit: An important software for molecular biology. *GERF Bull. Biosci.* **2011**, *2*, 60–61. [CrossRef]
75. Sievers, F.; Wilm, A.; Dineen, D.; Gibson, T.J.; Karplus, K.; Li, W.; Lopez, R.; McWilliam, H.; Remmert, M.; Soding, J.; et al. Fast, scalable generation of high-quality protein multiple sequence alignments using Clustal Omega. *Mol. Syst. Biol.* **2011**, *7*, 539. [CrossRef]

76. Librado, P.; Rozas, J. DnaSP v5: A software for comprehensive analysis of DNA polymorphism data. *Bioinformatics* **2009**, *25*, 1451–1452. [CrossRef]
77. Excoffier, L.; Lischer, H.E.L. Arlequin suite ver 3.5: A new series of programs to perform population genetics analyses under Linux and Windows. *Mol. Ecol. Resour.* **2010**, *10*, 564–567. [CrossRef]
78. Byers, J.E. The distribution of an introduced mollusc and its role in the long-term demise of a native confamilial species. *Biol. Invasions* **1999**, *1*, 339–352. [CrossRef]
79. Smouse, P.E.; Long, J.C.; Sokal, R.R. Multiple Regression and Correlation Extensions of the Mantel Test of Matrix Correspondence. *Syst. Zool.* **1986**, *35*, 627–632. [CrossRef]
80. Network v. 4.5.1.6 Software. Available online: http://www.fluxus-technology.com (accessed on 21 July 2020).
81. Darriba, D.; Taboada, G.L.; Doallo, R.; Posada, D. jModelTest 2: More models, new heuristics and parallel computing. *Nat. Methods* **2012**, *9*, 772. [CrossRef] [PubMed]
82. Hasegawa, M.; Kishino, H.; Yano, T. Dating of the human-ape splitting by a molecular clock of mitochondrial DNA. *J. Mol. Evol.* **1985**, *22*, 160–174. [CrossRef] [PubMed]
83. Kimura, M. A simple method for estimating evolutionary rates of base substitutions through comparative studies of nucleotide sequences. *J. Mol. Evol.* **1980**, *16*, 111–120. [CrossRef] [PubMed]
84. Kumar, S.; Stecher, G.; Tamura, K. MEGA7: Molecular Evolutionary Genetics Analysis Version 7.0 for Bigger Datasets. *Mol. Biol. Evol.* **2016**, *33*, 1870–1874. [CrossRef]
85. Voronoi, G. Nouvelles applications des paramètres continus à la théorie des formes quadratiques. *J. für die Reine und Angew. Math.* **1908**, *133*, 97–178. [CrossRef]
86. Brassel, K.E.; Reif, D. A Procedure to Generate Thiessen Polygons. *Geogr. Anal.* **1979**, *11*, 289–303. [CrossRef]
87. Monmonier, M.S. Maximum-Difference Barriers: An Alternative Numerical Regionalization Method*. *Geogr. Anal.* **1973**, *5*, 245–261. [CrossRef]
88. Drummond, A.J.; Suchard, M.A.; Xie, D.; Rambaut, A. Bayesian hylogenetics with BEAUti and the BEAST 1.7. *Mol. Biol. Evol.* **2012**, *29*, 1969–1973. [CrossRef]
89. Newton, M.A.; Raftery, A.E. Approximate Bayesian Inference with the Weighted Likelihood Bootstrap. *J. R. Stat. Soc. Ser. B* **1994**, *56*, 3–48. [CrossRef]
90. Suchard, M.A.; Weiss, R.E.; Sinsheimer, J.S. Bayesian selection of continous-time Markov chain evolutionary models. *Mol. Biol. Evol.* **2001**, *18*, 1001–1013. [CrossRef]
91. Zouros, E.; Ball, A.O.; Saavedra, C.; Freeman, K.R.; Skibinski, D.O.F.; Gallagher, C.; Beynon, C.M. Mitochondrial DNA inheritance. *Nature* **1994**, *368*, 818. [CrossRef] [PubMed]
92. Froufe, E.; Gonçalves, D.V.; Teixeira, A.; Sousa, R.; Varandas, S.; Ghamizi, M.; Zieritz, A.; Lopes-Lima, M. Who lives where? Molecular and morphometric analyses clarify which *Unio* species (Unionida, Mollusca) inhabit the southwestern Palearctic. *Org. Divers. Evol.* **2016**, *16*, 597–611. [CrossRef]
93. Nagel, K.-O. Testing hypotheses on the dispersal and evolutionary history of freshwater mussels (Mollusca: Bivalvia: Unionidae). *J. od Evol. Biol.* **2000**, *13*, 854–865. [CrossRef]
94. Patton, J.L.; Smith, M.F. Paraphyly, Polyphyly, and the Nature of Species Boundaries in Pocket Gophers (Genus *Thomomys*). *Syst. Biol.* **1994**, *43*, 11–26. [CrossRef]
95. Sperling, F.A.H.; Harrison, R.G. Mitochondrial DNA variation within and between species of the *Papilio machaon* group of swallowtail betterflies. *Evolution* **1994**, *48*, 408–422. [CrossRef] [PubMed]
96. Talbot, S.L.; Shields, G.F. Phylogeography of Brown Bears (*Ursus arctos*) of Alaska and Paraphyly within the Ursidae. *Mol. Phylogenet. Evol.* **1996**, *5*, 477–494. [CrossRef]
97. Avise, J.C.; Wollenberg, K. Phylogenetics and the origin of species. *Proc. Natl. Acad. Sci. USA.* **1997**, *94*, 7748–7755. [CrossRef]
98. Shaw, K.L. Conflict between nuclear and mitochondrial DNA phylogenies of a recent species radiation: What mtDNA reveals and conceals about modes of speciation in Hawaiian crickets. *Proc. Natl. Acad. Sci. USA* **2002**, *99*, 16122–16127. [CrossRef]
99. Griffiths, C.S.; Barrowclough, G.F.; Groth, J.G.; Mertz, L. Phylogeny of the Falconidae (Aves): A comparison of the efficacy of morphological, mitochondrial, and nuclear data. *Mol. Phylogenet. Evol.* **2004**, *32*, 101–109. [CrossRef]
100. Bowen, B.W.; Bass, A.L.; Soares, L.; Toonen, R.J. Conservation implications of complex population structure: Lessons from the loggerhead turtle (*Caretta caretta*). *Mol. Ecol.* **2005**, *14*, 2389–2402. [CrossRef]
101. Zanatta, D.T.; Fraley, S.J.; Murphy, R.W. Population structure and mantle display polymorphisms in the wavy-rayed lampmussel, *Lampsilis fasciola* (Bivalvia: Unionidae). *Can. J. Zool.* **2007**, *85*, 1169–1181. [CrossRef]

102. McCauley, D.E. Genetic consequences of local population extinction and recolonization. *Trends Ecol. Evol.* **1991**, *6*, 5–8. [CrossRef]
103. Błońska, D. Geneza słodkowodnej ichtiofauny polski. *Kosmos* **2012**, *2*, 261–270.
104. Ehlers, J.; Gibbard, P.L. *Quaternary Glaciations: Extent and Chronology. Part 1, Europe*; Elsevier: Amsterdam, The Netherlands, 2004; ISBN 9780080540146.
105. Witkowski, A. Analiza ichtiofauny baseny Biebrzy. Charakterystyka morfologiczno-systematyczna smoczkoustych i ryb. *Acta Univ. Wratislav.* **1984**, *14*, 3–100.
106. Stoeckl, K.; Taeubert, J.-E.; Geist, J. Fish species composition and host fish density in streams of the thick-shelled river mussel *(Unio crassus)* - implications for conservation. *Aquat. Conserv. Mar. Freshw. Ecosyst.* **2015**, *25*, 276–287. [CrossRef]
107. Lamand, F.; Roche, K.; Beisel, J.-N. Glochidial infestation by the endangered mollusc *Unio crassus* in rivers of north-eastern France: *Phoxinus phoxinus* and *Cottus gobio* as primary fish hosts. *Aquat. Conserv. Mar. Freshw. Ecosyst.* **2016**, *26*, 445–455. [CrossRef]
108. Lopes-Lima, M.; Froufe, E.; Do, V.T.; Ghamizi, M.; Mock, K.E.; Kebapçı, Ü.; Klishko, O.; Kovitvadhi, S.; Kovitvadhi, U.; Paulo, O.S.; et al. Phylogeny of the most species-rich freshwater bivalve family (Bivalvia: Unionida: Unionidae): Defining modern subfamilies and tribes. *Mol. Phylogenet. Evol.* **2017**, *106*, 174–191. [CrossRef]
109. Witkowski, A.; Kotusz, J.; Przybylski, M.; Marszał, L.; Heese, T.; Amirowicz, A.; Buras, P.; Kukuła, K. Pochodzenie, skład gatunkowy i aktualny stopień zagrożenia ichtiofauny w dorzeczu Wisły i Odry. *Arch. Polish Fish.* **2004**, *12*, 7–20.
110. Konopacka, A. Inwazyjne skorupiaki obunogie (Crustacea, Amphipoda) w wodach Polski. *Przegląd Zool.* **2004**, *48*, 141–162.
111. Durand, J.D.; Persat, H.; Bouvet, Y. Phylogeography and postglacial dispersion of the chub (*Leuciscus cephalus*) in Europe. *Mol. Ecol.* **1999**, *8*, 989–997. [CrossRef]
112. Nesbø, C.L.; Fossheim, T.; Vollestad, L.A.; Jakobsen, K.S. Genetic divergence and phylogeographic relationships among european perch (*Perca fluviatilis*) populations reflect glacial refugia and postglacial colonization. *Mol. Ecol.* **1999**, *8*, 1387–1404. [CrossRef] [PubMed]
113. Englbrecht, C.C.; Freyhof, J.; Nolte, A.; Rassmann, K.; Schliewen, U.; Tautz, D. Phylogeography of the bullhead *Cottus gobio* (Pisces: Teleostei: Cottidae) suggests a pre-Pleistocene origin of the major central European populations. *Mol. Ecol.* **2000**, *9*, 709–722. [CrossRef] [PubMed]
114. Šlechtová, V.; Bohlen, J.; Freyhof, J.; Persat, H.; Delmastro, G.B. The Alps as barrier to dispersal in cold-adapted freshwater fishes? Phylogeographic history and taxonomic status of the bullhead in the Adriatic freshwater drainage. *Mol. Phylogenet. Evol.* **2004**, *33*, 225–239. [CrossRef] [PubMed]
115. Mäkinen, H.S.; Merilä, J. Mitochondrial DNA phylogeography of the three-spined stickleback (*Gasterosteus aculeatus*) in Europe—Evidence for multiple glacial refugia. *Mol. Phylogenet. Evol.* **2008**, *46*, 167–182. [CrossRef]
116. Hänfling, B.; Brandl, R. Genetic variability, population size and isolation of distinct populations in the freshwater fish *Cottus gobio* L. *Mol. Ecol.* **1998**, *7*, 1625–1632. [CrossRef]
117. Weiss, S.; Schlötterer, C.; Waidbacher, H.; Jungwirth, M. Haplotype (mtDNA) diversity of brown trout *Salmo trutta* in tributaries of the Austrian Danube: Massive introgression of Atlantic basin fish-By man or nature? *Mol. Ecol.* **2001**, *10*, 1241–1246. [CrossRef]
118. Berg, D.J.; Cantonwine, E.G.; Hoeh, W.R.; Guttman, S.I. Genetic structure of *Quadrula quadrula* (Bivalvia: Unionidae): Little variation across large distances. *J. Shellfish Res.* **1998**, *17*, 1365–1373.
119. Hantke, R. *Flussgeschichte Mitteleuropas*.; Ferdinand Enke Verlag: Stuttgart, Germany, 1993.
120. Hewitt, G.M. Post-glacial re-colonization of European biota. *Biol. J. Linn. Soc.* **1999**, *68*, 87–112. [CrossRef]
121. Neve, G.; Verlaque, R. Relict species: Phylogeography and conservation biology. In *Relict Species: Phylogeography and Conservation Biology*; Habel, J.C., Assmann, T., Eds.; Springer: Berlin, Germany, 2010; pp. 277–294, ISBN 9783540921608.
122. Taberlet, P. Biodiversity at the intraspecific level: The comparative phylogeographic approach. *J. Biotechnol.* **1998**, *64*, 91–100. [CrossRef]
123. Hewitt, G.M. Genetic consequences of climatic oscillations in the Quaternary. *Philos. Trans. R. Soc. Lond. B. Biol. Sci.* **2004**, *359*, 183–195. [CrossRef]

124. McDevitt, A.D.; Zub, K.; Kawałko, A.; Oliver, M.K.; Herman, J.S.; Wójcik, J.M. Climate and refugial origin influence the mitochondrial lineage distribution of weasels (*Mustela nivalis*) in a phylogeographic suture zone. *Biol. J. Linn. Soc.* **2012**, *106*, 57–69. [CrossRef]
125. Haas, F. Superfamilia Unionacea. In *Das Tierreich*, 88th ed.; de Gruyter: Berlin, Germany, 1969.
126. Falkner, G. *Vorschlag für eine Neufassung der Roten Liste der in Bayern vorkommenden Mollusken (Weichtiere). Mit einem revidierten systematischen Verzeichnis der in Bayern nachgewiesenen Molluskenarten*; Schriftenr. Bayer. Landesamt für Umweltschutz: Bayer, Germany, 1990; pp. 61–112.
127. Nesemann, H. Zoogeographie und Taxonomie der Muschel-Gattungen *Unio* Philipsson, 1788, *Pseudanodonta* Bourguignat. 1877 und *Pseudunio* Haas, 1910 im oberen und mittleren Donausystem (Bivalvia: Unionidae, Margaritiferidae). *Nachrichtenblatt der Ersten Vor. Malakol. Gesellschaft* **1993**, *1*, 2–40.
128. Haas, F. Superfamilia Unionacea. In *Das Tierrich.*; Mertens, R., Hennig, W., Eds.; Walter de Gruyter: Berlin, Germany, 1969; p. 663.

 © 2020 by the authors. Licensee MDPI, Basel, Switzerland. This article is an open access article distributed under the terms and conditions of the Creative Commons Attribution (CC BY) license (http://creativecommons.org/licenses/by/4.0/).

Article

One in a Million: Genetic Diversity and Conservation of the Reference *Crassostrea angulata* Population in Europe from the Sado Estuary (Portugal)

Stefania Chiesa [1,2,*], Livia Lucentini [3], Paula Chainho [4,5], Federico Plazzi [6], Maria Manuel Angélico [7], Francisco Ruano [7], Rosa Freitas [8] and José Lino Costa [4]

1. Department of Molecular Sciences and Nanosystems, Ca' Foscari University of Venice, 30172 Venice, Italy
2. ISPRA—The Italian Institute for Environmental Protection and Research, 00144 Rome, Italy
3. Department of Chemistry, Biology and Biotechnologies, University of Perugia, 06123 Perugia, Italy; livia.lucentini@unipg.it
4. MARE—Marine and Environmental Sciences Centre, Faculdade de Ciências da Universidade de Lisboa, 1749-016 Lisboa, Portugal; pmchainho@fc.ul.pt (P.C.); jlcosta@fc.ul.pt (J.L.C.)
5. ESTSetubal-CINEA, Instituto Politécnico de Setúbal, Estefanilha, 2910-761 Setúbal, Portugal
6. Department of Biological, Geological and Environmental Sciences, University of Bologna, 40126 Bologna, Italy; federico.plazzi@unibo.it
7. Department of Sea and Marine Resources, IPMA—Portuguese Institute of Sea and Atmosphere, 1495-006 Lisboa, Portugal; mmangelico@ipma.pt (M.M.A.); fruano@ipma.pt (F.R.)
8. Department of Biology and CESAM, University of Aveiro, 3810-193 Aveiro, Portugal; rosafreitas@ua.pt
* Correspondence: stefania.chiesa@unive.it

Citation: Chiesa, S.; Lucentini, L.; Chainho, P.; Plazzi, F.; Angélico, M.M.; Ruano, F.; Freitas, R.; Costa, J.L. One in a Million: Genetic Diversity and Conservation of the Reference *Crassostrea angulata* Population in Europe from the Sado Estuary (Portugal). *Life* 2021, *11*, 1173. https://doi.org/10.3390/life11111173

Academic Editor: Edgar Lehr

Received: 7 September 2021
Accepted: 27 October 2021
Published: 3 November 2021

Publisher's Note: MDPI stays neutral with regard to jurisdictional claims in published maps and institutional affiliations.

Copyright: © 2021 by the authors. Licensee MDPI, Basel, Switzerland. This article is an open access article distributed under the terms and conditions of the Creative Commons Attribution (CC BY) license (https://creativecommons.org/licenses/by/4.0/).

Abstract: The production of cupped oysters is an important component of European aquaculture. Most of the production relies on the cultivation of the Pacific oyster *Crassostrea gigas*, although the Portuguese oyster *Crassostrea angulata* represents a valuable product with both cultural and economic relevance, especially in Portugal. The authors of the present study investigated the genetic diversity of Portuguese oyster populations of the Sado estuary, both from natural oyster beds and aquaculture facilities, through *cox1* gene fragment sequencing. Then, a comparison with a wide dataset of cupped oyster sequences obtained from GenBank (up to now the widest available dataset in literature for the Portuguese oyster) was performed. Genetic data obtained from this work confirmed that the Pacific oyster does not occur in the natural oyster beds of the Sado estuary but showed that the species occasionally occurs in the oyster hatcheries. Moreover, the results showed that despite the founder effect and the bottleneck events that the Sado populations have experienced, they still exhibit high haplotype diversity. Risks are arising for the conservation of the Portuguese oyster reference populations of the Sado estuary due to the occurrence of the Pacific oyster in the local hatcheries. Therefore, researchers, local authorities, and oyster producers should work together to avoid the loss of this valuable resource.

Keywords: *Crassostrea angulata*; Portuguese oyster; mtDNA; *cox1*; phylogeography; phylogenetics; haplotype diversity; oyster conservation; genetic diversity

1. Introduction

Oyster farming has a high economic relevance in the European economy, mainly relying on cupped oyster production. The European production of cupped oysters was 93,103 tons in 2014, approximately 2% of the worlds production, with France, Ireland, and the Netherlands being the main producers (96%) [1]. EU production reached 108,910 tons in 2008 before severe outbreaks of pathogens, including a *Herpes* virus, that struck French production and spread to all shellfish-producing European countries including Portugal. In 2015, production started to rise again, reaching 110,000 tones the following year [1]. The largest production increases have been observed in Ireland and Portugal, who target the French market. Half of the spat used for oyster farming is supplied by hatcheries; the remaining 50% is wild spat collected by farmers.

Concerning their taxonomy, cupped oysters belong to two identified sister species [2]: the Pacific oyster *Crassostrea gigas* (Thunberg, 1793) and the Portuguese oyster *Crassostrea angulata* (Lamarck 1819), both introduced in Europe from their native ranges in the Northwestern Pacific [2].

The taxonomic classification of cupped oysters has been debated for almost two decades, being identified as a single species or two depending on their cross-fertilization and the genetic variation estimated by different molecular markers—for a complete list of references, see [2]. Recent genomic studies [2,3] have reinforced the hypothesis of two genetically similar but differentiated species. Previous studies [2] have stated that the two species exhibit partial reproductive isolation but also genetic introgression as a result of secondary contacts in the areas where both species have been introduced.

Nowadays, the consensus is that the Portuguese oyster was the first to be introduced in Europe. It was accidentally introduced by Portuguese merchants during the 16th century, probably from Taiwan [2,4,5]; however, it is impossible to establish where the original stocks came from [2,6]. Following the first accidental introduction in Portugal, by the end of the 19th century, the species was already occurring in France, where it was voluntarily introduced for farming and exploitation—see [2] and references therein. Moreover, the Portuguese oyster was introduced in other European countries for shellfish farming, replacing the native flat oyster *Ostrea edulis* (Linnaeus, 1758) (see [2] and references therein). *C. angulata* had a high economic relevance in European aquaculture until the late 1970s, when it practically disappeared due to high mortality rates [7].

The main cause of such a massive mortality, which almost led to the extinction of the species, was associated with the rapid and severe degradation of the main oyster bed ecosystems. In Portuguese systems such as the Sado and Tagus estuaries, this was caused by an impressive development in industrial fabric on the Lisbon and Setúbal water fronts. A second cause was related to the occurrence of a severe pathology, a gill disease characterized by a severe lesion frame in the gills and mantle tissues. The disease was first described by Alderman in 1969 [8] and later associated with the presence of a pathogen identified by Comps in 1976 as an iridovirus [9].

The consequence of such events was the decline of an important oyster industry that permanently employed more than 5000 people, especially in the Sado and Tagus estuaries. The Pacific oyster was then introduced to replace Portuguese oyster cultivation, and it is currently the main species supporting oyster production in Europe (for details on historical and oyster production data in Portugal, see [10,11]). A few populations of *C. angulata*, mainly located in Southern Europe and Northern Africa, survived the massive mortality events of the 1970s and the introduction of *C. gigas* [12]; these populations nowadays occur in Portugal, Spain, and Morocco [2,13,14].

It is worth mentioning that very recent genomic data [2] highlight that the Portuguese population of the Sado estuary represents the reference *C. angulata* population in Europe due to its low level of genetic introgression with the Pacific oyster. Therefore, special attention should be paid to the management and conservation of this valuable Portuguese oyster population. Even if *C. angulata* cannot be strictly considered to be a native species, it has a relevant commercial and cultural value in Portugal, where it has been exploited for over a century and is now considered a valuable natural resource.

Despite this evidence, there has been a lot of pressure put on aquaculture producers to increase the production of the Pacific oyster in both the Sado and Mira estuaries, similarly to what happened in other areas of the Portuguese coast, such as the Ria Formosa and Ria de Aveiro lagoons [15]. The genetic introgression of the Portuguese oyster with Pacific oyster has already been confirmed in Ria Formosa [2], posing risks for the conservation of *C. angulata* wild populations.

Therefore, the specific genetic characterization of both natural and farmed cupped oyster populations of the Sado estuary was performed in this study, with the aim of contributing to better management and conservation plans for this valuable resource in Portugal and Europe in general.

2. Materials and Methods

2.1. Study Area

The Sado estuary is located in Southern Portugal, covering a total area of 180 km^2, with a mean river flow of 40 m$^3 \cdot$s^{-1} and a mean depth of 6 m [14,16]. The estuary hosts both a shipping port and recreational marinas 14], but it also represents one of the most important sites for aquaculture production, especially of oysters, in the country. Wetlands, intertidal mudflats, and saltmarshes are predominant habitats [14] (and references therein), mostly characterized by sandy bottoms [14,17]. Sandy and muddy bottom habitats have high invertebrate species richness, and NIS (non-indigenous species) have also been detected [14].

2.2. Oyster Collection

Oyster collection was carried out in 2015 from Portuguese natural oyster beds and aquaculture facilities. The collection from natural oyster beds was conducted at seven sites along the estuary salinity gradient, whilst the collection of farmed samples was conducted in seven aquaculture facilities located in the estuarine region (Figure 1).

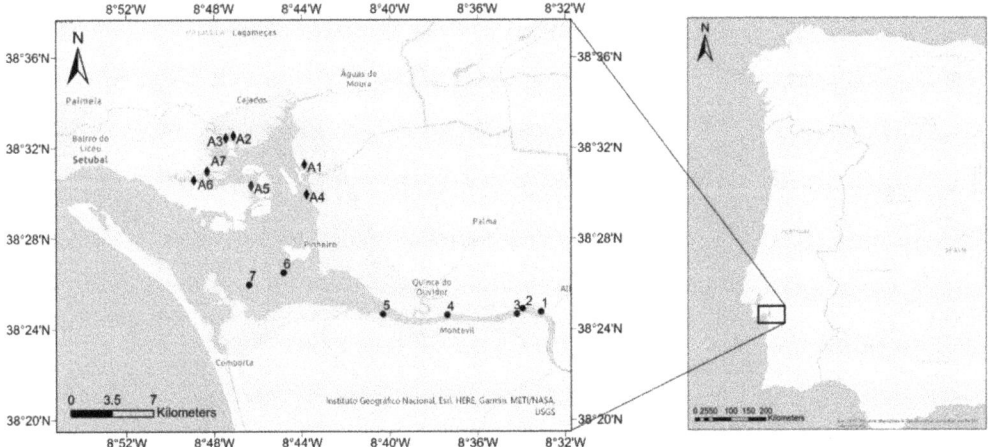

Figure 1. Sampling locations of *C. angulata* from seven natural oyster beds (black dots) and seven aquaculture facilities (black diamonds) in the Sado estuary, Portugal.

For each sampling site, whether from natural oyster beds or aquaculture, 20 individuals were collected. The adductor muscle was dissected in each specimen, individually fixed in absolute ethanol, and preserved at -20 °C until DNA extraction and purification. Ten additional samples of *C. gigas* were also collected in the Ria de Aveiro lagoon as reference material (40°69' N, 8°69' W).

2.3. HMW DNA Extraction and Purification

High molecular weight (HMW) total genomic DNA was extracted and purified for each sample from the adductor muscle fixed in absolute ethanol with the DNeasy Blood & Tissue Kit (Qiagen, Hilden, Germany) following the manufacturer's instructions, and its quality and quantity were verified via an electrophoretic run in 1% agarose gel and TAE buffer (1×).

2.4. Cox1 Gene Fragment Amplification and Sequencing

A fragment of the mitochondrial cytochrome coxidase subunit I (*cox1*) gene was amplified by PCR using the universal primers *LCO1490* (5′-GGTCAACAAATCATAAAGATATTGG-3′) and *HCO2198* (5′-TAAACTTCAGGGTGACCAAAAAATCA-3′) [18] and the specific PCR conditions developed for oysters [19].

The amplification reactions were performed in a total volume of 25 µL, including 20.375 µL of sterilized distilled water, 2.5 µL of a 5× colorless reaction buffer, 0.75 µL of $MgCl_2$ (50 mM), 0.25 µL of each primer (10 pmol/µL), 0.5 µL of dNTP mixture (10 mM), 0.125 µL of Taq polymerase (Enzytech, Roche Diagnostics, Mannheim, Germany), and 0.25 µL of DNA.

PCR was carried out for 4 min at a denaturation temperature of 95 °C, followed by 40 cycles of 1 min at 95 °C, 1 min at 45 °C, 2.5 min at 72 °C, and a final extension of 7 min at 72 °C.

The quality of PCR products was verified with an electrophoretic run on a 2.5% agarose gel and 1× TAE buffer, and they were visualized under UV light: amplification products exhibited a molecular weight of about 650 bp. The PCR products were then purified by Promega Wizard™ SV Gel and PCR Clean-Up System (Promega, Madison, WI USA), following the standard protocol; finally, Sanger sequencing was conducted by STAB Vida, Caparica, Portugal.

2.5. Phylogenetic Analyses and Haplotype Analysis

Electropherograms were visualized in Mega X (https://www.megasoftware.net/, accessed on 2 January 2021) and imported into a multiple sequence alignment [20]. The sequences obtained were compared with all *C. angulata* available in GenBank up to December 2020, and those of other species of the genus *Crassostrea*—*C. gigas*, *C. dianbaiensis*, *C. sikamea*, *C. nippona*, *C. virginica*, *C. ariakensis*—available on GenBank (Table S1); moreover, *Saccostrea glomerata* and *S. cucullata* were used as outgroups in the final alignment (Table S1).

Sequences were aligned using amino acids as a guide through the TranslatorX server [21] using the Muscle [22,23] algorithm and the invertebrate mitochondrial genetic code, with no alignment cleaning. Sites with low or noisy phylogenetic signal were masked using Gblocks 0.91b [24]: the minimum number of sequences for a flank position was set to 50% + 1, the maximum number of contiguous no conserved positions was set to 10, the minimum length of a block was set to 5, and all gap positions were allowed. The aligned *cox1* fragment was split into the three codon positions thanks to a custom-tailored Python script (available from FP upon request), which resulted in three datasets: cox1_1, cox1_2, and cox1_3; these datasets were concatenated into the final alignment. A phylogenetic tree was inferred using IQ-TREE 1.7-beta7 [25] with 1000 ultrafast bootstrap replicates [26]. ModelFinder [27] was used to select substitution models; the greedy strategy was chosen to select the best partitioning scheme [28,29].

In order to estimate the degree of saturation in our dataset, the substitution saturation test developed by Xia and colleagues [30,31] was applied. Eventually, the EMBOSS 6.6.0 distmat application [32] was used to compute pairwise (uncorrected) p-distances to be plotted over pairwise ML distances computed in RAxML 8.2.12 [33].

The PopART v 1.7 software [34] was used to draw the minimum spanning network by selecting the statistical parsimony criterion and setting $\varepsilon = 0$. The sequences were also analyzed using statistical parsimony performed in [35] tested through the TCS v.1.21 program [36], in which we set the network connection limit at 90% and gaps as "missing". TCS allowed for the identification of different haplotypes, desegregating them in haplogroups. TCS produced networks that clarified the relationships between different haplotypes/haplogroups, showing the significant number of substitutions connecting haplotypes. The network was visualized and plotted with tcsBU [37]. Spatial or demographic expansion was estimated through the Tajima D neutrality test [38] using the DNAsp 5.0 program [39]. Tajima's D statistic tested the departure from neutrality by measuring the differences between the average number of pairwise nucleotide differences and the number of segregating sites [38]. If both balancing or purifying selection were absent, only the population expansion significantly lowered Tajima's D to zero; the positive increase of this statistics may be related to a population bottleneck [38].

3. Results

The final alignment included 394 *cox1* sequences, 110 of which were the original sequences collected in this study in the Sado estuary (Table S2) from seven natural oyster beds and seven aquaculture facilities. The total alignment length was 543 base pairs (bps).

3.1. Phylogenetic Analysis

No significant saturation was detected across the three codon positions; therefore, they were all retained for subsequent analyses (Figure S1 and Table S3).

The *C. angulata* and *C. gigas* species resulted in a single monophyletic clade, with ultrafast bootstrap support (UFboot) equal to 100 (Figure 2). Moreover, *C. gigas* was recovered as monophyletic (UFboot = 100) within a wide polytomy of *C. angulata* OTUs, where the phylogenetic relationships were not completely resolved (Figure 2). The *C. gigas* clade was retrieved as the sister group of the *C. angulata* clade (UFboot = 82), which entirely comprised Pacific specimens, but the statistical support of the node was low (UFboot = 72).

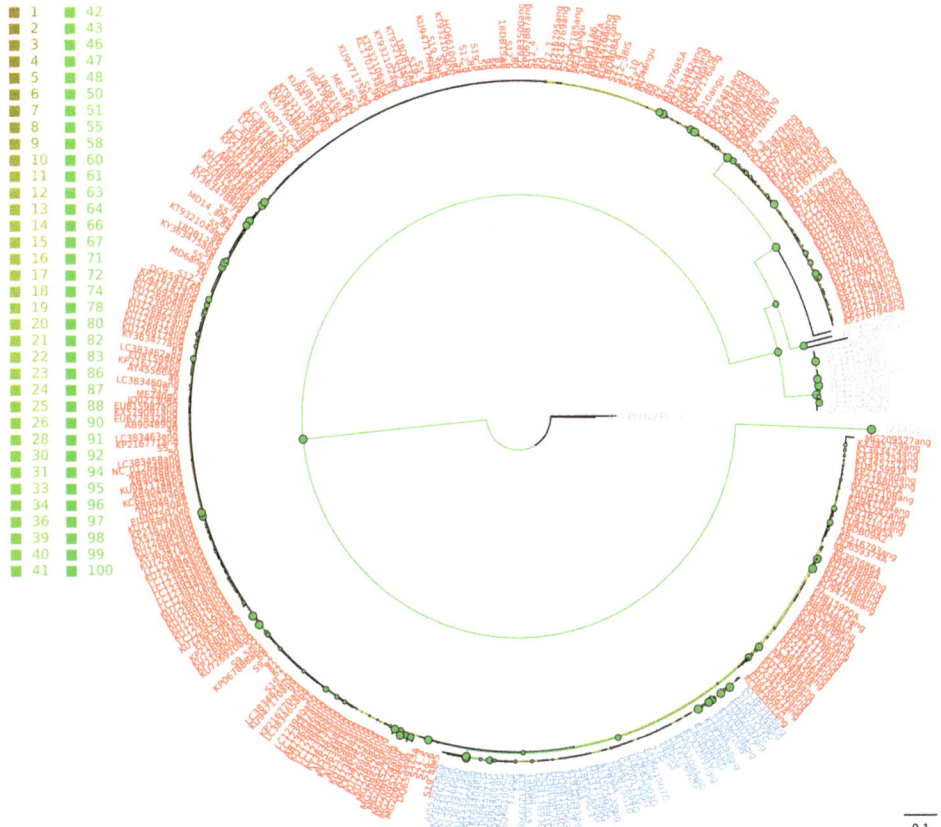

Figure 2. Maximum likelihood phylogenetic tree. Red, *C. angulata*; blue, *C. gigas*; grey, outgroup. The color of the dots depicted at nodes indicates ultrafast bootstrap values (light brown = 0%; solid green = 100%). Correspondence between bootstrap values and colors are reported as a legend.

All the original samples analyzed within this study from the Sado estuary were nested within *C. angulata*, except for twelve samples collected in one aquaculture facility (Figure 2). The reference material collected in the Ria de Aveiro lagoon was confirmed to be *C. gigas*. Moreover, the EU007507, EU007510 and EU007512 sequences [19] (which were previously deposited in GenBank as *C. gigas*) nested as *C. angulata* in the phylogenetic analysis. Therefore, they were considered to be *C. angulata* in the subsequent analyses.

Such separation was confirmed by the median joining network (Figure S2). This network clearly evidenced the separation between *C. angulata* and *C. gigas*, as well as a clear-cut divergence of these two species from the congeneric *C. dianbaiensis*, *C. sikamea*, *C. nippona*, *C. virginica*, *C. ariakensis*, and (obviously) the two outgroups of *S. glomerata* and *S. cucullata*.

3.2. Haplotype Analysis and Genetic Diversity of Portuguese Oyster Populations

In total, 134 haplotypes were identified (Figure 3; Table S4): 104 for *C. angulata*; 18 for *C. gigas*; 5 for *C. dianbaiensis*; and one each for *C. sikamea*, *C. nippona*, *C. virginica*, *C. ariakensis*, *S. glomerata*, and *S. cucullata*. These haplotypes segregated into nine haplogroups: one for *C. angulata*; one for *C. gigas*; one each for *C. dianbaiensis*, *C. sikamea*, *C. nippona*, *C. virginica*, and *C. ariakensis*; and one each for outgroups *S. glomerata* and *S. cucullata*.

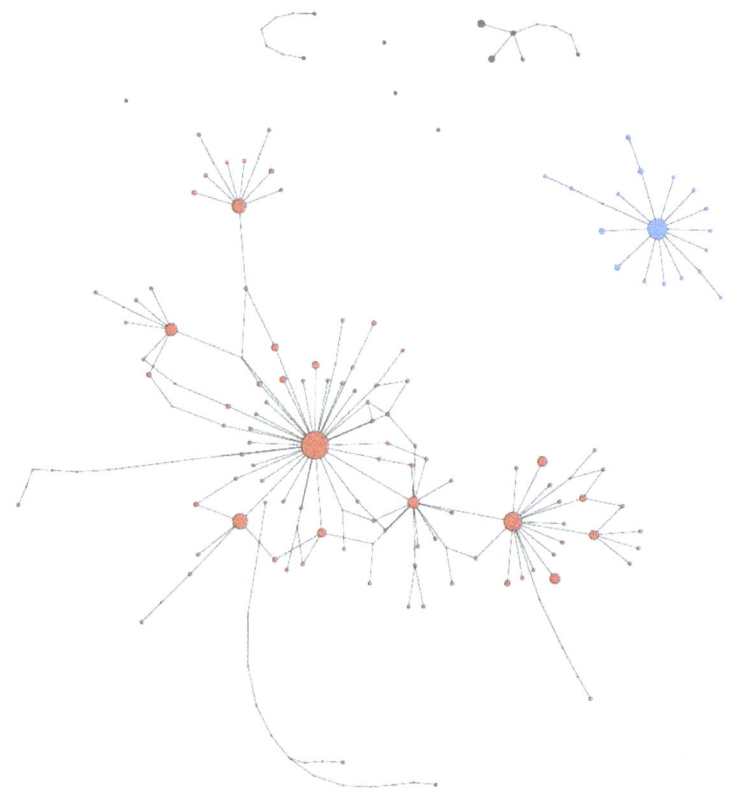

Figure 3. Haplotype minimum spanning network. Each circle represents a unique haplotype, each color represents each species as reported above (red, *C. angulata*; blue, *C. gigas*; grey, *C. dianbaiensis*, *C. sikamea*, *C. nippona*, *C. virginica*, *C. ariakensis*, and the two outgroups *S. glomerata* and *S. cucullata*), and the size of the circle is proportional to number of samples represented by each haplotype.

The obtained haplotype network confirmed the separation of the *C. angulata* and *C. gigas* haplogroups (Figure 3 and Table S4). Original samples of *C. gigas* collected in this study were included in five haplotypes, labelled as *Cr_gi01–Cr_gi05*, and deposited in GenBank under accession numbers OK021655–OK021659.

For *C. angulata*, 17 of the 104 described haplotypes included the original sequences collected from this study in the Sado estuary: haplotype sequences were labelled as *Cr_an01–Cr_an17* and deposited in GenBank under accession numbers OK030211–OK030227. Among them, five were newly identified (*Cr_an10, 14, 15, 16,* and *17*) since they were never detected before among all *C. angulata* sequences available in GenBank.

Considering the frequency of the 104 haplotypes obtained for the Portuguese oyster, four haplotypes were the most common ones, including sequences collected both from the Indo-Pacific and from Portuguese and Spanish populations. In detail, haplotype *Cr_an01* (reference haplotype sequence: OK030211) comprised 75 sequences (biggest haplotype probability: 0.076), haplotype *Cr_an03* (reference haplotype sequence: OK030213) comprised 32 sequences (haplotype probability 0.038), haplotype *Cr_an08* (reference haplotype sequence: OK030218) comprised 22 sequences (haplotype probability: 0.047), and haplotype *Cr_an07* (reference haplotype sequence: OK030217) comprised 18 sequences (haplotype probability: 0.012) (Table S4).

However, most of the identified haplotypes (89 out of 104) were characterized by a single sequence (singletons) or two sequences; mainly, 73 were distributed in the original area of species distribution (China, Taiwan, Korea, and Japan), but 16 were identified in Portuguese populations.

As for their geographic distribution, among the global 104 haplotypes identified for Portuguese oysters, 28 were identified in the European populations in Portugal and Spain (Cadiz) (Figure 4).

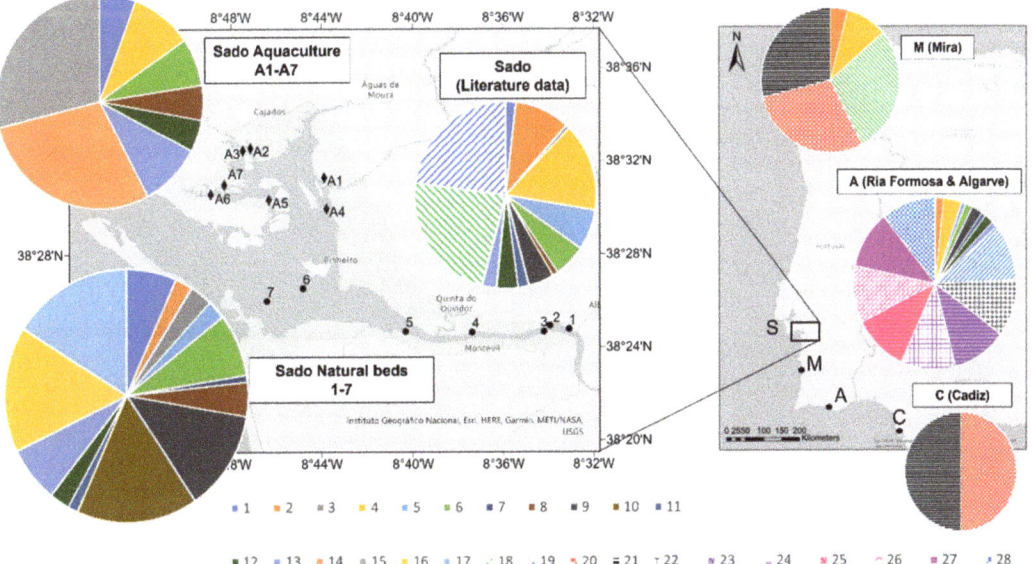

Figure 4. Distribution of the 28 haplotypes of *C. angulata* occurring in European populations. The original data of this study refer to the Sado aquaculture (A1–A7) and Sado natural oyster beds (1–7); all other data were collected from GenBank sequences. Haplotype numbers 1–17 (originally named *Cr_an01–Cr_an17*) refer to sequences OK030211–OK030227, haplotype 18 corresponds to reference sequence AY397686, haplotype 19 corresponds to reference sequence KY363483, haplotype 20 corresponds to reference sequence AJ553907, haplotype 21 corresponds to reference sequence AJ553908, and haplotypes 22–28 correspond to reference sequences MG209523-29.

Moreover, 19 of these haplotypes—21 if considering two additional haplotypes previously identified also in Spain (Cadiz)—only occurred in Portugal, namely in the Sado and Mira estuaries, Ria Formosa, and Algarve.

Eight haplotypes were shared among Indo-Pacific and Portuguese populations, whilst 75 haplotypes were never detected in Portuguese or Spanish populations (both original samples and previously deposited data).

Focusing on the original samples collected in this study from the Sado Estuary, the results obtained revealed some differences between the *C. angulata* samples belonging to natural oyster beds and aquaculture facilities (see Figure 4). The samples from the natural beds were characterized by the presence of 14 haplotypes: *Cr_an01–Cr_an03, Cr_an05– Cr_an13, Cr_an16*, and *Cr_an17*. Eight haplotypes were identified in farmed samples: *Cr_an01, Cr_an04, Cr_an06, Cr_an08, Cr_an12, Cr_an13, Cr_an14*, and *Cr_an15*. Five haplotypes were shared by both groups: *Cr_an01, Cr_an06, Cr_an08, Cr_an12*, and *Cr_an13*.

It is noteworthy that most of the described haplotypes were found to have a very low frequency, being represented only by one or a few sequences. A total of 45% of the sequences collected in the studied natural oyster beds and aquaculture facilities were grouped into haplotype *Cr_an01*, which was the most common one even when considering the entire *C. angulata* dataset (Figure 3 and Table S4). This could be useful to understand the Tajima Neutrality test, which showed a negative D parameter for all tested combinations: when applied to all *C. angulata* sequences ($D = -2.23$, ** $p < 0.01$); when only applied to the sequences obtained from Portuguese populations that also included GenBank data ($D = -1.87$, ** $p < 0.05$); and when limited to the sequences collected for the present study, although it was not significant in this case ($D = -1.32, p > 0.10$).

4. Discussion

This study allowed for the collection of *cox1* sequences of *C. angulata* from the Sado estuary from both natural oyster beds and aquaculture facilities, and their comparison with a wide dataset of cupped oyster sequences obtained from GenBank—to our knowledge the widest available for Portuguese oysters.

Our analysis confirmed that most of the original samples of this study could be taxonomically identified as *C. angulata*, except for a limited number of samples collected from an aquaculture facility that were identified as *C. gigas*. Moreover, results obtained from *cox1* sequence analyses confirmed the existence of a genetic distance between the two species, although it was lower than those occurring with all other species of the same genus.

The oyster populations of the Sado estuary showed high haplotypic variability and were characterized by five original haplotypes that were not previously described; however, most of them were found to have a very low frequency, being represented by only one or two sequences. Natural populations showed higher haplotypic variability compared to farmed ones. The sharing of five common haplotypes in both natural and cultivated oysters, two of which (*Cr_an01* and *Cr_an08*) showed higher frequencies, is compatible with the fishery activities that are carried out in the Sado estuary. A richness in singletons, i.e., pronounced/strong sweeps related to an excess of low frequency polymorphisms [40,41], and the high frequencies of very common haplotypes result in significantly negative values of the D Tajima's statistics [38]. A negative D value can indicate a possible recent population expansion in the Sado estuary that is compatible with a founder effect related to the non-native origin of *C. angulata* or with a genetic drift and a bottleneck caused by the strong demographic reduction in the 1960s and 1970s and the recent recovery that shaped their current genetic structure and diversity. As previously underlined [42], this test is frequently used by conservation biologists due to its advantages, including the fact that the Tajima test can be performed on sequences belonging to any coding or noncoding locus of any species and no outgroup is required [42].

The haplotypic variability of the samples collected in natural oyster beds was an example of in situ sustainable management that clearly demonstrated the importance of the integrated conservation of wild populations. Their natural genetic diversity will constitute a fundamental source of variation and greater adaptation to the naturally variable conditions of the Sado estuary, reinforcing the viewpoint that aquaculture activities must ensure the high genetic diversity and fitness of cultivated stocks in a global change scenario.

Moreover, the absence of *C. gigas* from the sampled natural oyster beds is good news from the viewpoint of Portuguese oyster conservation, especially considering that *C. angulata* beds in the Sado estuary are among the last existing populations in Europe and are considered the purest ones in terms of introgression with the Pacific oyster [2]. The conservation of the ancestral genetic traits of *C. angulata* may be due to the fact that the introduction of cultured *C. gigas* in the estuary has been forbidden by local authorities, a situation that has not taken place in the Ria Formosa and the Ria de Aveiro. In fact, recent genomic data [2] showed that the Ria Formosa population has higher level of introgression with genetic traits of *C. gigas* because of Pacific oyster cultivation that has been conducted for more than 15 years [2,43,44].

Nevertheless, the data obtained in this study showed that specimens of *C. gigas* are cultivated in the local aquaculture facilities. These results were corroborated by interviews with local oyster producers, who confirmed that some of them import oyster seeds from France [45]. The cultivation of the Pacific oyster represents a concrete risk to Portuguese oyster conservation due to its ability to hybridize. In fact, though the genetic divergence between these two species is low, they show phenotypic differences, including in their resistance to diseases, growth rates, and physiological behavior [2], that are particularly relevant in terms of biodiversity conservation and aquaculture production. The introduction of the Pacific oyster could also represent a risk for the spreading of new pathogens. Unfortunately, this last risk became real in several Portuguese systems where the Pacific oyster was introduced. The severe outbreaks of a *Herpes* virus that have occurred in France since 2012 reached the Portuguese populations in 2014, 2015, and 2016 with mortality rates close to 90%. Therefore, since 2008, the Portuguese authorities have forbidden the introduction of the Pacific oyster into the Natural Reserve of the Sado Estuary to preserve the Portuguese oyster beds. However, as highlighted by the results of this paper, the Pacific oyster occurred in at least one of the investigated aquaculture facilities.

It is therefore clear that the conservation of last "pure" populations of *C. angulata* from the Sado estuary should be considered a priority, especially for local authorities and oyster producers, due to their biological, ecological, cultural, and economic value. Efforts should be made to tightly regulate the introduction of *C. gigas* in both natural beds and aquaculture facilities in order to prevent the hybridization of the valuable Portuguese oysters with the invasive Pacific ones.

It is noteworthy that interviews conducted with the oyster producers in the Sado estuary have indicated that the "certified origin" of the product and the creation of a "Sado label" are two of the most important measures to improve oyster production and cultivation [45].

Additionally, the specific regulations regarding the use of NIS in aquaculture [46] and the specific restrictions and measures required for NIS of EU concern [47] should be implemented to effectively mitigate the risk [14]. Therefore, the monitoring and restoration of Portuguese oyster populations, with reference to the Sado estuary, should be regularly carried out, as previously suggested [2].

5. Conclusions

Although the Portuguese oyster cannot be strictly described as a native species of Portugal, it was introduced a long time ago and is an important component of the estuarine habitats, with a relevant cultural and economic value. Therefore, the conservation of the last reference populations of *C. angulata* should be considered a priority, both in the Natural

Reserve of the Sado estuary and in the Mira estuary, another estuarine system included in a protected area.

Genetic data obtained from this study confirmed that the Pacific oyster occasionally occurs in oyster aquaculture facilities but does not occur in the natural oyster beds of the Sado estuary. However, the presence of the hybridizing congeneric represents a concrete problem, and risks are arising for the conservation of the Portuguese oyster reference populations of the Sado estuary: therefore, researchers, local authorities, and oyster producers should work together to avoid the loss of this valuable resource.

Supplementary Materials: The following are available online at https://www.mdpi.com/article/10.3390/life11111173/s1, Figure S1: substitution saturation test: (a) first codon position; (b) second codon position; (c) third codon position. Figure S2: Median-joining network constructed using PopArt 1.7 based on 394 sequences. Each circle represents a unique genome/group of affine genomes, the colors represent the species, as reported above (red, *C. angulata*; blue, *C. gigas*; grey, *C. dianbaiensis*, *C. sikamea*, *C. nippona*, *C. virginica*, *C. ariakensis*, and the two outgroups *S. glomerata* and *S. cucullate*), and the size of the circle is proportional to number of genomes included in each circle. Figure legend represents dimension proportionality to sample number included in each circle and colour correspondence. Numbers of temples (one line per mutation) represent the number of single nucleotide variations (SNVs) between groups. Table S1: GenBank sequences included in the alignment and phylogenetic analyses. For each sequence the species, the GenBank accession number, the original source, and the sampling location (if available) are provided. The sequences marked with (*) A.N. EU007507, 510, and 512 were deposited in GenBank as *C. gigas*, but the phylogenetic analyses conducted here confirmed that they belong to *C. angulata*. Previous data collected from Sado Estuary are marked in bold. Table S2: List of original samples sequenced in this study. Table S3: Test of substitution saturation (Xia and Lemey, 2009, Xia et al., 2003). If Iss was significantly smaller than Iss.c, only a little saturation was observed. Iss.cSym was Iss.c when assuming a symmetrical topology; Iss.cAsym was Iss.c when assuming an asymmetrical topology. Given the large number of OTUs in the present dataset, only the results for the highest number of OTUs (32) are shown; P was estimated to be 0.0000 in all cases. Table S4: Haplotype analysis of all the sequences included in the final alignment. Haplogroup/species; haplotype name with reference sequence; specimens/sequences for each haplotype are indicated. Original haplotypes from this study are indicated in bold.

Author Contributions: Conceptualization, S.C., L.L., F.P., P.C., J.L.C., F.R., R.F. and M.M.A.; methodology, S.C., L.L., F.P., P.C. and M.M.A.; software, S.C., L.L. and F.P.; validation, S.C., L.L. and F.P.; formal analysis, S.C., L.L. and F.P.; investigation, S.C., L.L. and F.P.; resources, R.F., F.R., P.C. and J.L.C.; data curation, S.C., L.L. and F.P.; writing—original draft preparation, S.C., L.L. and F.P.; writing—review and editing, all authors.; visualization, S.C., L.L., F.P. and P.C.; supervision S.C., L.L., R.F., F.R., P.C. and J.L.C.; project administration, R.F., F.R., P.C., J.L.C. and M.M.A.; funding acquisition, R.F., F.R., P.C. and J.L.C. All authors have read and agreed to the published version of the manuscript.

Funding: This research was funded by CRASSOSADO project "Estado atual da ostra portuguesa (*Crassostrea angulata*) no estuário do Sado, ameaças e oportunidades para a sua exploração como recurso" funded by Instituto da Conservação da Natureza e das Florestas (ICNF) and the Navigator Company. Thanks are also due for the financial support to "Progetto Ricerca Di Base 2020", University of Perugia (Italy).

Institutional Review Board Statement: Ethical review and approval were waived for this study due to the use of commercial species.

Informed Consent Statement: Not applicable.

Data Availability Statement: Original data from this study have been deposited in GenBank under Accession Numbers OK021655–OK021659 and OK030211–OK030227.

Acknowledgments: The authors express thanks for financial support to MARE (UIDB/04292/2020), COASTNET (PINFRA/22128/2016) and to CESAM (UIDB/50017/2020 + UIDP/50017/2020), to FCT/MEC through national funds, and the co-funding by the FEDER, within the PT2020 Partnership Agreement and Compete 2020. P. Chainho was funded by the 2020.01797.CEECIND contract. Moreover, the authors express thanks for financial support to "Progetto Ricerca Di Base 2020", University of Perugia (Italy). The authors would like to thank the oyster farmers of the Sado estuary for their help with collecting the samples, as well as Carlos Silva (Instituto de Conservação da Natureza e Florestas). The authors would like to thank AU personnel (Adilia Pires and Anthony Moreira) for their help with DNA extraction and PCR procedures. The authors would like to thank MARE, IPMA, and ICNF personnel and especially to Drs. Narcisa Bandarra, Helena Lourenço, Maria de Jesus Fernandes, Ana Cristina Falcão, Ana Grade, Filipa Marques, and Teresa Portela, for the project development and logistic and personnel support during sampling surveys, as well as Sara Cabral for her help with the georeferenced mapping of sampling stations. The authors would like to thank Thomas Goulding for English editing. Finally, the authors would like to thank the two anonymous referees who revised the paper for their valuable suggestions, the Guest Editors of this Special Issue, and the Editorial office at LIFE journal.

Conflicts of Interest: The authors declare no conflict of interest. The funders had no role in the design of the study; in the collection, analyses, or interpretation of data; in the writing of the manuscript, or in the decision to publish the results.

References

1. EUMOFA. EUMOFA N. 2/2017. 2017. Available online: https://www.eumofa.eu/ (accessed on 10 January 2021).
2. Lapègue, S.; Heurtebise, S.; Cornette, F.; Guichoux, E.; Gagnaire, P.A. Genetic Characterization of Cupped Oyster Resources in Europe Using Informative Single Nucleotide Polymorphism (SNP) Panels. *Genes* **2020**, *11*, 451. [CrossRef] [PubMed]
3. Gagnaire, P.; Lamy, J.B.; Cornette, F.; Heurtebise, S.; Dégremont, L.; Flahauw, E.; Boudry, P.; Bierne, N.; Lapègue, S. Analysis of Genome-Wide Differentiation between Native and Introduced Populations of the Cupped Oysters *Crassostrea gigas* and *Crassostrea angulata*. *Genome Biol. Evol.* **2018**, *10*, 2518–2534. [CrossRef] [PubMed]
4. Boudry, P.; Heurtebise, S.; Collet, B.; Cornette, F.; Gérard, A. Differentiation between populations of the Portuguese oyster, *Crassostrea angulata* (Lamark) and the Pacific oyster, *Crassostrea gigas* (Thunberg), revealed by mtDNA RFLP analysis. *J. Exp. Mar. Biol. Ecol.* **1998**, *226*, 279–291. [CrossRef]
5. Huvet, A.; Lapègue, S.; Magoulas, A.; Boudry, P. Mitochondrial and nuclear DNA phylogeography of *Crassostrea angulata*, the Portuguese oyster endangered in Europe. *Conserv. Genet.* **2000**, *1*, 251–262. [CrossRef]
6. Grade, A.; Chairi, H.; Lallias, D.; Power, D.M.; Ruano, F.; Leitão, A.; Drago, T.; King, J.W.; Boudry, P.; Batista, F.M. New insights about the introduction of the Portuguese oyster, *Crassostrea angulata*, into the North East Atlantic from Asia based on a highly polymorphic mitochondrial region. *Aquat. Living Resour.* **2016**, *29*, 404. [CrossRef]
7. FAO. *Global Aquaculture Production 1950–2004*; Fisheries Global Information System: Rome, Italy, 2007.
8. Alderman, D.J.; Gras, P. "Gill Disease" of Portuguese oysters. *Nature* **1969**, *224*, 616–617. [CrossRef]
9. Comps, M.; Bonami, J.R.; Vago, C. Pathologie des invertébrés: Une virose de l'huître portugaise (*Crassostrea angulata* Lmk.). *Comptes Rendus De L'académie Des Sci. De Paris Série D* **1976**, *282*, 1991–1993. (In French)
10. Ruano, F.D.L. Fisheries and Farming of Important Marine Bivalves in Portugal. *Hist. Present Cond. Future Molluscan Fish. North Cent. Am. Eur.* **1997**, *3*, 191–2001.
11. INE/DGRM. 2017. Available online: https://www.dgrm.mm.gov.pt/web/guest (accessed on 10 January 2021).
12. Batista, F. Assessment of the Aquacultural Potential of the Portuguese Oyster *Crassostrea angulata*. Ph.D. Thesis, University of Porto, Porto, Portugal, 2007; 245p.
13. Fabioux, C.; Huvet, A.; Lapègue, S.; Heurtebise, S.; Boudry, P. Past and present geographical distribution of populations of Portuguese (*Crassostrea angulata*) and Pacific (*C. gigas*) oysters along the European and north African Atlantic coasts. *Haliotis* **2002**, *31*, 33–44.
14. Cabral, S.; Carvalho, F.; Gaspar, M.; Ramajal, J.; Sá, E.; Santos, C.; Silva, G.; Sousa, A.; Costa, J.L.; Chainho, P. Non-indigenous species in soft-sediments: Are some estuaries more invaded than others? *Ecol. Indic.* **2020**, *110*, 105640. [CrossRef]
15. CRASSOSADO. *Estado Atual da Ostra Portuguesa (Crassostrea angulata) no Estuário Do Sado, Ameaças e Oportunidades Para a Sua Exploração Como Recurso—CRASSOSADO*; Relatório Final de Projecto; ICNF, IPMA, MARE/FCUL, CESAM/UA: Lisboa, Portugal, 2016; 98p.
16. França, S.; Costa, M.J.; Cabral, H.N. Assessing habitat specific fish assemblages in estuaries along the Portuguese coast. *Estuar. Coast. Shelf Sci.* **2009**, *83*, 1–12. [CrossRef]
17. Rodrigues, A.M.J.; Quintino, V.M.S. Horizontal biosedimentary gradients across the Sado estuary. W. Portugal. Netherlands. *J. Aquat. Ecol.* **1993**, *27*, 449–464. [CrossRef]
18. Folmer, O.; Black, M.; Hoeh, W.; Lutz, R.; Vrijenhoek, R. DNA primers for amplification of mitochondrial cytochrome c oxidase sub-unit I from diverse metazoan invertebrates. *Mol. Mar. Biol. Biotech.* **1994**, *3*, 294–299.
19. Reece, K.S.; Cordes, J.F.; Stubbs, J.B.; Hudson, K.L.; Francis, E.A. Molecular phylogenies help resolve taxonomic confusion with Asian *Crassostrea* oyster species. *Mar. Biol.* **2008**, *153*, 709–721. [CrossRef]
20. Tamura, K.; Stecher, G.; Peterson, D.; Filipski, A.; Kumar, S. MEGA6: Molecular evolutionary genetics analysis version 6.0. *Mol. Biol. Evol.* **2013**, *30*, 2725–2729. [CrossRef]
21. Abascal, F.; Zardoya, R.; Telford, M.J. TranslatorX: Multiple alignment of nucleotide sequences guided by amino acid translations. *Nucleic Acids Res.* **2010**, *38*, W7–W13. [CrossRef] [PubMed]

22. Edgar, R.C. MUSCLE: A multiple sequence alignment method with reduced time and space complexity. *BMC Bioinform.* **2004**, *5*, 113. [CrossRef] [PubMed]
23. Edgar, R.C. MUSCLE: Multiple sequence alignment with high accuracy and high throughput. *Nucleic Acids Res.* **2004**, *32*, 1792–1797. [CrossRef] [PubMed]
24. Castresana, J. Selection of conserved blocks from multiple alignments for their use in phylogenetic analysis. *Mol. Biol. Evol.* **2000**, *17*, 540–552. [CrossRef] [PubMed]
25. Nguyen, L.T.; Schmidt, H.A.; von Haeseler, A.; Minh, B.Q. IQ-TREE: A fast and effective stochastic algorithm for estimating maximum likelihood phylogenies. *Mol. Biol. Evol.* **2015**, *32*, 268–274. [CrossRef]
26. Hoang, D.T.; Chernomor, O.; von Haeseler, A.; Minh, B.Q.; Vinh, L.S. UFBoot2: Improving the ultrafast bootstrap approximation. *Mol. Biol. Evol.* **2018**, *35*, 518–522. [CrossRef] [PubMed]
27. Kalyaanamoorthy, S.; Minh, B.Q.; Wong, T.K.F.; von Haeseler, A.; Jermiin, L.S. ModelFinder: Fast model selection for accurate phylogenetic estimates. *Nat. Methods* **2017**, *14*, 587–589. [CrossRef]
28. Chernomor, O.; von Haeseler, A.; Minh, B.Q. Terrace aware data structure for phylogenomic inference from supermatrices. *Syst. Biol.* **2016**, *65*, 997–1008. [CrossRef] [PubMed]
29. Lanfear, R.; Calcott, B.; Ho, S.Y.; Guindon, S. Partitionfinder: Combined selection of partitioning schemes and substitution models for phylogenetic analyses. *Mol. Biol. Evol.* **2012**, *29*, 1695–1701. [CrossRef] [PubMed]
30. Xia, X.; Lemey, P. Assessing substitution saturation with DAMBE. In *The Phylogenetic Handbook: A Practical Approach to DNA and Protein Phylogeny*, 2nd ed.; Lemey, P., Salemi, M., Vandamme, A.M., Eds.; Cambridge University Press: Cambridge, UK, 2009; pp. 615–630.
31. Xia, X.; Xie, Z.; Salemi, M.; Chen, L.; Wang, Y. An index of substitution saturation and its application. *Mol. Phylogenet. Evol.* **2003**, *26*, 1–7. [CrossRef]
32. Rice, P.; Longden, I.; Bleasby, A. EMBOSS: The European Molecular Biology Open Software Suite. *Trends Genet.* **2000**, *16*, 276–277. [CrossRef]
33. Stamatakis, A. RAxML version 8: A tool for phylogenetic analysis and post-analysis of large phylogenies. *Bioinformatics* **2014**, *30*, 1312–1313. [CrossRef]
34. Leigh, J.W.; Bryant, D. PopArt: Full-feature software for haplotype network construction. *Methods Ecol. Evol.* **2015**, *6*, 1110–1116. [CrossRef]
35. Templeton, A.R.; Crandall, K.A.; Sing, C.F. A cladistic analysis of phenotypic associations with haplotypes inferred from restriction endonuclease mapping and DNA sequence data. III. Cladogram estimation. *Genetics* **1992**, *132*, 619–633. [CrossRef]
36. Clement, M.; Posada, D.; Crandall, K.A. TCS: A computer program to estimate gene genealogies. *Mol. Ecol.* **2000**, *9*, 1657–1659. [CrossRef]
37. dos Santos, A.M.; Cabezas, M.P.; Tavares, A.I.; Xavier, R.; Branco, M. tcsBU: A tool to extend TCS network layout and visualization. *Bioinformatics* **2016**, *32*, 627–628. [CrossRef]
38. Tajima, F. Statistical methods to test for nucleotide mutation hypothesis by DNA polymorphism. *Genetics* **1989**, *123*, 585–595. [CrossRef] [PubMed]
39. Rozas, J.; Ferrer-Mata, A.; Sánchez-Del Barrio, J.C.; Guirao-Rico, S.; Librado, P.; Ramos-Onsins, S.E.; Sánchez-Gracia, A. DnaSP 6: DNA Sequence Polymorphism Analysis of Large Data Sets. *Mol. Biol. Evol.* **2017**, *34*, 3299–3302. [CrossRef] [PubMed]
40. Braverman, J.M.; Hudson, R.R.; Kaplan, N.L.; Langley, C.H.; Stephan, W. The hitchhiking effect on the site frequency spectrum of DNA polymorphisms. *Genetics* **1995**, *140*, 783–796. [CrossRef] [PubMed]
41. Wakeley, J. *Coalescent Theory: An Introduction*, 1st ed.; Roberts and Company: Greenwood Village, CO, USA, 2008.
42. Rand, D.M. Neutrality Tests of Molecular Markers and the Connection Between DNA Polymorphism, Demography, and Conservation Biology. *Conserv. Biol.* **1996**, *10*, 665–671. [CrossRef]
43. Huvet, A.; Fabioux, C.; McCombie, H.; Lapègue, S.; Boudry, P. Natural hybridization between genetically differentiated populations of *Crassostrea gigas* and *C. angulata* highlighted by sequence variation in flanking regions of a microsatellite locus. *Mar. Ecol. Prog. Ser.* **2004**, *272*, 141–152. [CrossRef]
44. Batista, F.M.; Fonseca, V.G.; Ruano, F.; Boudry, P. Asynchrony in settlement time between the closely related oysters *Crassostrea angulata* and *C. gigas* in Ria Formosa lagoon (Portugal). *Mar. Biol.* **2017**, *164*, 110. [CrossRef]
45. CRASSOSADO II. *Nova Contribuição Para o Conhecimento Do Estado Atual da Ostra Portuguesa (Crassostrea angulata) No Estuário Do Sado, Com Destaque Para a Determinação de Ameaças e Oportunidades Para a Sua Exploração–CRASSOSADO II*; Relatório Final de Projecto; ICNF, IPMA, MARE/FCUL, CESAM/UA: Lisboa, Portugal, 2017; 56p.
46. Council Regulation (EC) No 708/2007 of 11 June 2007 Concerning Use of Alien and Locally Absent Species in Aquaculture OJ L 168, 28.6.2007; pp. 1–17. Available online: https://eur-lex.europa.eu/eli/reg/2007/708/oj (accessed on 10 April 2021).
47. Regulation (EU) No 1143/2014 of the European Parliament and of the Council of 22 October 2014 on the Prevention and Management of the Introduction and Spread of Invasive Alien Species. Available online: http://data.europa.eu/eli/reg/2014/1143/oj (accessed on 10 April 2021).

Article

Molecular Correlation between Larval, Deutonymph and Adult Stages of the Water Mite *Arrenurus* (*Micruracarus*) *Novus*

Pedro María Alarcón-Elbal [1], Ricardo García-Jiménez [2], María Luisa Peláez [2], Jose Luis Horreo [3],* and Antonio G. Valdecasas [2]

[1] Instituto de Medicina Tropical & Salud Global (IMTSAG), Universidad Iberoamericana (UNIBE), 22333 Santo Domingo, Dominican Republic; Pedro.Alarcon@uv.es
[2] Museo Nacional de Ciencias Naturales, CSIC. C/José Gutiérrez Abascal, 2, 28006 Madrid, Spain; rgarcia@mncn.csic.es (R.G.-J.); maaller@yahoo.es (M.L.P.); valdeca@mncn.csic.es (A.G.V.)
[3] UMIB Research Unit of Biodiversity (UO, CSIC, PA), C/Gonzalo Gutiérrez de Quirós s/n, 33600 Mieres, Spain
* Correspondence: horreojose@gmail.com; Tel.: +34-985-10-30-00 (ext. 5943)

Received: 27 May 2020; Accepted: 7 July 2020; Published: 9 July 2020

Abstract: The systematics of many groups of organisms has been based on the adult stage. Morphological transformations that occur during development from the embryonic to the adult stage make it difficult (or impossible) to identify a juvenile (larval) stage in some species. Hydrachnidia (Acari, Actinotrichida, which inhabit mainly continental waters) are characterized by three main active stages—larval, deutonymph and adult—with intermediate dormant stages. Deutonymphs and adults may be identified through diagnostic morphological characters. Larvae that have not been tracked directly from a gravid female are difficult to identify to the species level. In this work, we compared the morphology of five water mite larvae and obtained the molecular sequences of that found on a pupa of the common mosquito *Culex* (*Culex*) *pipiens* with the sequences of 51 adults diagnosed as *Arrenurus* species and identified the undescribed larvae as *Arrenurus* (*Micruracarus*) *novus*. Further corroborating this finding, adult *A. novus* was found thriving in the same mosquito habitat. We established the identity of adult and deutonymph *A. novus* by morphology and by correlating COI and cytB sequences of the water mites at the larval, deutonymph and adult (both male and female) life stages in a particular case of 'reverse taxonomy'. In addition, we constructed the Arrenuridae phylogeny based on mitochondrial DNA, which supports the idea that three *Arrenurus* subgenera are 'natural': *Arrenurus*, *Megaluracarus* and *Micruracarus*, and the somewhat arbitrary distinction of the species assigned to the subgenus *Truncaturus*.

Keywords: Acari Actinotrichida; COI; cytochrome B; genetic identification; Hydrachnidia; Culicidae; reverse taxonomy; species identification

1. Introduction

The systematics of many groups of organisms has traditionally been based on the adult stage [1]. For species showing a deep morphological transformation from the embryonic to the adult stage, due to a change in habitat or different behavior (parasitic and free-living stages) it may be difficult, if not impossible, to identify the juvenile stage without a previous correlation with the corresponding adult.

Members of the clade Hydrachnidia (Acari Actinotrichida) mainly inhabit continental waters [2] and are characterized by three main active life stages—larval, deutonymph and adult—with intermediate dormant stages [3]. Deutonymphs and adults share a characteristic morphology and may be identified through diagnostic characters. Larval morphology is very different from that of the two other stages. The standard procedure to match the larvae of a species with its corresponding adult is to track them directly from a fertilized female. In short, a fertilized female is kept in a small vial with a strip of paper or another substrate until they oviposit eggs and larvae emerge [4,5]. It is hard to identify the species level of larvae that have not been tracked in this way with certainty.

The water mite genus *Arrenurus* has a worldwide distribution, except for the perpetual snow regions, but there are no cosmopolitan species; rather, each main area upholds its own set of species. There are 152 *Arrenurus* in Europe and, of these, about 30% have a wide distribution [6–8]. As in other Hydrachnidia genera, *Arrenurus* deutonymphs and adults share diagnostic characters, however, the larval stage has a distinct morphology and requires independent morphological characterization. Water mite larvae usually attach to insects at preimaginal stages before they hatch, using adults as vector to disperse [9]. Adults and larvae of the same species, therefore, may be found in very different locations in their life cycle. Some *Arrenurus* species have occasionally been recorded as an ectoparasite of aquatic insects at the juvenile stage, either nymphs or larvae [2]. Some have even been recorded in insect exuviae [6,10]. Given these challenges, species identification of isolated *Arrenurus* larvae is fraught with uncertainty.

In this work, we characterized *Arrenurus* sp. larvae that were attached to larvae and a pupa of the common mosquito *Culex* (*Culex*) *pipiens* and cross-correlate their sequences with those of 51 adult *Arrenurus* species. One of these species, *Arrenurus* (*Micruracarus*) *novus* George, 1884, was in the same mosquito habitat from which the *Arrenurus* larvae were isolated. We compare cytochrome C oxidase subunit I (COI) and cytochrome B (cytB) DNA sequences of the water mite larvae with those of *A. novus* deutonymphs and adults (male and female) and other *Arrenurus* species, in order to establish the identity of the water mite larvae in a particular case of 'reverse taxonomy' [11].

2. Materials and Methods

2.1. Taxon Sampling

Water mite and mosquito specimens were both collected during a survey of mosquitoes at preimaginal stages carried out every two weeks between May and October of 2012 at Acequia del Caminàs (UTM 30S 741855.92 E 4324431.48), a semiartificial irrigation channel within the marsh waters of the Xeraco and Xeresa wetlands (Valencia, Spain; Figure 1). For more details on the characteristics of the area, see Alarcón-Elbal et al. [12]. Further collection efforts were carried out in May 2019 without success.

We used a quick dipping technique [13] with a standard 500-mL mosquito dipper (BioQuip Products Inc., Rancho Dominguez, CA, USA). Culicids at preimaginal stages were sorted manually at the laboratory and then transferred to a 60 °C water bath for a minute and then to absolute ethanol. Culicid specimens with ectoparasites and free-living water mites in the same sample were both preserved in absolute ethanol and stored at −20 °C.

Preservation in absolute ethanol ensures that both microscopic and molecular techniques can be used on any water mite specimen collected from potential hosts or directly from the field [14,15]. Both the culicids and the water mite specimens were morphologically identified, and subsequently used for DNA extraction.

Figure 1. Map of the sampling site (Xeraco, Valencia, Spain).

2.2. Morphological Identification of Culicids

Morphological identification of culicids at preimaginal stages were made to the species level based on external morphology and diagnostic characters related with chaetotaxy, as described by Schaffner et al. [16]. Specimens were examined under a Nikon SMZ-1B Stereozoom microscope.

2.3. Morphological Identification of Water Mites

2.3.1. Rearing of Adult Female Water Mites

In an attempt to rear larvae, six live adult *Arrenurus* sp. females were collected in the field and then kept under laboratory conditions in individual transparent poliestirene plastic containers (diameter 10–15 mm, height 15 mm) with a piece of paper and mineral water at room temperature for 14–15 days [4]. Following the monitoring period, the females were preserved in absolute ethanol for morphological identification of the species level.

2.3.2. Morphological Identification of Water Mite Adults, Deutonymphs and Larvae

Morphological identification of the adult and deutonymph water mite specimens collected from the field site was performed using the diagnostic characters originally described by George [17] and subsequently revised by Viets [18] and Gerecke [2].

The study of larval water mite specimens was based on four specimens found anchored on the body of a culicid larva (L1–L4 in Table 1) and one larva found on a culicid pupa (L5). The culicid larva, pupa and attached water mite larvae were examined first under a Bausch and Lomb stereomicroscope and then with laser scanning confocal microscopy (LSCM, Leica TCS SPE). The host attachment site, where the water mite larval mouthparts were embedded, was recorded and the larvae were individually and carefully detached using fine tweezers and transferred to a drop of glycerol on a microscope slide for LSCM stack acquisition. After imaging, the five larvae were then processed for molecular analyses.

Table 1. Standard measurements (see [8]) of the larval body morphology (in μm; only an approximation due to specimen orientation on the slide).

	L1	L2	L3	L4	L5
Body length	197	182	190	204	200
Body width	175	168	168	164	156
Dorsal plate length	190	186	190	185	185
Dorsal plate width	171	152	160	148	147
CpI medial margin length	57	55	61	57	62
CpII medial margin length	27	30	30	27	26
CpIII medial margin length	38	42	44	40	42
Distance between C1 and CpI median margin	21	19	23	18	20
Distance between C4 and CpIII median margin	28	27	24	27	29
Distance between C1 and C2	40	36	40	38	39
Excretory pore plate length	25	27	26	22	29
Excretory pore plate width	32	30	32	30	31
Distance between Exp and Expp posterior margin	11	15	15	12	13
PIII length	27	21	26	31	31
Length of PIV claw	27	23	14	16	17

2.3.3. Laser Scanning Confocal Microscopy Imaging

Laser scanning confocal microscopy (LSCM, Leica TCS SPE) was employed to acquire z-stack images with the following objectives: 10×/0.30 NA, 20×/0.70 NA, 40×/1.25 NA oil immersion and 63×/1.30 NA glycerol immersion. The samples were subjected to an excitation wavelength of 488 nm and an emission range between 520 and 660 nm. For further information on the stack acquisition procedure, see Valdecasas and Abad [14]. Serial images were processed with Amira (ver. 5.4.3) and ImageJ/Fiji to obtain maximum 2D projections and Voltex volume renderings. Different parameters of visualization were used with Voltex to best highlight the characters of interest. Fiji/ImageJ were used to take all measurements directly from the image stacks. As all the material was used for the subsequent molecular analyses, no morphological vouchers could be kept. Therefore, a photographic voucher comprised of the full set of original unaltered stacks is stored in the confocal collection at the National Museum of Natural Sciences of Spain (CSIC–MNCN).

2.4. Molecular Analyses

A total of nine water mite samples—three adults, one deutonymph and five larvae—were processed for molecular analyses. DNA extraction, PCR amplification and sequencing were performed as previously described) [15]. The COI gene was amplified using the primer pair LCO1490 and HC02198 [19]. For the larval water mite samples ($n = 5$), the internal primer COI-AV-F (5′-ATAAGATTTTGACTTCTYCC-3′) and HC02198 (360 basepairs length) were used because the complete amplification of the gene was not successful. CytB was amplified using the primer pair COB-F/R [20]. Sequencing reactions were performed using ABI BigDye® v3.1 Cycle Sequencing Kit (Applied Biosystems) and run on an ABI 3100 Genetic Analyzer. Sequence chromatograms were checked for accuracy and edited using Sequencher® version 5.0 (Gene Codes Corporation, Ann Arbor, MI, USA). DNA sequences were deposited in GenBank (accession numbers MT598102 to MT598106 for COI and MT607634 to MT607637 for cytB).

The sequences of both gene fragments were analyzed for all three life cycle stages in order to molecularly correlate the undescribed larvae with the diagnosed adults and deutonymphs. Genetic distances were estimated using the Kimura-2-Parameter (K2P) distance model [21], as implemented in MEGA v.7 [22].

Additionally, COI and cytB sequences of *Arrenurus* species were searched in GenBank (https://www.ncbi.nlm.nih.gov/genbank/; accessed 21 May 2020) in order to estimate the genetic distances among all available sequences, including the new ones of this study. The distances of the GenBank sequences were calculated using MEGA. A Maximum-Likelihood phylogenetic tree of COI sequences was constructed with the IQ-TREE web server (http://iqtree.cibiv.univie.ac.at) with the following settings: automatic detection of substitution model, perturbation strength of 0.5, IQ-TREE stopping rule of 100, ultrafast bootstrap analysis (1000 replicates) and SH-aLRT single branch tests (1000 replicates). For this tree, two outgroup sequences were used: *Torrenticola lundbladi* (GenBank accession number JX629050) and *Lebertia maderigena* (KX421869).

3. Results

3.1. Morphological Identification of Culicids

All of the culicid specimens were identified as *Culex* (*Culex*) *pipiens* Linnaeus, 1758. The principal and reliable characters for diagnosis at the larval stage are the syphon index, the syphon/saddle index, the branch number of seta one on abdominal segments III–IV, 1a–S tuft, 1b–S tuft and syphon shape [23]. One culicid larva and one pupa were parasitized with four and one water mite larvae, respectively (Figures 2–4).

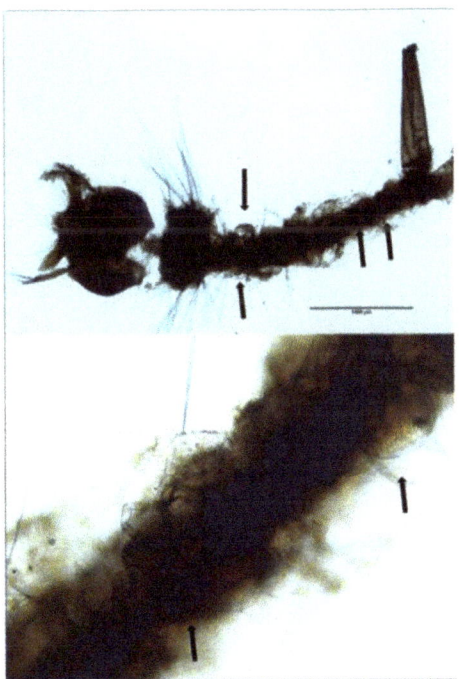

Figure 2. *Culex pipiens* larva with *Arrenurus* sp. larvae attached (arrows) and detail.

Figure 3. *Culex pipiens* pupa with *Arrenurus* sp. larva attached (white dot).

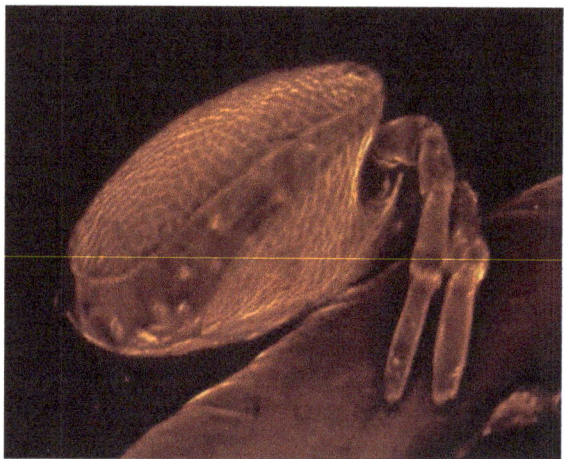

Figure 4. *Culex pipiens* pupa with *Arrenurus* sp. larva attached (laser scanning confocal microscopy (LSCM) image).

3.2. Morphological and Molecular Identification of Water Mites

3.2.1. Rearing of Adult Female Water Mites

Six *Arrenurus* females were maintained for 14–15 days, but no eggs or larvae were observed during this period. We were not able to ascertain whether the females were gravid and if the lack of eggs and larvae was due to unsuitable laboratory conditions.

3.2.2. Larval Description

The description of the water mite larvae is preliminary: confocal microscopy could not resolve some of the extremely fine morphological details, an issue faced by other researchers of water mite larvae [7]. Therefore, scanning electron microscopy analyses are necessary for a complete description. The terminology used here follows that of Zawal [7]. Figures 5 and 6 show a Voltex projection of the ventral and the dorsal habitus of a larva, respectively. The projections do not indicate morphometric distances, and the scales are only an approximation of size. Tables 1 and 2 include some standard morphometric data taken independently of the Voltex projections. Figure 7 shows a schematic of the approximate distribution of setae on the leg segments.

Arrenurus novus larvae (see molecular analysis below) may be distinguished from other previously described *Micruracarus* larvae [7] by the following set of characters: dorsal plate ovoid, narrowed anteriorly; coxas distinct, ratio lateral length of coxal plate I, II and III: 5:3:4; a hexagonal excretory plate perimeter and the excretory pore in line or slightly above E2 setae (Figure 8). Total length of Leg1 < Leg II = Leg III.

Table 2. Standard measurements of the larval leg morphology in μm.

Larvae		Trochanter	Femur	Genu	Tibia	Tarsus
L1	leg1	25	21	29	42	55
	leg2	30	25	30	46	57
	leg3	30	25	30	46	57
L2	leg1	23	21	29	40	49
	leg2	23	25	30	44	59
	leg3	27	23	25	48	57
L3	leg1	30	23	30	38	49
	leg2	30	32	27	42	57
	leg3	34	34	27	44	57
L4	leg1	29	23	27	44	55
	leg2	21	25	29	46	53
	leg3	32	29	29	46	59
L5	leg1	24	23	27	40	41
	leg2	29	22	26	45	60
	leg3	30	25	30	40	31

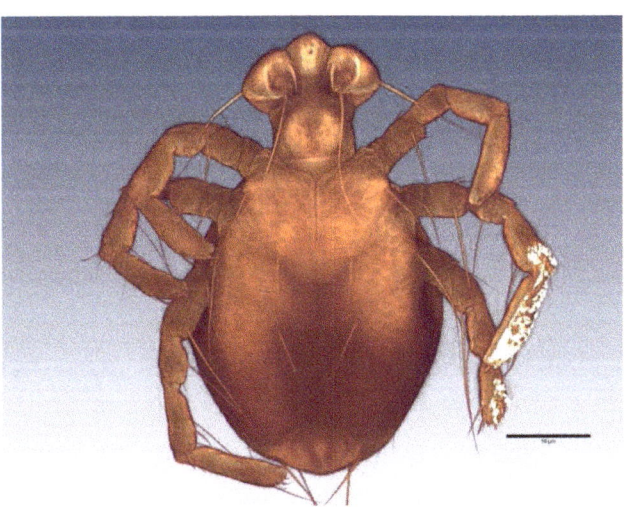

Figure 5. *Arrenurus (Micruracarus) novus* larva, ventral view (LSCM).

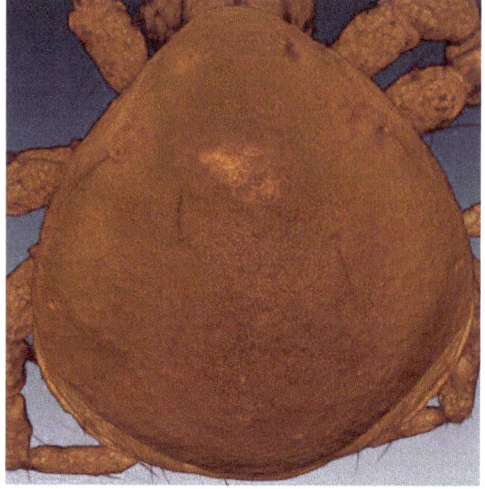

Figure 6. *Arrenurus (Micruracarus) novus* larva, dorsal view (LSCM).

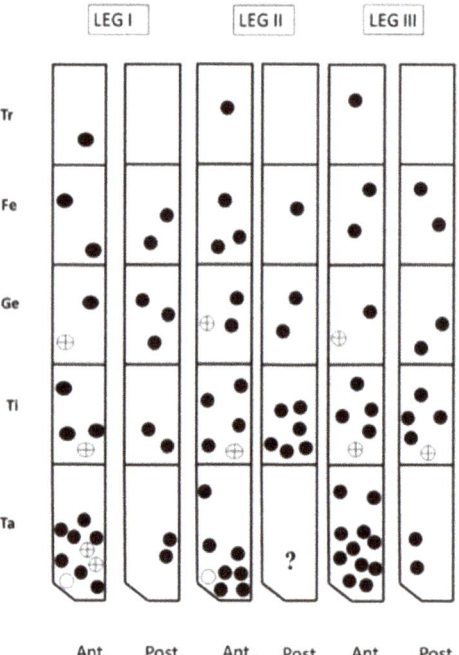

Figure 7. ASchematic showing the distribution of setae on the leg segments of an Arrenurus (Micruracarus) novus larva. Tr: trochanter; Fe: femur; Ge: genu; Ti: tibia; Ta: tarsus. ● Round setae; ○ solenidum; ⊕ other setae.

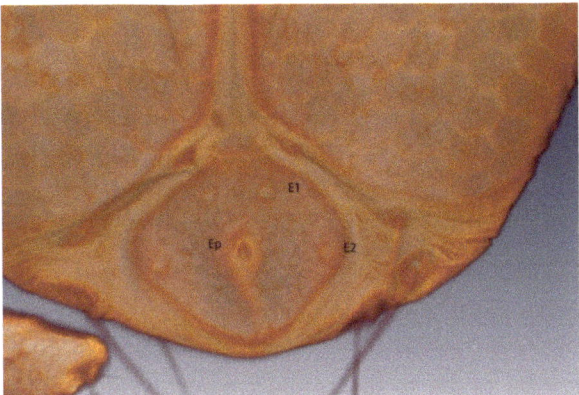

Figure 8. *Arrenurus (Micruracarus) novus* larva, excretory plate (LSCM).

The number of setae on the pedipalp segments are in agreement with Zawal's description of *Arrenurus* larvae [7]: none on P-I; one on P-II; two on P-III, one long and thick and the other short; four on P-IV, three thin and one thick; and a solenoid and seven setae on P-V (Figure 9).

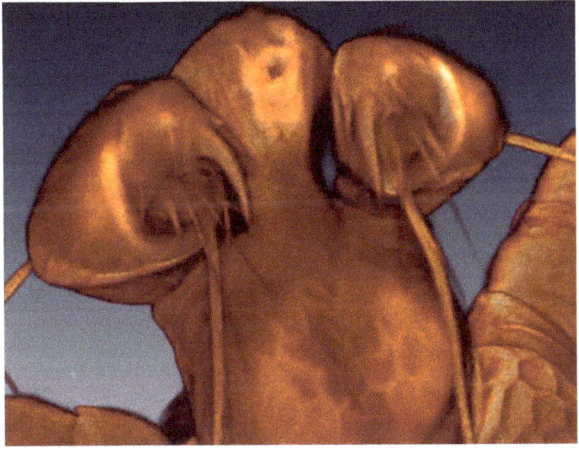

Figure 9. *Arrenurus (Micruracarus) novus* larva, pedipalp (LSCM).

3.2.3. Morphological Identification of Adult and Deutonymph Water Mites

Seven male and six female adults plus two deutonymphs were identified as *A. novus*, a species found in standing waters throughout Europe [2]. Among the main diagnostic characters identifying *A. novus* adults are the petiole in the male (see arrow in Figure 10A), the setation of male and female palps, and the shape and length of cauda (Figure 10). The deutonymph diagnosis was based on palp shape and setation and habitat co-occurrence (Figure 11): approximate body length: 526 µm; body width: 467 µm; P-1 without setae; P-II with three medial setae; P-III with one medial setae; PIV uncate and a strong distal setae and PV with a basal fine setae.

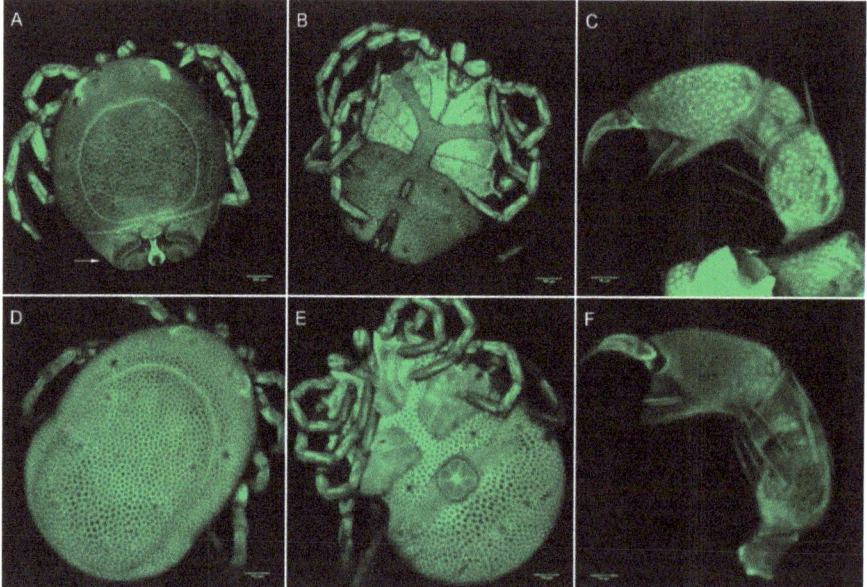

Figure 10. *Arrenurus (Micruracarus) novus* adults (LSCM). (**A**) male, dorsal; (**B**) male, ventral; (**C**) male, palp; (**D**) female, dorsal; (**E**) female, ventral; (**F**) female, palp.

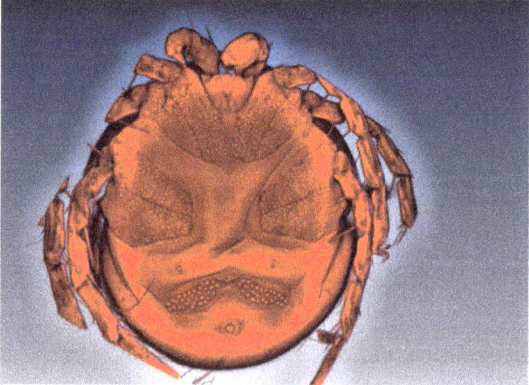

Figure 11. Ventral view of *Arrenurus (Micruracarus) novus* deutonymph.

3.3. Molecular Analyses

GenBank identified 990 *Arrenurus* barcoding COI sequences but no *Arrenurus* cytB sequences (probably because cytB is not generally used for species delimitation). These sequences belonged to 51 species of the genus plus *A. novus*, whose sequence is reported here for the first time (genetic distances among species are shown in Table S1 and phylogenetic tree in Figure 12). The mean genetic distances among the 52 *Arrenurus* species was 16.4% (standard deviation = 4.1). The here-amplified COI sequences from all the different *A. novus* life-cycle stages were all the same.

Very few phylogenetic information about Arrenurus currently exist [24]. The here-constructed COI phylogenetic tree (Figure 12, Material S1) may be divided into three main arms: the upper branch is dominated by species belonging to the subgenus *Micruracarus*, with occasional incursions of species belonging to the subgenera *Arrenurus*, *Truncaturus*, *Megaluracarus* and *Micrarrenurus*. In this upper branch is located *Arrenurus* (*Micruracarus*) *novus*. The intermediate branch of the tree is composed exclusively of species of the subgenus *Arrenurus*, except for a sequence belonging to subgenus *Truncaturus*. The lower section of the tree consists exclusively of species belonging to the sub-genus *Megaluracarus*, with European and American representatives.

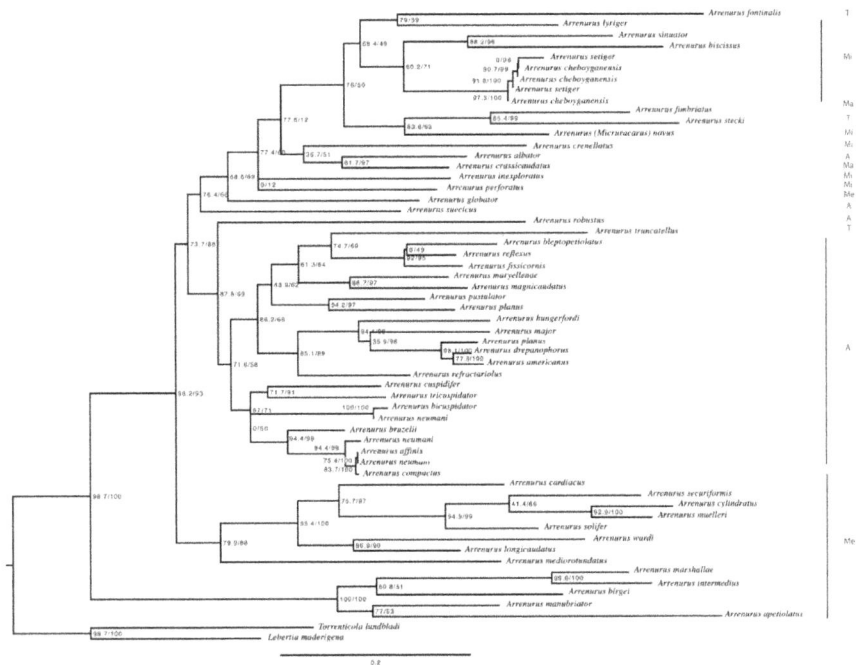

Figure 12. Maximum Likelihood phylogenetic tree of the *Arrenurus* C oxidase subunit I (COI) sequences employed in this study and the outgroups (*Torrenticola lundbladi* and *Lebertia maderigena*). Node numbers show SH-aLRT (left) and bootstrap (right) support values (1000 replicates). Letters at right show subgenus acronyms: *Arrenurus* (A), *Micruracarus* (Mi), *Truncaturus* (T), *Megaluracarus* (Me) and *Micrarrenurus* (Ma).

The 360-bp COI fragment was successfully amplified only from one larvae; however, no genetic variation was found among the sequences of this region (two adults, one deutonymph and one larvae). Genetic differences were, however, observed in the 330-bp cytB fragment (Table 3). One of the adults and all five larvae shared the same cytB haplotype, whereas the deutonymph and the other adult shared a different haplotype (Table 3).

Table 3. K2P distances among the adult, deutonymph and larval *Arrenurus* sp. cytB sequences analyzed in this study (alignment: 330 bp).

K2P	Adult 1	Adult 2	Nymph	Larvae
Adult 1	-			
Adult 2	0.006	-		
Nymph	0.006	0.000	-	
Larvae	0.000	0.006	0.006	-

4. Discussion

The number of molecular sequences available for Hydrachnidia has increased substantially in recent years. GenBank list 6450 Hydrachnidia nucleotide sequences (as of 21 May 2020). We did not search the BOLD Systems database for two main reasons: many of the *Arrenurus* sequences are privately held, and it was recently shown that some water mite sequences do not agree with their species assignment [25]. Of the GenBank sequences, almost 70% are identified to the generic level. Some new species have been described with morphological and molecular data [15,26]. For the genus *Arrenurus*, there are 1071 sequences (990 of them belonging to COI gene) but only 198, representing 51 taxa, are at the species level. These were employed to the genetic distance calculations. All of the *Arrenurus* sequences in GenBank identified to the species level are based on the adult stage; sequences obtained from larvae are only given a generic rank and are frequently singled out as operational taxonomic units (e.g. [27]). There is a need for more species-level sequences based on different life cycle stages to widen the usefulness of this type of data to address ecological, biogeographical and evolutionary questions. We identified the adult and deutonymph water mite specimens as *A. novus*, a broadly distributed species found from the Western Palearctic to the Afrotropic, commonly in standing waters [2,25]. However, according to Gerecke et al. [2], it has been "little reported in Europe".

In the present study, we combined morphological and molecular techniques to associate undescribed larvae with corresponding adults and deutonymphs for the first time in water mites. Associations among life cycle stages are especially relevant for this genus given that less than 20% of arrenurid larvae are known [28]. To link undescribed larval specimens with other life stages (i.e., deutonymph and adult), we employed a DNA-based identification method in which sequences are considered as standardized comparative characters that can be used to support and integrate morphological datasets [29]. Even in cases with very small genetic distances, these data can be useful for species identification [30]. The results of our analyses of two mitochondrial genes (COI and cytB), which proved to be informative, clearly support the same conclusion: the larvae belong to the species *A. novus*. The larval COI sequences were compared with 52 adult *Arrenurus* species sequences (51 were from GenBank; the *A. novus* sequences, from one adult male, two adult females, one deutonymphs and one larvae, are provided for the first time here). Genetic distances among the three *A. novus* life-cycle stages were 0.000 (the same DNA sequence), confirming the co-specificity of the larvae, deutonymph and adults. The distance of *A. novus* with other *Arrenurus* species ranged between 9.8 with *A. setiger* and 24.3 with *A. megalurus* (see their phylogenetic relationships in Figure S1). The cytB sequences of the larvae and one of the adults were identical and highly similar to those of the other adult and the deutonymph (which were identical; see Table 3), strongly supporting the conclusion that they belong to the same species. It is important to note here that we are studying only mitochondrial DNA, thus, in the case of hybrids, we would be only describing the maternal species of the larvae. In addition, the phylogenetic tree based on COI mitochondrial information (Figure 12) supports the idea that three *Arrenurus* subgenera are 'natural'—*Arrenurus*, *Megaluracarus* and *Micruracarus*—and the somewhat arbitrary distinction of the species assigned to the subgenus *Truncaturus*. This topology differs from the only other *Arrenurus* phylogeny published [24] and points to the need for more molecular data in order to clarify their phylogenetic relationships.

This study demonstrates the usefulness of an integrative approach to resolve the taxonomic uncertainty of water mites at the larval stage. Furthermore, it appears to be a promising approach for identifying larvae at the species level in terms of both reliability and speed, in comparison with rearing or hatching approaches. With these data, we can also begin to elucidate host–parasite interactions between a specific water mite species and its host(s) with greater detail. To date, a host of *A. novus* larvae was unknown [2]. Indeed, only a few studies have described a host-parasite interaction between a *Culex* species and an *Arrenurus* species (e.g. [6,10,31,32]). *Arrenurus* is the genus most reported to parasitize mosquitoes, evidence of its flexibility regarding host specificity [33]. Despite this, recent findings suggest that these mites prefer *Culex* species [34]. Aside from reducing host fitness by piercing their exoskeleton to feed on hemolymph, each host–parasite association has its own characteristics, which are determined mainly by the size of the partners, intrinsic defence mechanisms and environmental conditions [35]. Parasitic mites may play a significant role in the biological control of mosquitos in wetlands, especially of adult populations. Therefore, more observations and experimental data of water mite species are needed to better understand their host–parasite interactions, as well as to incorporate the use of molecular techniques in their identification, particularly at the larval stage.

Supplementary Materials: The following are available online at http://www.mdpi.com/2075-1729/10/7/108/s1, Table S1: K2P distances among the COI sequences of the 52 *Arrenurus* species included in the analysis. The final alignment was 659 bp. *Arrenurus novus* is indicated in bold. Material S1: Maximum Likelihood phylogenetic tree of the *Arrenurus* COI sequences employed in this study and the outgroups (*Torrenticola lundbladi* and *Lebertia maderigena*) in newick format.

Author Contributions: Field data sampling and culicids identification, P.M.A.-E.; Bright field microscopy and Laser Scanning Confocal Microscopy imaging, and identification of water mites, A.G.V.; Molecular extraction, amplification and sequencing R.G.-J.; Genbank searches and downloading and molecular analysis, M.L.P. and J.L.H.; all the authors contributed to the writing and reviewed the last version of the manuscript. All authors have read and agreed to the published version of the manuscript.

Funding: J.L.H. was supported by the Regional Government of Asturias (reference SV-16-UMB-1).

Acknowledgments: We thank Annie Machordom for access to the Systematics Molecular Laboratory of the MNCN.

Conflicts of Interest: The authors declare no conflict of interest. The funders had no role in the design of the study; in the collection, analyses, or interpretation of data; in the writing of the manuscript, or in the decision to publish the results.

References

1. Thorp, J.H.; Rogers, D.C.; Dimmick, W.W. Introduction to invertebrates of inland waters. In *Freshwater Invertebrates*, 4th ed.; Elsevier BV: Amsterdam, The Netherlands, 2015; pp. 1–19.
2. Gerecke, R.; Gledhill, T.; Pešić, V.; Smit, H. *Süßwasserfauna von Mitteleuropa, Bd. 7/2-3 Chelicerata*; Springer Spektrum: Heidelberg, Germany, 2016.
3. Smith, I.M.; Cook, D.R.; Smith, B.P. Water mites (Hydrachnidiae) and other arachnids. In *Ecology and Classification of North American Freshwater Invertebrates*; Elsevier BV: Amsterdam, The Netherlands, 2010; pp. 485–586.
4. Cook, W.J.; Smith, B.P.; Brooks, R.J. Allocation of reproductive effort in female *Arrenurus* spp. water mites (Acari: Hydrachnidia; Arrenuridae). *Oecologia* **1989**, *79*, 184–188. [CrossRef]
5. Tuzovsky, P. A new water mite species of the genus *Arrenurus* Dugès, 1834 (Acariformes: Hydrachnidia: Arrenuridae) from Eastern Palaearctic. *Acarina* **2012**, *20*, 173–179.
6. Zawal, A. Phoresy and parasitism: Water mite larvae of the genus *Arrenurus* (Acari: Hydrachnidia) on Odonata from lake Binowskie (NW Poland). *Biol. Lett.* **2006**, *43*, 257–276.
7. Zawal, A. Morphological characteristics of water mite larvae of the genus *Arrenurus* Duges, 1834, with notes on the phylogeny of the genus and an identification key. *Zootaxa* **2008**, *1765*, 1–75. [CrossRef]
8. Zawal, A. Morphology of the larval stages of *Arrenurus affinis* Koenike, 1887, A. neumani Piersig, 1895, and A. vietsi Koenike, 1911 (Acari: Hydrachnidia). *Genus* **2008**, *19*, 161–169.
9. Bohonak, A.J.; Smith, B.P.; Thornton, M. Distributional, morphological and genetic consequences of dispersal for temporary pond water mites. *Freshw. Boil.* **2004**, *49*, 170–180. [CrossRef]
10. Lanciani, C.A. Sexual bias in host selection by parasitic mites of the mosquito anopheles crucians (Diptera: Culicidae). *J. Parasitol.* **1988**, *74*, 768. [CrossRef]
11. Markmann, M.; Tautz, D. Reverse taxonomy: An approach towards determining the diversity of meiobenthic organisms based on ribosomal RNA signature sequences. *Philos. Trans. R. Soc. B: Boil. Sci.* **2005**, *360*, 1917–1924. [CrossRef]
12. Alarcón-Elbal, P.M.; Murillo, J.M.S.; Estrella, S.D.; Ruiz-Arrondo, I.; Prieto, R.P.; Curdi, J.L. Asociación de vector del VNO e hidrófito invasor: *Culex pipiens* Linnaeus, 1758 y Ludwigia grandiflora (Michaux) Greuter and Burdet en el marjal de Xeraco-Xeresa, Valencia. *Anales Biol.* **2013**, *35*, 17–27. [CrossRef]
13. Service, M.W. A critical review of procedures for sampling populations of adult mosquitoes. *Bull. Entomol. Res.* **1977**, *67*, 343–382. [CrossRef]
14. Valdecasas, A.G.; Abad, A. Morphological confocal microscopy in arthropods and the enhancement of autofluorescence after proteinase K extraction. *Microsc. Microanal.* **2010**, *17*, 109–113. [CrossRef] [PubMed]
15. Pešić, V.; Valdecasas, A.G.; García-Jimenez, R. Simultaneous evidence for a new species of *Torrenticola* Piersig, 1896 (Acari, Hydrachnidia) from Montenegro. *Zootaxa* **2012**, *3515*, 38–50. [CrossRef]

16. Schaffner, E.; Angel, G.; Geoffroy, B.; Hervy, J.P.; Rhaiem, A.; Brunhes, J. *Les Moustiques d'Europe: Logiciel d'Identification et d'Enseignement = The Mosquitoes of Europe: An Identification and Training Programme*; IRD Editions & EID Méditerranée: Montpellier, France, 2001.
17. George, C.F. The British fresh-water mites. *Hardwicke's Sci.* **1884**, *20*, 80–81.
18. Viets, K. Zur kenntnis dert Hydracarinen-Fauna von Spanien. *Arch. Hydrobiol.* **1930**.
19. Folmer, O.; Black, M.; Hoeh, W.; Lutz, R.; Vrijenhoek, R. DNA primers for amplification of mitochondrial cytochrome c oxidase subunit I from diverse metazoan invertebrates. *Mol. Mar. Biol. Biotechnol.* **1994**, *3*, 294–299. [CrossRef] [PubMed]
20. Ernsting, B.R.; Edwards, D.D.; Aldred, K.J.; Fites, J.S.; Neff, C.R. Mitochondrial genome sequence of Unionicola foili (Acari: Unionicolidae): A unique gene order with implications for phylogenetic inference. *Exp. Appl. Acarol.* **2009**, *49*, 305–316. [CrossRef]
21. Kimura, M. A simple method for estimating evolutionary rates of base substitutions through comparative studies of nucleotide sequences. *J. Mol. Evol.* **1980**, *16*, 111–120. [CrossRef]
22. Kumar, S.; Stecher, G.; Tamura, K. MEGA7: Molecular evolutionary genetics analysis version 7.0 for bigger datasets. *Mol. Boil. Evol.* **2016**, *33*, 1870–1874. [CrossRef]
23. Azari-Hamidian, S.; Harbach, R.E. Keys to the adult females and fourth-instar larvae of the mosquitoes of Iran (Diptera: Culicidae). *Zootaxa* **2009**, *2078*, 1–33. [CrossRef]
24. Więcel, M. Effects of the Evolution of Intromission on Courtship Complexity and Male and Female Morphology: Water Mites of the Genus *Arrenurus* (Acari; Hydrachnida) from Europe and North America. Ph.D. Thesis, Adam Mickiewicz University, Poznan, Poland, 2016.
25. Valdecasas, A.G.; García-Jiménez, R.; Marín, M. Two rare water mite species (Acari, Parasitengona, Hydrachnidia) new for the Iberian Peninsula. *Rev. Ibérica Aracnol.* **2019**, *35*, 33–37.
26. Pešić, V.; Smit, H. Neumania kyrgyzica sp. nov. a new water mite from Kyrgyzstan based on morphological and molecular data (Acari, Hydrachnidia: Unionicolidae). *Syst. Appl. Acarol.* **2017**, *22*, 885. [CrossRef]
27. Mlynarek, J.J.; Knee, W.; Forbes, M.R. Explaining susceptibility and resistance to a multi-host parasite. *Evol. Boil.* **2013**, *41*, 115–122. [CrossRef]
28. Martin, P. Wassermilben (Hydrachnidia, Acari) und Insekten: Ein Überblick über eine selten betrachtete Beziehung. *Entomol. Heute* **2008**, *20*, 45–75.
29. Scorrano, S.; Aglieri, G.; Boero, F.; Dawson, M.N.; Piraino, S. Unmasking Aurelia species in the Mediterranean Sea: An integrative morphometric and molecular approach. *Zool. J. Linn. Soc.* **2016**, *180*, 243–267. [CrossRef]
30. Horreo, J.L.; Ardura, A.; Pola, I.G.; Martinez, J.L.; Garcia-Vazquez, E. Universal primers for species authentication of animal foodstuff in a single polymerase chain reaction. *J. Sci. Food Agric.* **2012**, *93*, 354–361. [CrossRef] [PubMed]
31. Mullen, G.R. Water mites of the subgenus Truncaturus (Arrenuridae, *Arrenurus*) in North America. *Ithaca Agric. Entomol.* **1976**, *6*, 1–35.
32. Kirkhoff, C.J.; Simmons, T.W.; Hutchinson, M.; Simmons, T.W. Adult mosquitoes parasitized by larval water mites in Pennsylvania. *J. Parasitol.* **2013**, *99*, 31–39. [CrossRef] [PubMed]
33. Dos Santos, E.B.; Favretto, M.A.; Dos Santos Costa, S.G.; Navarro-Silva, M.A. Mites (Acari: Trombidiformes) parasitizing mosquitoes (Diptera: Culicidae) in an Atlantic Forest area in southern Brazil with a new mite genus country record. *Exp. Appl. Acarol.* **2016**, *69*, 323–333. [CrossRef] [PubMed]
34. Atwa, A.A.; Bilgrami, A.L.; Al-Saggaf, A.I. Host–parasite interaction and impact of mite infection on mosquito population. *Rev. Bras. Entomol.* **2017**, *61*, 101–106. [CrossRef]
35. Gerson, U.; Smiley, R.L.; Ochoa, R. *Mites (Acari) for Pest Control*; Blackewell: Malden, MA, USA, 2003.

© 2020 by the authors. Licensee MDPI, Basel, Switzerland. This article is an open access article distributed under the terms and conditions of the Creative Commons Attribution (CC BY) license (http://creativecommons.org/licenses/by/4.0/).

Article

Inferring the Phylogenetic Positions of Two Fig Wasp Subfamilies of Epichrysomallinae and Sycophaginae Using Transcriptomes and Mitochondrial Data

Dan Zhao, Zhaozhe Xin, Hongxia Hou, Yi Zhou, Jianxia Wang, Jinhua Xiao * and Dawei Huang *

Institute of Entomology, College of Life Sciences, Nankai University, Tianjin 300071, China;
2120181038@mail.nankai.edu.cn (D.Z.); 1120180392@mail.nankai.edu.cn (Z.X.);
1120180393@mail.nankai.edu.cn (H.H.); 1120170366@mail.nankai.edu.cn (Y.Z.);
1120170365@mail.nankai.edu.cn (J.W.)
* Correspondence: xiaojh@nankai.edu.cn (J.X.); huangdw@nankai.edu.cn (D.H.); Tel.: +86-185-2245-2108 (J.X.); +86-139-1025-6670 (D.H.)

Abstract: Fig wasps are a group of insects (Hymenoptera: Chalcidoidea) that live in the compact syconia of fig trees (Moraceae: *Ficus*). Accurate classification and phylogenetic results are very important for studies of fig wasps, but the taxonomic statuses of some fig wasps, especially the non-pollinating subfamilies are difficult to determine, such as Epichrysomallinae and Sycophaginae. To resolve the taxonomic statuses of Epichrysomallinae and Sycophaginae, we obtained transcriptomes and mitochondrial genome (mitogenome) data for four species of fig wasps. These newly added data were combined with the data of 13 wasps (data on 11 fig wasp species were from our laboratory and two wasp species were download from NCBI). Based on the transcriptome and genome data, we obtained 145 single-copy orthologous (SCO) genes in 17 wasp species, and based on mitogenome data, we obtained 13 mitochondrial protein-coding genes (PCGs) for each of the 17 wasp species. Ultimately, we used 145 SCO genes, 13 mitochondrial PCGs and combined SCO genes and mitochondrial genes data to reconstruct the phylogenies of fig wasps using both maximum likelihood (ML) and Bayesian inference (BI) analyses. Our results suggest that both Epichrysomallinae and Sycophaginae are more closely related to Agaonidae with a high statistical support.

Keywords: fig wasps; classification; phylogeny; mitochondrial gene; transcriptome

1. Introduction

Fig wasps (Hymenoptera: Chalcidoidea) refer to all wasps that must rely on the syconia of fig trees (Moraceae: *Ficus*) to complete their life histories. According to whether they pollinate the figs, fig wasps are broadly classified into two categories of pollinating fig wasps and non-pollinating fig wasps [1]. Among them, the symbiosis of figs–pollinating fig wasps is a classical model of the mutualistic system, which originated about 75 million years ago [2]. The plant–insect interaction system between figs and fig wasps provides an ideal model for the study of the co-evolution of species and symbiotic relationship among organisms [3], and these studies are inseparable from the correct identification and phylogenetic history reconstruction of fig wasps.

The current taxonomic information listed on the fig web (http://www.figweb.org/Fig_wasps/Classification/index.htm) is mainly based on the taxonomy system proposed by Heraty [4], in which the fig wasps are classified into five families (Agaonidae, Eurytomidae, Ormyridae, Pteromalidae, Torymidae) and ten subfamilies (Agaoninae, Kradibiinae, Sycophaginae, Tetrapusiinae, Colotrechinae, Pteromalinae, Epichrysomallinae, Otiteselinae, Sycoecinae, and Sycoryctinae) [4]. The taxonomic statuses of some subfamilies are difficult to determine, such as the non-pollinating wasps of Epichrysomallinae and Sycophaginae [5–8].

The taxonomic history of Epichrysomallinae and Sycophaginae has undergone multiple changes since the beginning of the taxonomy of fig wasps in the 19th century. Sycophaginae was founded by Walker in 1875 [9], then it was included in the family of Torymidae based on cleptoparasitic habits by Joseph in 1964 [10]. Epichrysomallinae was established and classified in Torymidae by Hill in 1967, who also agreed to classify Sycophaginae in Torymidae [5]. In 1981, Bouček revised and adjusted Epichrysomallinae into Pteromalidae on the basis of morphological characteristics [6]. In 1988, Bouček further considered the whole Agaonidae as a monophyletic group based on the reproductive characteristics, of which the Agaoninae was only the most specific group with the behaviors of pollination, and thus both Epichrysomallinae and Sycophaginae were classified in Agaonidae [7]. In 1998, Rasplus et al. used the D1 and D2 domains of nuclear 28S rRNA to infer the phylogenetic relationships of six subfamilies proposed by Bouček, and they indicated that Agaonidae that Bouček (1988) referred to was not a monophyletic group [8]; they further revised Agaonidae and pointed out that Agaonidae only contained the subfamily of Agaoninae, and the taxonomic statuses of Epichrysomallinae and Sycophaginae in Chalcidoidea were undetermined. In 2013, Heraty et al. used 233 morphological and two molecular datasets (nuclear ribosomal 18S and 28S D2-D5 expansion regions) to make phylogenetic inference of a variety of wasps in Chalcidoidea and the results showed that the two subfamilies of Epichrysomallinae and Sycophaginae previously undetermined were included in Pteromalidae and Agaonidae, respectively [4]. In 2018, Peters et al. constructed a phylogenetic tree using 3239 homologous genes based on transcriptomic data from 62 species of Chalcidoidea, which showed that Epichrysomallinae was more closely related to Agaonidae than Pteromalidae, but Sycophaginae was not involved in their studies [11]. Therefore, due to different sources of taxonomic evidences (morphology, biological characteristics or molecular evidences), the taxonomy and phylogenetic positions of the two subfamilies of Epichrysomallinae and Sycophaginae, especially the latter, are still unclear, and more data are needed to clarify whether they belong to Agaonidae or Pteromalidae.

The second-generation high-throughput sequencing technology represented by transcriptome RNA sequencing (RNA-seq) can obtain a large-scale sequencing of the transcripts of specific tissues of a certain species. By using bioinformatics tools for splicing assembly, we can quickly obtain almost all the gene coding sequences (CDS) of the specific tissue of the species at a time point. With the advantages of low cost, large data volume, high efficiency and high accuracy, RNA-seq has shown great potential in the field of molecular phylogenetic research and become an effective means for molecular biology research of non-model animals [12–14]. The mitochondrial genome (mitogenome) sequences are also widely used in molecular evolution, phylogeny, phylogeography and population genetics because of their advantages, such as small genome size, maternal inheritance, no intron, relatively high evolutionary rate, simple structure, conserved gene content, and rare recombination [15–17].

In this study, we obtained conserved single-copy orthologous (SCO) genes in the nucleus based on transcriptomes (*Odontofroggatia galili*, *Walkerella microcarpae*, *Micranisa ralianga*, *Platyneura mayri* and *Encarisa Formosa*) and genomes (another 12 wasp species) from 17 wasp species and mitogenome sequences based on second-generation genome sequencing. Phylogenies of fig wasps were reconstructed based on three different molecular datasets, conserved SCO genes in the nucleus, 13 mitochondrial protein-coding genes (PCGs) and combined genes (SCO genes combined with 13 mitochondrial PCGs) to obtain new evidences to explore the taxonomic status of Epichrysomallinae and Sycophaginae.

2. Materials and Methods

2.1. Taxon Sampling and Data Collection

A total of 17 wasp species were included in this study (Table 1), including 15 fig wasp species (representing three families and six subfamilies). Four fig wasp species (*O. galili*, *W. microcarpae*, *M. ralianga* and *P. mayri*) were newly sequenced in this study. Data on 11 fig wasp species were from our laboratory and 2 wasp species were download from

NCBI. Among the four species, *O. galili*, *W. microcarpae*, and *M. ralianga* were collected from Qinzhou, Guangxi, China (N21°57′, E108°37′), and *P. mayri* was collected from Xishuangbanna, Yunnan, China (N21°41′, E101°25′). The information on the host fig trees was as follows: *O. galili* and *W. microcarpae* were associated to *Ficus microcarpa*, *P. mayri* was associated to *Ficus racemosa*, and *M. ralianga* was associated to *Ficus altissima*. We collected the figs in their natural state of maturity (but fig wasps have not left the fig yet) in the wild. All the fig wasps were collected after they came out from the figs, and then identified by using the SMZ-168 microscope (Motic, China). The identification of fig wasps was based mainly on descriptions and pictures in the literatures [7,18], and also combined with photographs of fig wasps left by Rasplus in Yunnan province. The alive fig wasps were immediately stored in RNA Hold (TransGen, Beijing, China) at −80 °C (for RNA extraction) or in 95% ethanol at −20 °C (for DNA extraction).

Table 1. List of the 17 species analyzed in this study.

Species	Subfamily	Family	Accession No. (Mitochondrial Genome)	Accession No. (Genomes or Transcriptomes)
Platyneura mayri *	Sycophaginae *		MW167114	PRJNA672045
Odontofroggatia galili *	Epichrysomallinae *		MW167113	PRJNA671819
Walkerella microcarpae *	Otitesellinae *	Pteromalidae	MW167116	PRJNA672219
Micranisa ralianga *	Otitesellinae *	Pteromalidae	MW167115	PRJNA672141
Euprisitina koningsbergeri #	Agaoninae #	Agaonidae	MT947597	PRJNA641212
Platyscapa corneri #	Agaoninae #	Agaonidae	MT947604	PRJNA641212
Dolichoris vasculosae #	Agaoninae #	Agaonidae	MT947596	PRJNA641212
Wiebesia pumilae #	Agaoninae #	Agaonidae	MT947601	PRJNA641212
Kradibia gibbosae #	Kradibiinae #	Agaonidae	MT947598	PRJNA641212
Ceratosolen fusciceps #	Kradibiinae #	Agaonidae	MT916179	PRJNA494992
Sycophaga agreansis *	Sycophaginae *		MT947599	PRJNA641212
Sycophila sp.2 *	-	Eurytomidae	MT947603	PRJNA641212
Sycobia sp.2 *	Epichrysomallinae *		MT947600	PRJNA641212
Apocrypta bakeri *	Sycoryctinae *	Pteromalidae	MT906648	PRJNA641212
Philotrypesis tridentata *	Sycoryctinae *	Pteromalidae	MT947602	PRJNA641212
Nasonia vitripennis	Pteromalinae	Pteromalidae	EU746609.1, EU746613.1	PRJNA594415
Encarisa formosa	Coccophaginae	Aphelinidae	MG813797.1	PRJNA252167

The species in bold represent the fig wasps sequenced in this study. * non-pollinating fig wasps. # pollinating fig wasps.

2.2. RNA Extraction, Transcriptome Sequencing and Assembly

For the RNA extraction of the four fig wasps, we set up two sequencing samples (one female and one male, with 20 to 30 wasps included in each sample) for each wasp species. We picked well-preserved fig wasps in RNA hold and washed them with RNase-free water. We used TransZolUp Plus RNA Kit (TransGen, Beijing, China) to extract RNA for each sample according to the manufacturer's instructions and finally dissolved the RNA into 40 uL RNase-free water. The concentration and purity of RNA were examined according to the OD values by Thermo Scientific NanoDrop One (ThermoFisher, Waltham, MA, USA). The sequencing libraries were constructed with NEBNext® Ultra™ RNA Library Prep Kit (NEB, Ipswich, MA, USA) and sequenced by second-generation Illumina HiSeq TM2000 platform (Novogen, Tianjin, China).

We obtained at least 6 Gb of raw data for each sample (at least 12 Gb of total data per species). The raw reads were quality controlled by Fastp software [19] and yielded clean reads. For each species, all clean reads were used for *de novo* assembly by using Trinity v2.5.1 [20], with a spliced result file called "Trinity. fasta" generated. Based on the Trinity splicing, Corset program [21] (with default parameters) was employed to cluster the transcripts to obtain the "cluster _all. fasta" file. We selected the longest transcript of each gene in the file of "cluster _all. fasta" as the unique sequence of that gene (also called Unigene).

2.3. Prediction of CDS and Identification of SCO Genes

For wasps with non-reference transcriptomes (*O. galili*, *W. microcarpae*, *M. ralianga*, *P. mayri*, and *Encarisa Formosa*), TransDecoder v5.5.0 was used to identify the open reading frames (ORFs) of the unigene to obtain the CDS sequences [22]. They were subsequently translated into amino acid sequences (https://web.expasy.org/cgi-bin/translate/dna2aa.cgi), while the CDS and protein sequences of the other species used for analysis were from NCBI or our laboratory.

OrthoMCL v2.0 [23] (with default parameters) was used to construct a local protein database of protein sequences of all species and to perform all-vs-all BLASTP matching to obtain similarity results between protein sequences. The best matched pairs between species (Orthologous pairs) were found via the OrthoMCL Pairs module in OrthoMCL. Finally, all the orthologous homologous proteins were classified and numbered by using Markov Cluster algorithm (MCL) [23]. Eventually, we removed multi-copy homologous proteins and selected families of orthologous proteins with one-to-one relationships (SCO proteins) for subsequent phylogenetic analysis.

2.4. DNA Extraction, Library Construction and Sequencing; Mitogenome Assembly and Annotation of 13 PCGs

We picked well-preserved wasps in 95% ethanol and washed them with sterile double distilled water. For each species, we used 30–40 female wasps to extract DNA by using the LiCl/KAc method and dissolved the DNA into 25 uL sterile double distilled water. The concentration and purity of DNA were examined according to the OD values by Thermo Scientific Nano Drop One (ThermoFisher, USA) and DNA integrity was monitored by 1% agarose electrophoresis. Sequencing libraries with an average of insert size of 350 bp were constructed with NEBNext® Ultra™ DNA Library Prep Kit (NEB, USA) and sequenced by second-generation Illumina HiSeq TM2000 platform (Novogen, Tianjin, China). The amount of sequencing data per species was set up at least 4 Gb. Fastp software [19] was used to quality-filtered to obtain the high-quality clean reads.

Mira v4.0.2 [24] and MITObim v1.9.1 [25] were used to assemble the mitogenomes of the four fig wasps, each with a reference mitogenome. The reference species used for *P. mayri* was *Sycophaga agreansis*; the reference species used for *O. galili* was *Sycobia* sp. 2; the reference species used for *W. microcarpae* and *M. ralianga* was *Apocrypta bakeri*. We used Mira v4.0.2 [24] to map clean reads to the reference genome, and MITObim v1.9.1 [25] to assemble the clean reads according to the overlapping regions between the sequences. Then the Mitos web server (http://mitos.bioinf.uni-leipzig.de/index.py) and NCBI ORFfinder (https://www.ncbi.nlm.nih.gov/orffinder/) were used to view and annotate the assembly results. Geneious v2020 [26] was used to assist the poorly assembled fragments of MITObim. For the poorly assembled genomes of *P. mayri* and *M. ralianga*, specific primers were designed to fill the gap regions of the CDS by PCR amplification (Table 2).

Table 2. Specific primers used for PCR amplification of partial mitochondrial regions in this study.

Species	Primers	Sequence (5′–3′)	Annealing Temperature	Targeted CDS Region
Platyneura mayri	PF1 PR1	ctatataaatttatgaaactatgattaatatctactaatcataaatatattgg gataatctaggaggtaataatcaaaatcttatattatttattcgtgg	54 °C	The middle part of the *cox1*
Micranisa ralianga	MF1 MR1	caattaaagttaaacaaattaataagtaaataattgaaattaatattg caatttaataataatcattgattttcttatattatatttttaatcatagtag	50 °C	The middle part of the *nad6*

2.5. Multiple Sequences Alignment, Model Selection and Construction of Phylogeny

For each species, the SCO gene sequences, 13 mitochondrial PCGs sequences, or the combination of these sequences (SCO + mitochondrial PCGs sequences) were respectively concatenated in a specific order into a supergene sequence. The supergene sequences were translated into amino acid sequences using translation software (https://web.expasy.org/cgi-bin/translate/dna2aa.cgi). MAFFT v7.313 [27] was used for multiple sequence alignment. Gblocks v0.91b [28] was used to identify conserved regions and remove unreliably aligned sequences within the datasets. The processed sequences were then used to select the best amino acid substitution model according to the Akaike information criterion with ProtTest v3.4.2 [29]. Setting *Encarisa formosa* as an outgroup, we then performed a phylogeny reconstruction using both Maximum Likelihood (ML) and Bayesian Inference (BI) methods. ML analyses were performed with raxmlGUI v1.5b2 [30], with statistical support for each node estimated using bootstrap, and the rapid bootstrap replicates set to 1000. BI analyses were performed with MrBayes v3.2.6 [31] under the following conditions: 100,000,000 generations, sampled every 1000 generations, a burn-in step for the first 5000 generations. Convergence was deduced for BI phylogenetic tree based on the following metrics: on the one hand, the median standard deviation of split frequencies value was less than 0.01; on the other hand, Tracer v1.7.1 [32] was used to ensure that the effective sample sizes (ESS) value was more than 200. The resulting phylogenetic trees were visualized in FigTree v1.4.4.

3. Results

3.1. Transcriptomes and Mitogenomes of the Four Newly Sequenced Fig Wasp Species Obtained from High-Throughput Sequencing

In the transcriptome sequencing of *P. mayri*, *O. galili*, *W. microcarpae* and *M. ralianga*, we obtained 103,205,280, 110,176,152, 88,651,962, and 93,048,384 clean reads, respectively. The sequence length distribution of the transcripts and unigenes were showed in Table 3. By using second-generation genome sequencing, for the four species of *P. mayri*, *O. galili*, *W. microcarpae* and *M. ralianga*, we obtained 14,546,375, 25,413,085, 19,783,035, and 19,625,937 clean reads, respectively. Based on these data, we assembled and annotated the mitogenome of the four fig wasps (see Supplementary Tables S1–S4 for detailed annotation).

Table 3. Length distribution of the transcripts and unigenes clustered from the de novo assembly.

Length Range	*Platyneura Mayri*		*Odontofroggatia Galili*		*Walkerella Microcarpae*		*Micranisa Ralianga*	
	Transcript	Unigene	Transcript	Unigene	Transcript	Unigene	Transcript	Unigene
200 bp–500 bp	14,720	4950	38,315	13,104	21,455	6249	24,187	7148
500 bp–1000 bp	11,155	6015	27,822	14,868	13,405	8393	13,659	8435
1000 bp–2000 bp	9368	3799	20,849	8211	11,040	4716	9517	4391
>2000 bp	16,581	5705	19,191	6501	22,073	6541	19,922	6411
Total Number	51,824	20,469	106,177	42,684	67,973	25,899	67,285	26,385
Total Length	99,849,096	35,330,681	130,935,638	49,857,034	133,605,390	42,636,711	127,749,958	42,751,535
Mean Length	1927	1726	1233	1168	1966	1646	1899	1620

3.2. Identification of SCO Genes and Phylogenetic Analysis

The most important challenges of phylogenomic studies involve different methods for phylogenetic tree reconstruction that can influence the validity of phylogenetic trees [33]. In this study, we identified a total of 145 SCO genes in 17 wasp species. Then we used different molecular datasets including 145 SCO genes, 13 mitochondrial PCGs and combination of these sequences (SCO + mitochondrial PCGs sequences) to reconstruct the phylogenies of fig wasps based on two phylogenetic tree reconstruction methods of ML and BI. In the selection of the model, for the data of the SCO genes and the combined genes, the best-fit model for the amino acid sequences was the JTT + I + G + F model, and the closest substitution model (LG + I + G + F) was selected because the best-fit model was not available in MrBayes v3.2.6. For the 13 mitochondrial PCGs, the MtArt + I + G + F model was the best-fit model for the amino acid sequences, and the closest substitution model (MtREV + I + G + F) was selected because the best-fit model was not available in MrBayes v3.2.6.

The ML and BI phylogenetic trees constructed with 145 SCO genes were shown in Figure 1; the ML and BI phylogenetic trees constructed with combined genes (the nuclear genes combined with the mitochondrial genes) were shown in Figure 2; the BI tree constructed with 13 mitochondrial PCGs was shown in Figure 3; the ML tree constructed with 13 mitochondrial PCGs was shown in Figure 4. Further analyzing the topology of phylogenetic trees, we are convinced that with the addition of the four newly sequenced species data in this study, the genus affiliations of fig wasps were very clear in all phylogenetic trees we have constructed. *O. galili* and *Sycobia* sp.2 were clustered in one clade of the phylogenetic trees with high node support values (Bayesian posterior probability = 1, Bootstrap value = 100), and these two species belonged to Epichrysomallinae. Similarly, *P. mayri* and *S. agreansis* were clustered in one clade with high nodal support values (Bayesian posterior probability = 1, Bootstrap value = 100), and they belonged to Sycophaginae; *W. microcarpae* and *M. ralianga* belonged to Otitesellinae were clustered in one clade with high nodal support values (Bayesian posterior probability = 1, Bootstrap values = 100 and 94).

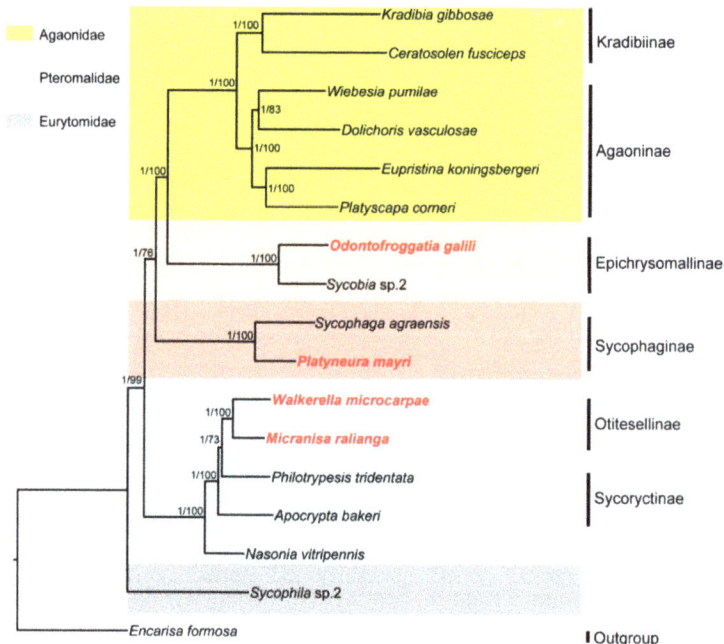

Figure 1. Inferred phylogenetic relationships of 17 chalcidoids based on amino acid (AA) datasets of 145 nuclear SCO genes using ML and BI analyses. *Encarisa Formosa* was used as the outgroup. Bayesian posterior probabilities (BPP) and ML bootstrap values (BP) for each node are shown as: BPP based on AA dataset/BP based on AA dataset, with maxima of 1.00/100.

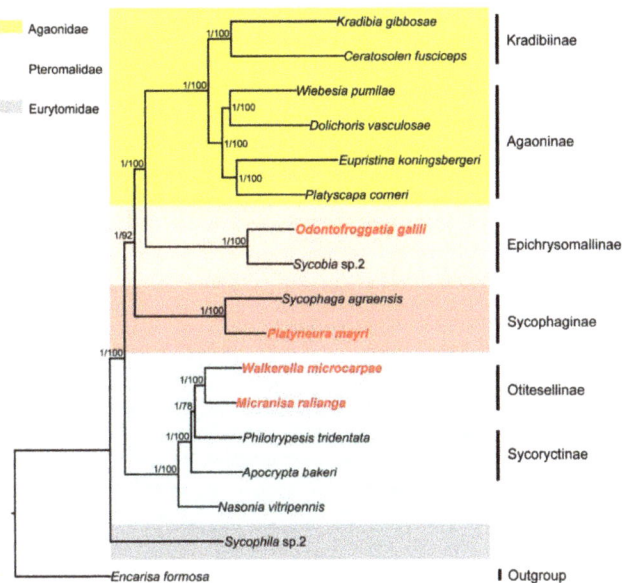

Figure 2. Inferred phylogenetic relationships of 17 chalcidoids based on amino acid (AA) datasets of combined genes (the 145 nuclear SCO genes combined with the 13 mitochondrial PCGs) using ML and BI analyses. *Encarisa Formosa* was used as the outgroup. Bayesian posterior probabilities (BPP) and ML bootstrap values (BP) for each node are shown as: BPP based on AA dataset/BP based on AA dataset, with maxima of 1.00/100.

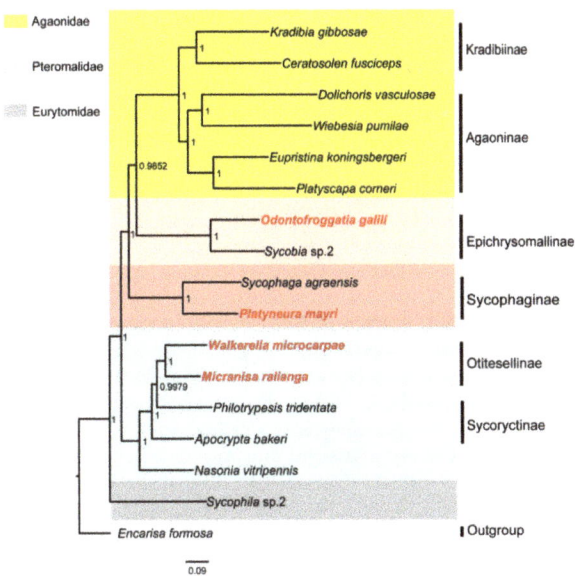

Figure 3. Inferred phylogenetic relationships of 17 chalcidoids based on amino acid (AA) datasets of 13 mitochondrial PCGs using BI analyses. *Encarisa Formosa* was used as the outgroup. Bayesian posterior probabilities (BPP) of each node are shown as BPP based on AA dataset 1.00.

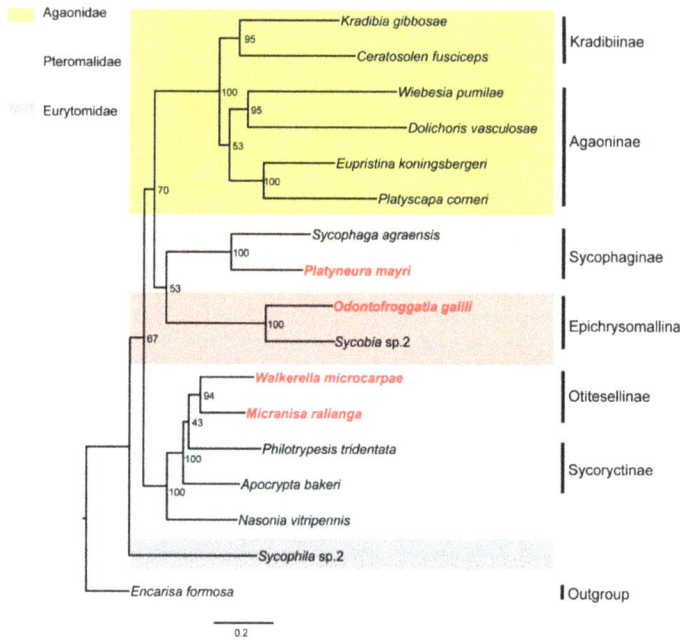

Figure 4. Inferred phylogenetic relationships of 17 chalcidoids based on amino acid (AA) datasets of 13 mitochondrial PCGs using ML analyses. *Encarisa Formosa* was used as the outgroup. Bootstrap values (BP) of each node are shown as BP based on AA dataset 100.

When considering the relationships between the subfamilies, our results showed that the ML and BI trees constructed with 145 SCO genes (Figure 1), the ML and BI trees constructed with the combined genes (Figure 2), and the BI tree constructed with 13 mitochondrial PCGs (Figure 3) all displayed the same topology: Epichrysomallinae and Agaonidae formed a clade, which subsequently clustered with Sycophaginae with high statistical support. However, the topology of the ML tree constructed with 13 mitochondrial PCGs (Figure 4) was somewhat divergent, Epichrysomallinae and Sycophaginae clustered together (Bootstrap value = 53), and then this clade was clustered with Agaonidae (Bootstrap value = 70), which was not necessarily reliable given its low statistical support.

4. Discussion

Transcriptome sequencing technology can economically and rapidly obtain all RNA information of organisms at a time point and plays an important role in finding molecular datasets for biological research [34]. Mitogenome sequences are also ideal molecular datasets for solving biological phylogeny due to their characteristics of genes without introns, and high evolutionary rate [35]. In this study of fig wasps, focusing on the undetermined taxonomic statuses of Epichrysomallinae and Sycophaginae, we used the SCO genes and mitochondrial PCGs of 17 species based on transcriptome, genome, and mitogenome data to construct phylogenetic trees using ML and BI methods. The final results suggested that both Epichrysomallinae and Sycophaginae were more closely related to Agaonidae, which updated information on the previously unclear phylogenetic statuses of both subfamilies. Compared with previous studies, this is the first time to use transcriptomes, genomes and mitogenome datasets to study the phylogenetic relationship reconstruction of the fig wasps.

Before the emergence of molecular data, the study of taxonomic statuses and phylogenetic relationships primarily relied on morphological characteristics. However, the evolutionary change of morphological characteristics is extremely complicated (even for a short evolutionary time) and phylogenetic trees derived from morphological data are often controversial [36]. Molecular phylogeny is the study of the interrelationships among various groups of organisms in their genealogical and evolutionary processes, by evolutionary study of the structure and function of biological macromolecules (proteins and nucleic acids) [37]. Since the evolutionary changes of DNA and amino acids follow a traceable pattern, it is possible to use a model to compare DNA or protein sequences among different organisms, so molecular phylogeny is expected to clarify problems that has been difficult to be resolved by the classical morphological approaches [38].

However, the incongruences of molecular data have also been frequently observed in phylogenetic trees, possibly due to lack of sufficient phylogenetic information from a single or a few genes [39]. Taking the phylogeny of the two subfamilies (Epichrysomallinae and Sycophaginae) of fig wasps studied in this study as an example, previous results for phylogenetic trees constructed with morphological data indicated that the taxonomic statuses of these two subfamilies had been changing, with Epichrysomallinae firstly being classified to Torymidae, then to Pteromalidae, and finally to Agaonidae by Bouček in 1988 [5–7], while Sycophaginae firstly being classified to Torymidae and finally to Agaonidae by Bouček in 1988 [7–10]. With the use of molecular data, the taxonomic statuses of these two subfamilies became clearer gradually. Initially, the phylogenetic tree constructed with D1 and D2 domains of nuclear 28S rRNA by Rasplus was not able to clarify the taxonomic statuses of these two subfamilies [8]. Subsequently, Heraty et al.'s results based on nucleotide sequences of nuclear ribosomal 18S and 28S D2-D5 expansion regions and 233 morphological data supported that Epichrysomallinae was closely related to Pteromalidae and Sycophaginae was closely related to Agaonidae [4]. Recently, Peters and his colleagues used transcriptomic data to show that Epichrysomallinae are closely related to Agaonidae, but the study did not include data for Sycophaginae [11]. In our study, we construct phylogenetic trees based on 145 SCO genes and 13 mitochondrial PCGs. Compared to the studies of Peters et al. [11], our results not only support their conclusions about Epichrysomallinae that it is closely related to Agaonidae, but our newly added species data about Sycophaginae further confirm that Sycophaginae is closely related to Agaonidae. Our results of Sycophaginae closely related to Agaonidae is consistent with the phylogenetic tree constructed by Heraty [4], even though our conclusions about the status of Epichrysomallinae is inconsistent. Therefore, according to the results of the above analyses, our study supports that both Epichrysomallinae and Sycophaginae are more closely related to Agaonidae. Our results suggest that the phylogenetic relationships of fig wasps cannot be well resolved by mitochondrial data alone, and the combination of nuclear genes, mitogenomes and morphological data will promote the reliability of fig wasp phylogeny. In the future, the addition of more species data will enable us to better understand the phylogenetic relationships of fig wasps.

5. Conclusions

In the taxonomic and phylogenetic studies of fig wasps, the taxonomic statuses of the two non-pollinating fig wasp subfamilies of Epichrysomallinae and Sycophaginae are still unclear. We here construct phylogenetic trees with nuclear conserved SCO genes and 13 mitochondrial PCGs genes from 17 wasp species by using the ML and BI methods. Our results show that both Epichrysomallinae and Sycophaginae are closely related to Agaonidae.

Supplementary Materials: The following are available online at https://www.mdpi.com/2075-1729/11/1/40/s1, Table S1: Summary of the mitogenome of *Platyneura mayri*. Table S2: Summary of the mitogenome of *Odontofroggatia galili*. Table S3: Summary of the mitogenome of *Walkerella microcarpae*. Table S4: Summary of the mitogenome of *Micranisa ralianga*.

Author Contributions: Conceptualization, J.X. and D.H.; methodology, D.Z., Z.X., H.H., and J.W.; validation, D.Z., Z.X., and Y.Z.; formal analysis, D.Z.; investigation, D.Z.; resources, D.Z. and H.H.; data curation, D.Z., Z.X., and Y.Z.; writing—original draft preparation, D.Z.; writing—review and editing, Z.X., H.H., Y.Z., J.X., and D.H.; visualization, D.Z.; supervision, J.X. and D.H.; project administration, J.X. and D.H.; funding acquisition, J.X. and D.H. All authors have read and agreed to the published version of the manuscript.

Funding: This research was funded by the National Natural Science Foundation of China (Nos. of 31830084, 31970440 & 32070466), and also supported by the construction funds for the "Double First-Class" initiative for Nankai University (Nos. 96172158, 96173250 & 91822294).

Acknowledgments: We thank Wenquan Zhen for help in collecting fig wasps.

Conflicts of Interest: The authors declare no conflict of interest.

References

1. Weiblen, G.D. How to be a fig wasp. *Annu. Rev. Entomol.* **2002**, *47*, 299–330. [CrossRef] [PubMed]
2. Machado, C.A.; Jousselin, E.; Kjellberg, F.; Compton, S.G.; Herre, E.A. Phylogenetic relationships, historical biogeography and character evolution of fig-pollinating wasps. *Proc. Biol. Sci.* **2001**, *268*, 685–694. [CrossRef] [PubMed]
3. Machado, C.A.; Robbins, N.; Gilbert, M.T.P.; Herre, E.A. Critical review of host specificity and its coevolutionary implications in the fig/fig-wasp mutualism. *Proc. Natl. Acad. Sci. USA* **2005**, *102* (Suppl. S1), 6558–6565. [CrossRef]
4. Heraty, J.M.; Burks, R.A.; Cruaud, A.; Gibson, G.A.P.; Liljeblad, J.; Munro, J.; Rasplus, J.-Y.; Delvare, G.; Janšta, P.; Gumovsky, A.; et al. A phylogenetic analysis of the megadiverse chalcidoida. *Cladistics* **2013**, *29*, 466–542. [CrossRef]
5. Hill, D.S. Figs (*Ficus* spp.) and fig-wasps (Chalcidoidea). *J. Nat. Hist.* **1967**, *1*, 413–434. [CrossRef]
6. Bouček, Z.; Watsham, A.; Wiebes, J.T. The fig wasp fauna of the receptacles of *Ficus thonningii* (Hymenoptera, Chalcidoidea). *Tijdschr. Entomol.* **1981**, *124*, 149–233.
7. Bouček, Z. *Australasian Chalcidoidea (Hymenoptera): A Biosystematic Revision of Genera of Fourteen Families, with a Reclassification of Species*; CAB International: Wallingford, UK, 1988.
8. Rasplus, J.-Y.; Kerdelhue, C.; Le Clainche, I.; Mondor, G. Molecular phylogeny of fig wasps. Agaonidae are not monophyletic. *C. R. Acad. Sci. III* **1998**, *321*, 517–526. [CrossRef]
9. Walker, F. Descriptions of new gengera and species of parasites, belonging to the families Proctotrupidae and Chalcididae, which attack insects destructive to the fig in India. *Entomologist* **1875**, *8*, 15–18.
10. Joseph, K.J. A proposed revision of the classification of the fig insects of the families Agaonidae and Torymidae (Hymenoptera). *Proc. R. Entomol. Soc. Lon. B* **1964**, *33*, 63–66.
11. Peters, R.S.; Niehuis, O.; Gunkel, S.; Blaser, M.; Mayer, C.; Podsiadlowski, L.; Kozlov, A.; Donath, A.; van Noort, S.; Liu, S.; et al. Transcriptome sequence-based phylogeny of chalcidoid wasps (Hymenoptera: Chalcidoidea) reveals a history of rapid radiations, convergence, and evolutionary success. *Mol. Phylogenet. Evol.* **2018**, *120*, 286–296. [CrossRef]
12. Delsuc, F.; Brinkmann, H.; Philippe, H. Phylogenomics and the reconstruction of the tree of life. *Nat. Rev. Genet.* **2005**, *6*, 361–375. [CrossRef] [PubMed]
13. Jeffroy, O.; Brinkmann, H.; Delsuc, F.; Philippe, H. Phylogenomics: The beginning of incongruence? *Trends Genet.* **2006**, *22*, 225–231. [CrossRef] [PubMed]
14. Zhao, Y.J.; Cao, Y.; Wang, J.; Xiong, Z. Transcriptome sequencing of *Pinus kesiya* var. *langbianensis* and comparative analysis in the *Pinus* phylogeny. *BMC Genom.* **2018**, *19*, 725. [CrossRef] [PubMed]
15. Cameron, S.L. Insect mitochondrial genomics: Implications for evolution and phylogeny. *Annu. Rev. Entomol.* **2014**, *59*, 95–117. [CrossRef]
16. Simon, C.; Buckley, T.R.; Frati, F.; Stewart, J.B.; Beckenbach, A.T. Incorporating molecular evolution into phylogenetic analysis, and a new compilation of conserved polymerase chain reaction primers for animal mitochondrial DNA. *Annu. Rev. Ecol. Evol. Syst.* **2006**, *37*, 545–579. [CrossRef]
17. Wang, R.Q.; Wang, D.Z.; Li, C.T.; Yang, X.R. Mitochondrial genome of the shorthead catfish (*Pelteobagrus eupogon*): Structure, phylogeny, and intraspecific variation. *Genet. Mol. Res.* **2016**, *15*. [CrossRef]
18. Bouček, Z. The genera of chalcidoid wasps from *Ficus* fruit in the New World. *J. Nat. Hist.* **1993**, *27*, 173–217. [CrossRef]
19. Yan, L.; Yang, M.; Guo, H.; Yang, L.; Wu, J.; Li, R.; Liu, P.; Lian, Y.; Zheng, X.; Yan, J.; et al. Single-cell RNA-Seq profiling of human preimplantation embryos and embryonic stem cells. *Nat. Struct. Mol. Biol.* **2013**, *20*, 1131–1139. [CrossRef]
20. Grabherr, M.G.; Haas, B.J.; Yassour, M.; Levin, J.Z.; Thompson, D.A.; Amit, I.; Adiconis, X.; Fan, L.; Raychowdhury, R.; Zeng, Q.; et al. Full-length transcriptome assembly from RNA-Seq data without a reference genome. *Nat. Biotechnol.* **2011**, *29*, 644–652. [CrossRef]
21. Davidson, N.M.; Oshlack, A. Corset: Enabling differential gene expression analysis for de novo assembled transcriptomes. *Genome Biol.* **2014**, *15*, 410. [CrossRef]
22. Haas, B.J.; Papanicolaou, A.; Yassour, M.; Grabherr, M.; Blood, P.D.; Bowden, J.; Couger, M.B.; Eccles, D.; Li, B.; Lieber, M.; et al. De novo transcript sequence reconstruction from RNA-seq using the Trinity platform for reference generation and analysis. *Nat. Protoc.* **2013**, *8*, 1494–1512. [CrossRef] [PubMed]
23. Li, L.; Stoeckert, C.J., Jr.; Roos, D.S. OrthoMCL: Identification of ortholog groups for eukaryotic genomes. *Genome Res.* **2003**, *13*, 2178–2189. [CrossRef] [PubMed]

24. Chevreux, B.; Wetter, T.; Suhai, S. Genome sequence assembly using Trace signals and additional sequence Information. *J. Comput. Sci. Syst. Biol.* **1999**, *99*, 45–56.
25. Hahn, C.; Bachmann, L.; Chevreux, B. Reconstructing mitochondrial genomes directly from genomic next-generation sequencing reads-a baiting and iterative mapping approach. *Nucleic Acids Res.* **2013**, *41*, e129. [CrossRef] [PubMed]
26. Kearse, M.; Moir, R.; Wilson, A.; Stones-Havas, S.; Cheung, M.; Sturrock, S.; Buxton, S.; Cooper, A.; Markowitz, S.; Duran, C.; et al. Geneious Basic: An integrated and extendable desktop software platform for the organization and analysis of sequence data. *Bioinformatics* **2012**, *28*, 1647–1649. [CrossRef]
27. Katoh, K.; Standley, D.M. MAFFT multiple sequence alignment software version 7: Improvements in performance and usability. *Mol. Biol. Evol.* **2013**, *30*, 772–780. [CrossRef]
28. Talavera, G.; Castresana, J. Improvement of phylogenies after removing divergent and ambiguously aligned blocks from protein sequence alignments. *Syst. Biol.* **2007**, *56*, 564–577. [CrossRef]
29. Darriba, D.; Taboada, G.L.; Doallo, R.; Posada, D. ProtTest 3: Fast selection of best-fit models of protein evolution. *Bioinformatics* **2011**, *27*, 1164–1165. [CrossRef]
30. Stamatakis, A. RAxML version 8: A tool for phylogenetic analysis and post-analysis of large phylogenies. *Bioinformatics* **2014**, *30*, 1312–1313. [CrossRef]
31. Altekar, G.; Dwarkadas, S.; Huelsenbeck, J.P.; Ronquist, F. Parallel Metropolis coupled Markov chain Monte Carlo for Bayesian phylogenetic inference. *Bioinformatics* **2004**, *20*, 407–415. [CrossRef]
32. Rambaut, A.; Drummond, A.J. Tracer v1.4. *Encyclopedia Atmos. Sci.* **2007**, *141*, 2297–2305. [CrossRef]
33. Philippe, H.; Delsuc, F.; Brinkmann, H.; Lartillot, N. Phylogenomics. *Annu. Rev. Ecol. Evol. S.* **2005**, *36*, 541–562. [CrossRef]
34. Dong, S.; Xiao, Y.; Kong, H.; Feng, C.; Harris, A.J.; Yan, Y.; Kang, M. Nuclear loci developed from multiple transcriptomes yield high resolution in phylogeny of scaly tree ferns (Cyatheaceae) from China and Vietnam. *Mol. Phylogenet. Evol.* **2019**, *139*, 106567. [CrossRef]
35. Wang, G.; Lin, J.; Shi, Y.; Chang, X.; Wang, Y.; Guo, L.; Wang, W.; Dou, M.; Deng, Y.; Ming, R.; et al. Mitochondrial genome in *Hypsizygus marmoreus* and its evolution in Dikarya. *BMC Genom.* **2019**, *20*, 765. [CrossRef]
36. Nei, M.; Kumar, S. *Molecular Evolution and Phylogenetics*; Oxford University Press: New York, NY, USA, 2000.
37. Maddison, W.P. Molecular approaches and the growth of phylogenetic biology. In *Molecular Zoology: Advances, Strategies, and Protocols*; Ferraris, J.D., Palumbi, S.R., Eds.; Wiley-Liss: New York, NY, USA, 1996; pp. 47–63.
38. Lin, G.-H.; Wang, K.; Deng, X.-G.; Nevo, E.; Zhao, F.; Su, J.-P.; Guo, S.-C.; Zhang, T.-Z.; Zhao, H. Transcriptome sequencing and phylogenomic resolution within Spalacidae (Rodentia). *BMC Genom.* **2014**, *15*, 32. [CrossRef] [PubMed]
39. Wang, K.; Hong, W.; Jiao, H.; Zhao, H. Transcriptome sequencing and phylogenetic analysis of four species of luminescent beetles. *Sci. Rep.* **2017**, *7*, 1814. [CrossRef] [PubMed]

Article

Chronological Incongruences between Mitochondrial and Nuclear Phylogenies of *Aedes* Mosquitoes

Nicola Zadra [1,2], Annapaola Rizzoli [1] and Omar Rota-Stabelli [1,2,3,*]

1 Research and Innovation Centre, Fondazione Edmund Mach, 38010 San Michele all Adige (TN), Italy; nicola.zadra@fmach.it (N.Z.); annapaola.rizzoli@fmach.it (A.R.)
2 Department of Cellular, Computational and Integrative Biology—CIBIO, University of Trento, 38123 Povo (TN), Italy
3 Center Agriculture Food Environment—C3A, University of Trento, 38010 San Michele all Adige (TN), Italy
* Correspondence: omar.rotastabelli@unitn.it

Citation: Zadra, N.; Rizzoli, A.; Rota-Stabelli, O. Chronological Incongruences between Mitochondrial and Nuclear Phylogenies of *Aedes* Mosquitoes. *Life* 2021, *11*, 181. https://doi.org/10.3390/life11030181

Academic Editors: Koichiro Tamura, Pedro Martinez, Federico Plazzi and Andrea Luchetti

Received: 15 January 2021
Accepted: 22 February 2021
Published: 25 February 2021

Publisher's Note: MDPI stays neutral with regard to jurisdictional claims in published maps and institutional affiliations.

Copyright: © 2021 by the authors. Licensee MDPI, Basel, Switzerland. This article is an open access article distributed under the terms and conditions of the Creative Commons Attribution (CC BY) license (https://creativecommons.org/licenses/by/4.0/).

Abstract: One-third of all mosquitoes belong to the Aedini, a tribe comprising common vectors of viral zoonoses such as *Aedes aegypti* and *Aedes albopictus*. To improve our understanding of their evolution, we present an updated multigene estimate of Aedini phylogeny and divergence, focusing on the disentanglement between nuclear and mitochondrial phylogenetic signals. We first show that there are some phylogenetic discrepancies between nuclear and mitochondrial markers which may be caused by wrong taxa assignment in samples collections or by some stochastic effect due to small gene samples. We indeed show that the concatenated dataset is model and framework dependent, indicating a general paucity of signal. Our Bayesian calibrated divergence estimates point toward a mosquito radiation in the mid-Jurassic and an *Aedes* radiation from the mid-Cretaceous on. We observe, however a strong chronological incongruence between mitochondrial and nuclear data, the latter providing divergence times within the Aedini significantly younger than the former. We show that this incongruence is consistent over different datasets and taxon sampling and that may be explained by either peculiar evolutionary event such as different levels of saturation in certain lineages or a past history of hybridization throughout the genus. Overall, our updated picture of Aedini phylogeny, reveal a strong nuclear-mitochondrial incongruence which may be of help in setting the research agenda for future phylogenomic studies of Aedini mosquitoes.

Keywords: divergence; mtDNA; Diptera; phylogeny; saturation; rates

1. Introduction

Mosquitoes (Culicidae) are one of the most successful Diptera radiation. They include more than 3600 species classified in two subfamilies and 44 genera and 145 subgenera [1–3]. Because they vector a variety of disease, mosquitoes are still the largest indirect cause of mortality among humans than any other group of organisms. Approximately one-third of mosquito species belong to the tribe Aedini, including 1261 species classified in 10 genera [3]. Aedini species are globally distributed and are vectors of many zoonosis of human and animals including filarial nematodes [4] and many arboviruses such as Chikungunya, Dengue, Zika, Yellow Fever, West Nile [5–7]. Aedini species include some of the most invasive and medically relevant mosquitoes: *Aedes aegypti* and *Aedes albopictus* [8–12]. *Aedes aegypti* has mainly spread outside its original African range, although it does not seem capable of settling stable populations in continental climates, such as the European one. *Aedes albopictus*, originally from South East Asia, is instead now reported from every continent and has quickly settled in Europe, China, and other temperate zones [7,13]. Genome resources exist for only these two species of *Aedes* [14–16], while whole genome data for other invasive *Aedes* is still lacking. These include *Aedes japonicus* and *Aedes koreicus*, which are quickly invading and establishing, respectively, in central Europe [17] and North

Italy [18,19] showing competence for the transmission of many arboviruses such as West Nile virus and Zika virus [8,19,20].

Knowledge of the reciprocal affinities of these and other invasive *Aedes* species and the timing of their evolution is important for various reasons. First, a robust phylogeny is essential to polarize key behavioral and ecological traits, as recently shown by Soghigian et al. [21]. In particular, a phylogeny can identify the sister-species of invasive *Aedes* of health concern. The sister-species shares a common ancestor with the species of interest (is the closest related in the phylogenetic tree) and is very useful for correctly polarizing evolutionary novelties, such as new genes in phylogenomics and transcriptomics studies [22,23]. Second, phylogenies may help to define taxonomy and classification. A recent classification [1] has raised the number of genera from 10 to 79; the genera, however, have been later reduced to 10 [3]. Molecular investigations of Aedini relationships can help to clarify these taxonomical issues. Third, dated phylogenies help to characterize the paleoecological scenario in which mosquito radiations happen, thus providing evidence with clues about their pre-adaptations as it has been shown, for example, in *Drosophila* [24,25]. Molecular studies have addressed Aedini evolution by studying their phylogeny using both mitochondrial and nuclear markers. While relationships within the Aedini group has been studied in detail using a multimarker approach [21], the origin of the family and their reciprocal affinity with other Culicinae are not well studied, or have not been addressed because datasets were centered only on Aedini [21]. Furthermore, the stability of clades within the Aedini has never been addressed by comparing different statistical frameworks (e.g., maximum likelihood versus Bayesian), or by employing a different model of replacement (e.g., homogeneous versus heterogeneous [26]).

One key aspect so far neglected in Aedini phylogenetic studies is the direct comparison of the phylogenetic signal from the DNA of the two cellular compartments: nuclear (nDNA) and mitochondrial (mtDNA). It has been shown that the nDNA and mtDNA may carry different phylogenetic signal and produce conflicting phylogenies, in some cases, because of hybridization events affecting mtDNA [27]. MtDNA substitution rate is typically faster compared to nuclear one; this can lead toward homoplasy caused by site saturation, which in turns may affect the topology and may underestimate the correct inference of substitution rates [27].

Little in general is known about how mtDNA and nDNA conflict for what concern estimation of divergence times. In some case, chronological signal can be consistent between nuclear and mitochondrial genes, as in fish [28] and amphibians [29], with discrepancy just in the shallow time part of the tree. In *Drosophila*, the two types of markers recover similar divergences with mtDNA supporting slightly younger estimates than nDNA [24]. In butterflies, the chronological conflict between nDNA and mtDNA is more marked, although it seems to be restricted to a few species experiencing hybridization [30]. In the above cases, the confidence interval of the divergence estimates using the two type of markers largely overlap. Therefore, the conflict is not statistically significant. The molecular clocks of mtDNA and nDNA have never been systematically compared in Aedini.

A systematic comparison of chronological signal in mosquitoes has never been undertaken. An effort to date the Aedini mosquito using nDNA data in a Bayesian framework [31] recovered the origin of Aedini at 157 (Credibility Intervals, CI: 187–124) millions of years ago (MYA) and a Culicidae radiation at 216 (229–192) MYA. A recent effort using a multigene (nDNA + mtDNA) strategy in a maximum likelihood framework [20] recovered an Aedini origin at circa 125 MYA. In the latter, the diversification of *A. albopictus* from its sister species *A. flavopictus* is circa 25 MYA, while *A. albopictus* and *A. aegypti* common ancestor was set at approximately 55 MYA, a time compatible estimate, but quite distant from that based on whole genomes 71 (44–107) MYA [14]. A recent divergence estimate of Culicinae using complete mtDNA [32] recovered the origin of Aedini at 130 (CI: 101–168) MYA and an *A. albopictus-A. aegypti* split at circa 67 (CI: 55–94) MYA. There are, therefore, certain discrepancies in available literature for what concerns the timing of Aedini radiation.

This work aims at providing an updated picture of Aedini phylogeny and divergence, by disentangling the phylogenetic signal in available genetic markers. We used four nuclear and four mitochondrial genes in a Bayesian framework to study the evolutionary history of the Aedini, their relationship with other Culicinae, and the timing of their origin and diversification. Our results revealed previously under looked incongruences between nuclear and mitochondrial data, for what concerns both their rate of evolution and their posterior divergence estimates. This has an important implication for our understanding of Aedini evolution and more generally for the long-lasting issue of incongruences between mitochondrial and nuclear data in inferring species phylogeny and divergences.

2. Materials and Methods

2.1. Genes and Taxa Selection

In our study, we employed four mitochondrial coded genes: Cytochrome c oxidase I (COI), Cytochrome c oxidase II (COII), NADH dehydrogenase subunit 4 (NAD4) and 16S, and four nuclear-coded genes: Enolase, Arginine Kinase, 18S, and 28S. We choose these genes after various rounds of literature and blast searches because they were the most evenly distributed through the Aedini tribe and the outgroup. Similarly to other recent Aedes phylogenetic studies [21], the current availability of genes in the database did not allow us to sample more genes. The number of annotated genes in Genebank for Aedes and other mosquitoes species is low, in general no more than 5 or 6 markers per species; many species are characterized by many variants of the same marker for example COI. Annotated genome and transcriptome data was present only for the model organisms *Aedes albopictus* and *Aedes aegypti*. We had to exclude from our gene list the genes encoding for white, hunchback, and Carbomoylphosphate synthase (CAD) because poorly sampled within the Culicidae family. We did not employ Internal Transcribed Spacer (ITS) because of poor and ambiguous alignment between Aedini and its outgroups. Since we were interested in studying the origin of Aedini, poor alignment of Aedini ITS sequence with those of the outgroups could have affected the correct inference of their phylogeny. We sampled genes from the same specimen whenever it was possible; in most cases we concatenate genes from different specimens of the same species. This is the common procedure when concatenating genes for inter-specific phylogenetic studies [21,26]. We followed the nomenclature as in [3,33]. For more clarity, in some of our phylogenetic trees, we displayed in brackets the proposed subgenera. Each chosen gene for all the available Aedini taxa was downloaded from GenBank. To reduce missing data and promote a direct comparison between nuclear and mitochondrial data, we selected a species only if at least two genes represented it in each of the two types of markers (nuclear and mitochondrial). Moreover, we excluded species that seemed to be ambiguously labelled. The final dataset finally was represented by 34 evenly phylogenetically distributed Aedini, plus 10 outgroups sampled from other Culicinae, Anopheline, and other Diptera samples (see Table S1 for the species list). The outgroup sequences were essential to root the Aedini phylogeny and to generate nodes for calibrating the molecular clock. For each of the eight markers, we filtered out ambiguous sequence. We used a fast bootstrap RaxML (see details below) to preliminarily check if sequences were clustering within their expected group (e.g., sequences for Aedini to form a monophyletic Aedini group). Sequence clustering to a different group considered unreliable and was excluded from downstream analysis.

2.2. Alignments and Phylogenetic Analyses

We aligned each of the eight genes independently. We aligned protein-coding genes using MAFFT through TranslatorX [34], and non-coding genes using MAFFT directly [35]. Finally, the genes were concatenated using FASconCAT [36] and manually edited to detect a few misaligned sites. We generated three aligned datasets: nuclear, mitochondrial, and concatenated. The nuclear dataset (nDNA) is composed of the concatenation of Enolase, Arginine Kinase, 18S, and 28S; it is 3270 nucleotides (nt) in length. The mitochondrial dataset (mtDNA) is composed of the concatenation of COI, COII, NAD4, and 16S; it is

4224 nt in length. The third dataset (concatenated) is the concatenation of nDNA and mtDNA and is 7494 nt in length. To further study Aedini relationships, we generated a fifth dataset based on the original 6298 nt alignment of [21], increasing site occupancy by using Gblocks at default parameters. The final dataset (named Soghigian) was composed of 71 sequences and 3815 nt with 8% of missing data. Although there is some overlap of genes between concatenated and Soghigian datasets, they substantially differ because of the presence of 4 genes (COII, NAD4, and 16S present in concatenated, while ITS is absent from concatenated) and mostly because of a very different taxon sampling. The concatenated dataset contains various outgroup to the Aedini because the aim was to set the origin of Aedini. Phylogenetic analyses were performed mainly at the nucleotide level using both maximum likelihood (ML) and Bayesian statistical frameworks, using, respectively, RAxML [37] and PhyloBayes [38] or BEAST (see below). The RAxML analyses were performed on all datasets using the General Time Reversible (GTR) replacement model plus four discrete rate categories of gamma (G) and employing 100 bootstrap replicates. PhyloBayes analyses were performed using the same model and repeated using the heterogeneous CAT (plus G) replacement model.

2.3. Divergence Estimates

BEAST v2.5 was used to reconstruct phylogenies and to estimate divergence times [39]. We use BEAUti to set the analyses using the following prior information to calibrate the clock. We employed a root prior based on the fruit fly-mosquito split using a normal distribution with mean 260 MYA and a 95% prior distribution to be between 296 and 238 MYA, as indicated by [40]. We employed three minimum calibration points for the diversification of Anophelinae, Culicinae, and Culicidae, using, respectively, 34 MYA, 34 MYA, and 99 MYA, according to the three oldest fossils known for each of these groups [41,42]. These calibrations were used for the mtDNA, nDNA, and concatenated datasets. We run BEAST using a different set of (model) priors and choose the most fitting combination of priors using the Harmonic mean, the Akaike Information Criterion (AICm), the stepping stone (SS), and the Path sampling (PS); for the latter, we used the Path sampler package and set the analysis at 50% burn-in with 40 steps of 500,000 chain length. All other chains were run for 100,000,000 generations until Beast log files indicated proper convergences of all posteriors and the likelihood using tracer1.7 [43]. Divergence estimates in PhyloBayes [38] where done using the same calibration priors described above, a CAT plus Gamma replacement model and a LogNormal relaxed clock.

3. Results

3.1. Conflicts between Nuclear and Mitochondrial Phylogenies

Bayesian inference of the eight concatenated genes dataset under a homogeneous replacement model (GTR, Figure 1A) reveals a generally well-supported tree with the *Aedes* genus divided into two distinct clades as in [21]: Clade A (in pink, Posterior Probability (PP): 1.00) comprises various species including *A. albopictus* and *A. aegypti*; Clade B (in orange, PP: 1.00) comprises various species often regarded as *Ochlerotatus* plus others referred to as *Aedes* such as *A. koreicus*. Species of the *Psorophora* genus are the sister of Clade A + Clade B (PP: 1.00). Within Clade A we observed two groups (dark pink), one consisting of species attributed to *Stegomya* + *Armigeres* (clade A1, PP: 1:00), the second containing four genera (*Aedimorphus*, *Catageiomyia*, *Diceromyia*, and *Scutomyia*; PP: 0.92). The mutual relationship of no-Aedini Culicinae is instead unresolved (PP: 0.48); four genera (*Sabethes*, *Wyeomyia*, *Malaya*, and *Toxorhynchites*) form however a robustly supported (PP: 1.00) group, which we have provisionally named Group C.

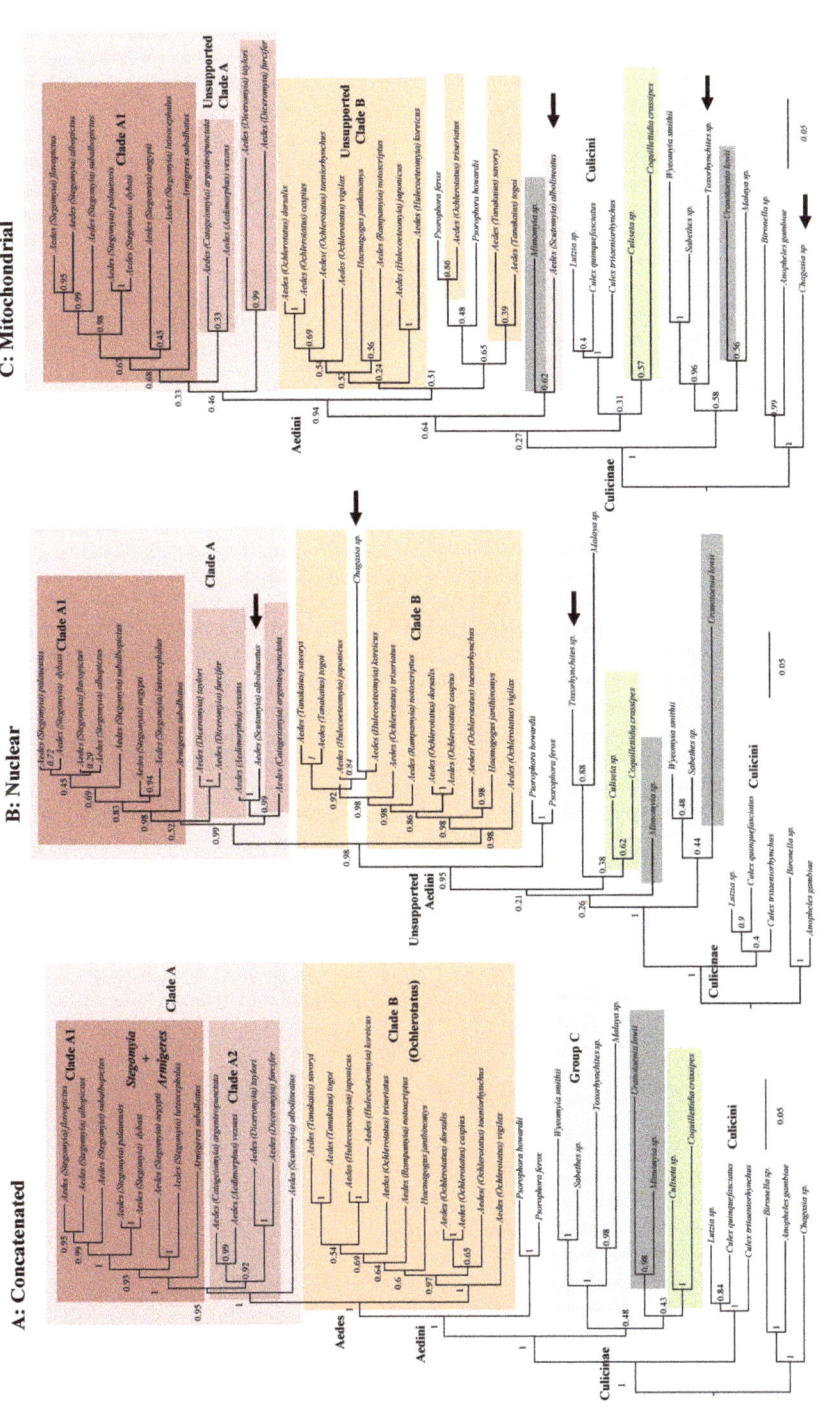

Figure 1. Topological incongruence between mitochondrial and nuclear data. (**A**): Bayesian consensus tree of concatenated dataset. (**B**): Bayesian consensus tree of nuclear (nDNA) dataset. (**C**): Bayesian consensus tree of mitochondrial (mtDNA) dataset. All analyses have been performed using a GTR+G model in PhyloBayes. Numbers at nodes are posterior probabilities (PP). Groups identified using the concatenated dataset have been colored. Arrows indicate highly supported incongruences between the nDNA and mtDNA datasets. The corresponding Maximum Likelihood trees are in Supplementary Figure S1.

Our concatenated analysis of Figure 1A recovers, at least for most nodes, a robust topology. One of the aims of our study was, however, to disentangle the phylogenetic signal for the Aedini by exploring its consistency over different data types and methodological treatments. We, therefore, analyzed the nDNA and mtDNA datasets separately (respectively, Figure 1B,C), and reveal various instances of mitochondrial-nuclear incongruence. Overall, both trees are less resolved than the concatenated tree (for example, they both do not support Group C nor Group A2), pointing toward the utility of concatenating genes. The nuclear tree is, however, markedly more resolved (it has overall higher supports at nodes) than the mitochondrial one. It does support, for example, the monophyly of *Aedes* and both Groups A and B (all with PP > 0.9), while mtDNA dataset does not support them. These differences may be explained by less phylogenetic signal in the mtDNA dataset. This is, however, not related to fewer nucleotide positions as the mtDNA alignment is larger than the nDNA one (4224 nt vs. 3270 nt). We identify some interesting cases of well-supported incongruences between the nDNA and the mtDNA trees involving *A. albolineatus*, *A. subalbopictus*, and *Toxorhynchites* sp (depicted by arrows in Figure 1B,C). There are various topological incongruences, for example for the position of *A. subalbopictus*, the two *Psorophora* and *Uranotaenia lowii*, but their affinities did not receive high PP in at least one of the two trees, therefore they are not considered statistically significant.

3.2. A Conservative Picture of Aedini and Other Culicinae Phylogeny

To explore in more detail the phylogenetic signal behind our Bayesian trees of Figure 1, we further performed phylogenetic analyses employing different statistical frameworks, different model of replacement, and type of datasets (Figure 2). In panel A we depict the result of a Maximum Likelihood (ML) analysis of the concatenated dataset. In panel B is the same dataset analyzed in a Bayesian framework using an among-site heterogeneous CAT model more suitable for ancient radiations and saturated datasets [44].

In panel C is the ML analysis of a dataset (named Soghigian) centered on Aedini and derived from [21]. To provide a conservative picture of Aedini phylogeny, we have collapsed a node if its bootstrap support (BS) from the Maximum Likelihood (ML) analysis was lower than 75% and if its posterior probability (PP) from Bayesian analysis was lower than 0.9. We found a consistent signal (compare Figure 1A with Figure 2A,B) for a group of *Sabethes*, *Wyeomyia*, and *Malaya* (Sabethini tribe), plus *Toxorhynchites* (Toxorhynchitini tribe) which we have provisionally named Group C. This group is monophyletic using both homogeneous and heterogeneous models of evolution, but its internal relationships, as well as its relative affinity with other Aedini, is inconsistent over different analyses and in general not significantly supported. This group is not consistent with a previous multigene phylogeny, which supports *Toxorhynchites* as closely related to *Mymoyia* than to the Sabethini [31]. Although highly supported in all our concatenated analyses, we advocate caution in considering the validity of Group C, as our analyses may have been biased by an unfortunate combination of reduced taxon and gene sampling; indeed, in most analyses, *Toxorhynchites* is the sister taxa of *Mymoyia*, therefore disrupting the monophyly of Sabethini. Our investigations are instead congruent with previous studies [20] in supporting Group A, and to a lesser extent Group A1 (*Stegomya + Armigeres*). Group B is instead supported only by the homogeneous GTR model of evolution. Site heterogeneous models of replacement such as CAT have been repeatedly shown as being capable of reducing systematic errors [45]; we cannot, therefore, exclude that the signal responsible for Group B is artefactual and we advocate care in considering it as monophyletic. From a systematic point of view, our phylogenies support the classical 10 genera classification of Aedini [2,20]. Overall, while some nodes are robustly supported in all analyses of Figure 2 (for example, Group A), other nodes are poorly supported or are supported only in one analysis.

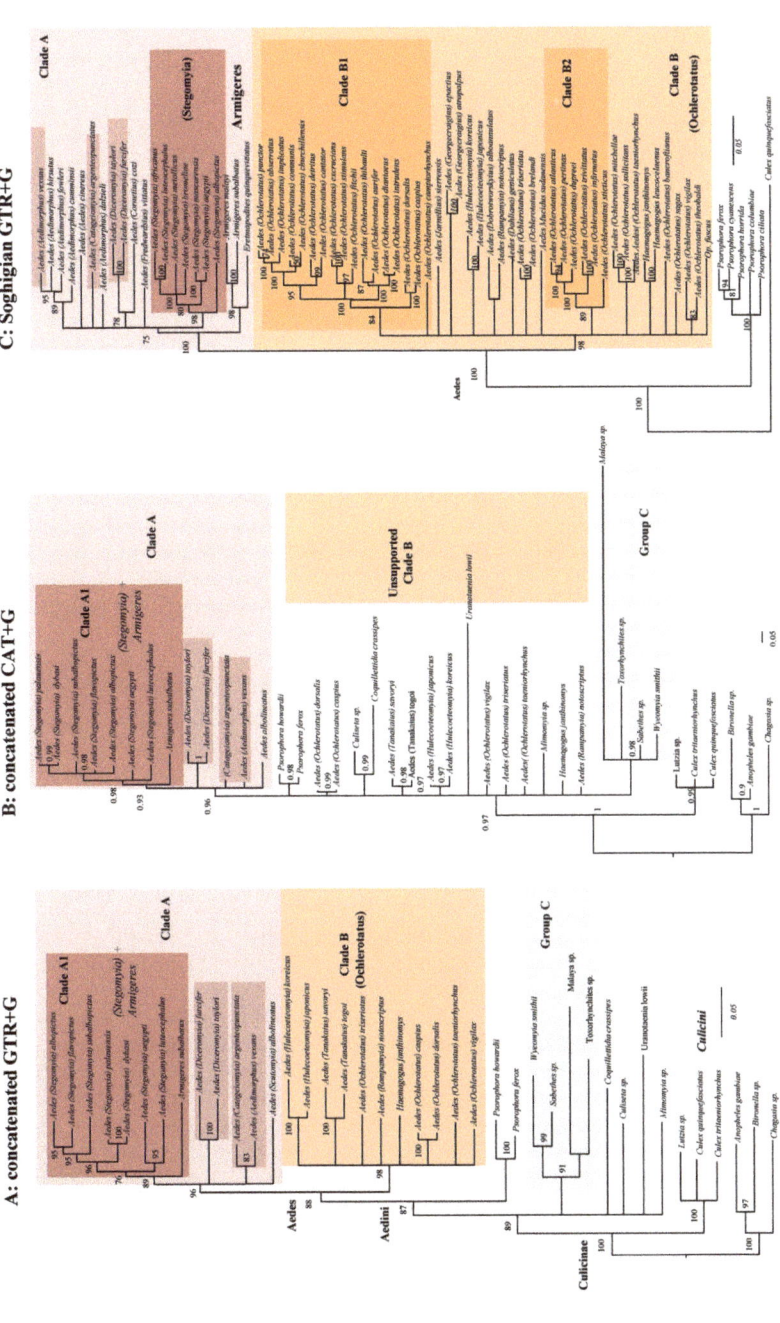

Figure 2. A conservative picture of Culicinae phylogeny using different models and datasets. To highlight lack of phylogenetic signal, all nodes below PP 0.90 and BS 75 have been collapsed. (**A**): Maximum likelihood tree of the concatenated dataset using the GTR+G model in RaXml. (**B**): Bayesian consensus tree of the concatenated dataset under the CAT+G model using PhyloBayes. (**C**): Maximum likelihood tree of a modified [21] dataset under the GTR+G model using RaXml. The number at nodes are bootstrap supports (BS) in panels A and C, and posterior probabilities (PP) in panel B. The backbone of the tree and the monophyly of Clade B (orange) are strongly supported using GTR, but not using CAT. Many relationships within Clade B are poorly supported in all analyses. Full trees with all nods and supports are in Supplementary Figures S2–S4.

3.3. Divergence Estimates of the Aedini

To define which evolutionary models better describe the radiation of mosquito in our concatenated dataset, we contrasted the strict clock versus the log-normal relaxed clock models, the coalescent versus the speciation model, and the HKY versus the GTR replacement model (Table 1). A relaxed clock is favoured over the strict clock. Furthermore, under the relaxed clock, the coefficient of variation rate was approximately 0.4 for the mitochondrial data and roughly 1 for the nuclear, further indicating that an uncorrelated clock hypothesis suits better our datasets than a strict clock. This is because if a lognormal clock has a coefficient of variation close to 0, it could be considered clock-like, so comparable with a strict clock [46]. Demographic speciation models are more supported than Coalescent model, but the stepping stone and path sampling could not discriminate between a Yule and a Birth Death model; we chose the Yule model because it was favoured by the AICm, which penalizes based on the number of free parameters. The two models provided nevertheless with similar results (Table 1). Therefore, we used a combination of GTR+G, relaxed log-normal, and Yule models for our clock analyses.

Table 1. Model tested with divergence estimates for two nodes.

Clock Model	Substitution Model	Tree Prior	logLikelihood	AICm	Harmonic Mean	PS/SS	Culicidae	Aedini
Strict	GTR	Yule	54,908.6	109,892.3	−54,926.7	4	156 (114–204)	90 (75–104)
Relaxed (LogN)	HKY	Yule	54,624.8	109,467.9	−54,666.3	5	166 (119–215)	96 (68–125)
	GTR	Yule	54,362.7	109,108.1	−54,424.3	1	180 (137–228)	113 (83–143)
		Birth Death	54,363.6	109,115	−54,414.9	1	180 (135–227)	112 (82–142)
		Coalescent Constant	54,370.2	109,208.7	−54,415.7	3	173 (123–225)	100 (66–132)

Our analysis of the concatenated dataset using the most fitting models, allows us to obtain a picture of Aedini evolutionary history, which we have contrasted with the appearance of some major vertebrate lineages and flowering plants in Figure 3. According to our posterior estimates, the mosquito family (Culicidae) diversified in its two subfamilies—Culicinae and Anophelinae—approximately 180 MYA (95% High Posterior Densities, HPD 137–228 MYA) in the lower Jurassic. The earliest fossil of a Chaoboridae, the Culicidae sister group, is 187 MYA [42]. This would suggest a very rapid diversification of Culicomorpha. Our estimates tend to match the proposed origin of angiosperm [47]; however their evolutionary history is not clear yet, and the origin of angiosperm could be older than expected [48]. Culicinae diversified in two clades (Culicini and the clade leading to Aedini) between the end of the Jurassic and the early Cretaceous, at 146 (108–182) MYA, while the Aedini tribe diversify at 113 (83–143) MYA with the split of *Aedes* from *Psorophora* genus. Within Aedini, Clade A, and Clade B originated circa 106 (77–133) MYA. Within Clade A, the subgenus *Stegomya* (which includes model organisms *A. albopictus* and *A. aegypti*) originated 84 (58–109) MYA, concomitantly with the diversification of Clade B (which include the subgenus *Ochlerotatus*) at 86 (61–111) MYA in the late Cretaceous. To test for the effect of outgroup on our dated phylogenies, we repeated the analysis of our concatenated alignment, excluding Brachycera outgroup. This additional analysis shows that the calibration point drives our divergence estimates at the root. The median height is younger without outgroups, although the two analyses are compatible for what concerns their (overlapping) 95% HPD (Table 2). The rooted tree provided more precise estimates. The 95% HPD is smaller in the root-calibrated phylogeny then in the unrooted one. Overall, our date estimates tend to be slightly younger than the ones provided previously [14,21,31] for what concern the origin of Culicinae, but slightly older for what concern the origin of Aedini (see Table 2).

3.4. Chronological Incongruences between Nuclear and Mitochondrial Data

Clock analysis using separately nuclear and mitochondrial genes (Figure 4) revealed unexpected strong incongruences. The estimates for the origin of the main mosquito clades (deep nodes of the phylogeny) are similar using the two datasets and reinforce our findings using the concatenated data of Figure 3. For example, Culicinae originated in the early Jurassic and Aedini in the Cretaceous both in the mtDNA and nDNA. However, there is a strong discrepancy for what concerns the diversifications within the Aedini lineages. For example, Group A diversified during the Cretaceous using mitochondrial (and also concatenated) data, but is much younger (Paleogene) using nuclear data (Table 2 for details). Even more discrepant is the origin of *Aedes* species: *A. aegypti* and *A. albopictus* split ranges from 81 (61–101) MYA using mitochondrial data, to 30 (15–45) MYA, using nuclear data; the split between *A. albopictus* and *A. flavopictus* is 32 (20–47) MYA using mitochondrial data and just 4 (0.5–11) MYA using nuclear data. Worryingly, those estimates do not overlap at their confidence interval. From a statistical point of view, this indicates that the two datasets reject each other. Overall, the estimates from the concatenated dataset are more similar to those of the mtDNA dataset then the nDNA dataset (Figure 4C–E).

Table 2. Divergence estimates of selected nodes from Figure 3 and other analyses. For each node, we provide the mean and the 95% high posterior density. On the right column, we provide estimates from previous studies.

Node	Taxonomic Level	Concatenated Dataset Figure 3	No Outgroup (Concatenated)	Nuclear Data Figure 4A	Mitochondrial Data Figure 4B	Others: Reidenbach09; Soghigian17 *; Da Silva 20 #; Chen 15 ^
a	Diptera	257 (223–294)		261 (225–296)	258 (224–293)	260 (239–295) ^
b	Culicidae split (Culicinae origin)	180 (137–228)	100 (50–185)	178 (113–245)	182 (143–223)	216 (229–192) 182 # 218 (181–260) ^
c	Culicinae split	146 (108–182)	92 (41–139)	139 (92–194)	150 (118–184)	204 (226–172) 130 # 179 (148–217) ^
d		137 (103–173)	86 (38–127)	123 (79–171)	135 (104–164)	
e	Aedini split (Aedes origin)	113 (83–143)	64 (34–122)	92 (55–137)	111 (95–150)	123 (155–90) 125 * 102 #
f	*Aedes* split (Clades A-B split)	105 (77–133)	57 (28–110)	69 (42–103)	107 (85–133)	92 (123–61) 102 *
g	Clade A split	99 (72–126)	51 (24–100)	49 (29–76)	96 (73–118)	
h		83 (59–109)	50 (22–93)	36 (20–57)	92 (71–116)	
i	*Stegomya* (*A. aegypti–A. albopictus*) split	73 (50–96)	36 (14–70)	27 (15–45)	81 (61–102)	55 * 67 # 71 (44–107) ^
j	*A. albopictus–A. flavopictus* split	28 (14–43)	36 (14–70)	3.7 (0.1–11.2)	33 (20–46)	25 *
l	*A. koreicus–A. japonicas* split	32 (15–51)	14 (3–31)	3.6 (0.2–10.9)	46 (24–71)	20 *

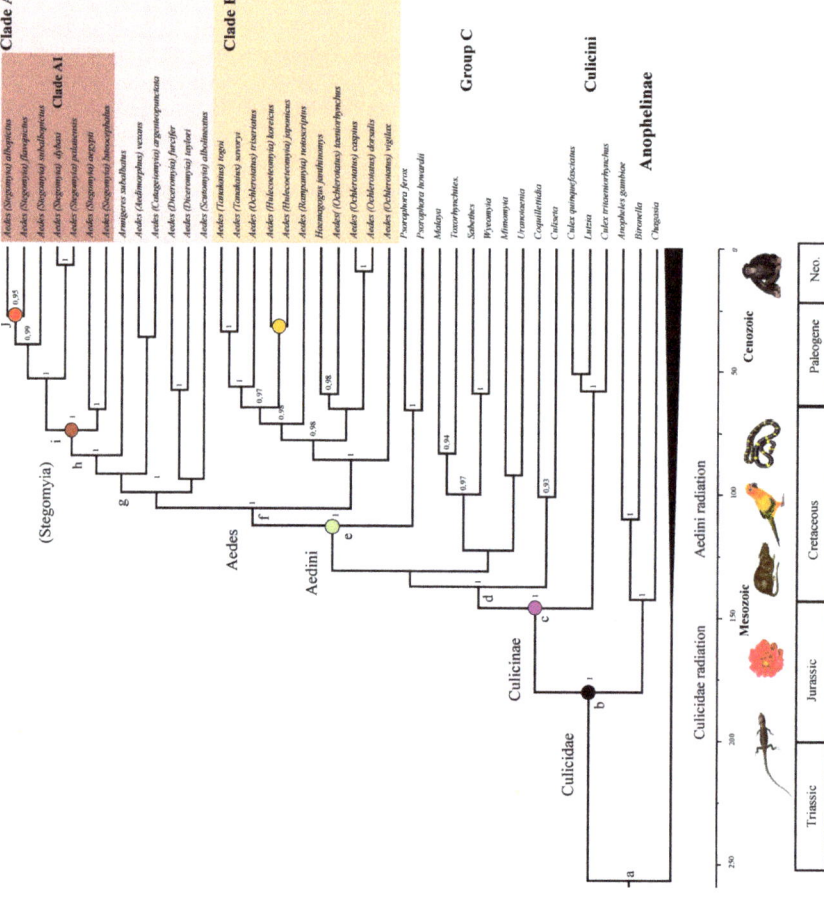

Figure 3. A Bayesian estimates of Aedini divergence. Posterior consensus tree from the analysis of the concatenated dataset. The two shaded distributions highlight the distribution of the 95% HPD for the origin of the Culicinae (b node) and the split of the Aedini (e node): for precise estimates and the 95% HPD see Table 2. Supports at nodes are posterior probabilities higher than 0.95. Time is in millions of years before the present.

We further inspected the posterior rates of both the nDNA and the mtDNA trees (Figure 5A,B, respectively; tree topologies are similar to those of Figure 1B,C). Additionally, in the case of rates there are various discrepancies between the two datasets: *A. furcifer* and *A. taylori* are, for example, fast-evolving according to mtDNA, but slow evolving using nDNA. These high rates are likely responsible for the dubious position of these two species in the mtDNA tree of Figure 5B. Our clock analyses returned mean posterior evolutionary rate (calculated over the whole tree) of 1.01×10^{-3} (sd = 1.12×10^{-4}) mutation per site per millions of years (msm) for the mtDNA and of 9.93×10^{-4} msm (sd = 1.8×10^{-4}) for the nDNA.

3.5. Mitochondrial-Nuclear Chronological Incongruences Are Consistent over Different Analytical Condition

We tested the robustness of the chronological incongruence observed between mitochondrial and nuclear data (Figure 4) by verifying if the results are biased by the taxon sampling and the number of gaps in our alignments. We first repeated the clock analyses using a reduced version of our dataset. We excluded three species (*Chagasia*, *Uranotenia*, *Aedes albolineatus*) which had an extremely different branching position in the phylogeny of the nDNA and mtDNA analysis. Results are very similar compared to when using the full dataset for what concerns both the divergence estimates (Figure S5) and the average mutation rate at branches (Figure S6). This indicates that the chronological discrepancy is not due to the presence of rough taxa in the dataset. We then tested if the pattern we observe is due to a particular taxon and site sampling by repeating the analyses using a different dataset. We inferred divergence estimates using separately the nuclear and mitochondrial partitions of the Soghigian alignment, derived from [21] and previously used for Figure 1C. This dataset is characterized by a different taxon representation compared to our dataset (it is centred on *Aedes* and contains few outgroups) and by a higher site representation (contains a lower amount of missing data, see methods for details). Results (Figure S7) provide a similar picture to when using our nDNA and mtDNA. Divergences closer to the root are similar, but those within *Aedes*, including the diversification of Clade A and Clade B are very different. This indicates that the chronological discrepancy is not due to peculiar taxon or gene sampling nor is affected by the amount of missing data in the datasets.

Because of its higher mutation rate, MtDNA is, in general, more prone to saturation than the nuclear genome [49]. Accordingly, we would expect to underestimate the number of observed mutations in mtDNA dataset compared to the nDNA one with the consequences that nodes using mtDNA dataset should appear younger than they are. We observe, however, exactly the opposite. Saturation and heterogeneity of the replacement pattern may have nevertheless played a certain role in overestimating the mitochondrial age in our mitochondrial phylogeny. We therefore tested our datasets for saturation by inferring divergences under the CAT model, a mixture model known to be less sensitive to systematic error in the presence of site-specific saturations [45]. The CAT trees are indeed slightly different than those obtained using homogeneous models of replacement (Figure S8). The divergences become more similar between the two datasets, but the nDNA dataset consistently return younger age for recent nodes compared to the mtDNA dataset. We conclude that site heterogeneity is only partially responsible for the mitochondrial–nuclear chronological discrepancy.

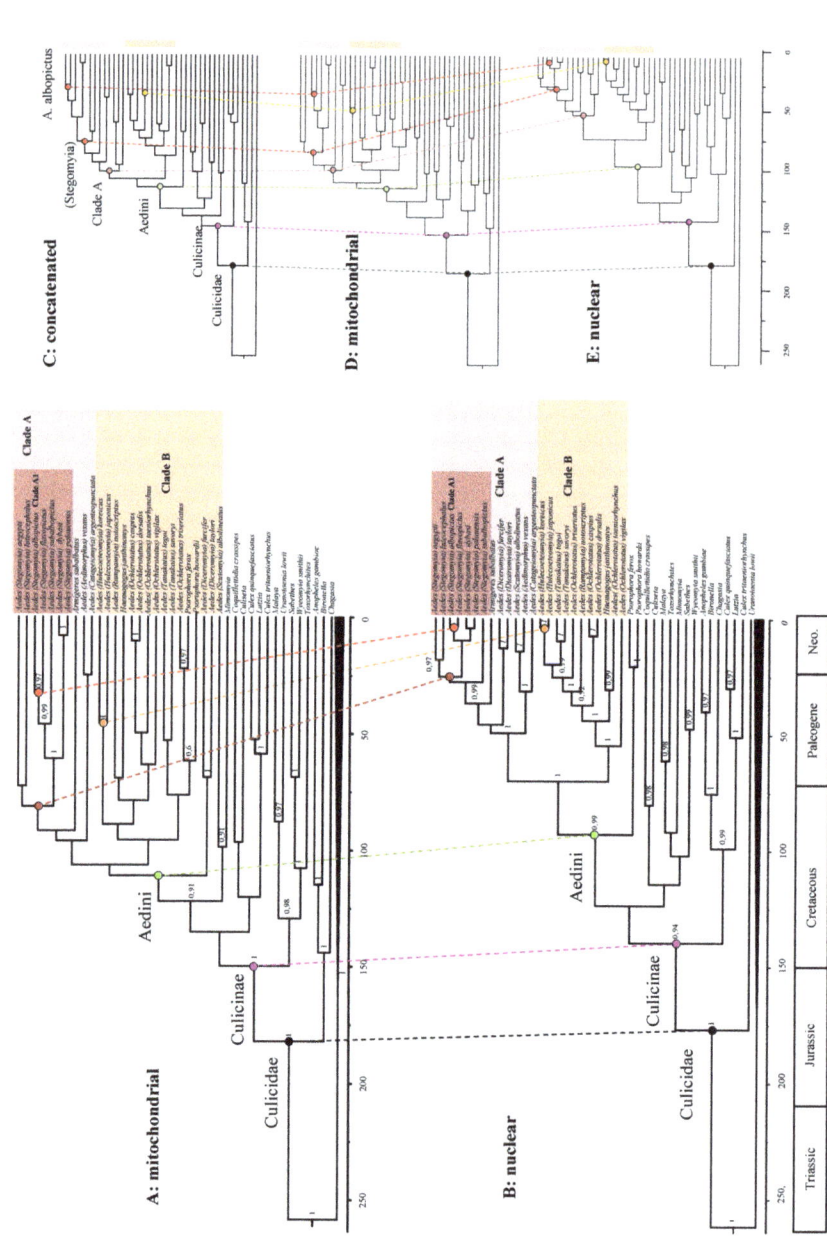

Figure 4. Chronological incongruence between mitochondrial and nuclear data. Note that while posterior estimates are similar for ancient nodes, there are strong incongruences for recent nodes. (**A**): posterior consensus tree from the analysis of the mtDNA dataset. (**B**): posterior consensus tree from the analysis of the nDNA dataset. (**C**–**E**): The concatenated, the mitochondrial, and the nuclear trees simplified for comparison. Supports at nodes are posterior probabilities higher than 0.95. Time is in millions of years before the present.

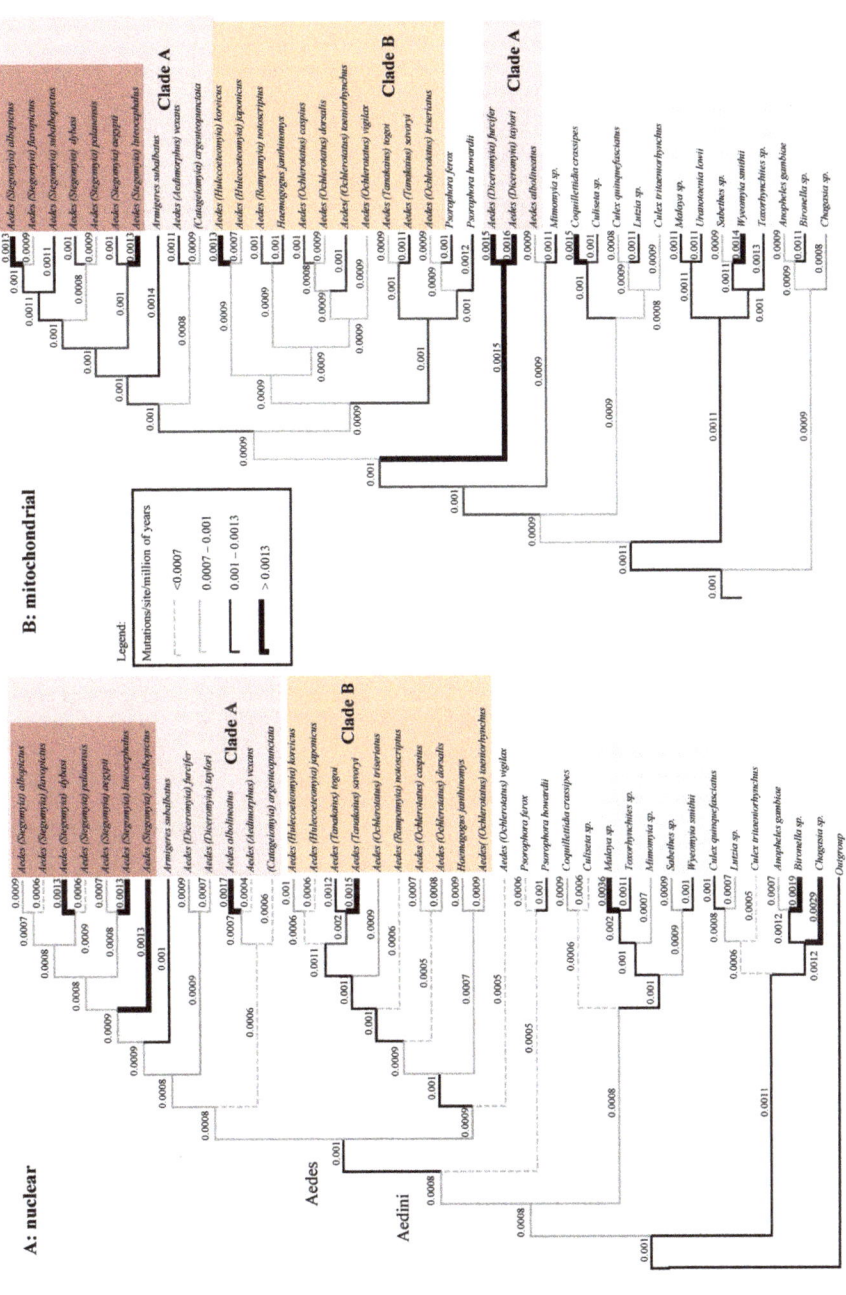

Figure 5. High degree of rate heterogeneity between nuclear and mitochondrial data. (**A**): the nDNA Bayesian phylogeny with mean posterior rates plotted on branches. (**B**): the Bayesian mtDNA phylogeny with mean posterior rates plotted on branches. Note that there are local accelerations of rate (bold lines) in certain taxa only in one of the two data types.

4. Discussion

Our phylogeny of Aedini using a concatenated dataset of eight mtDNA and nDNA markers (Figure 1A) recovers, at least for most nodes, a robustly supported tree topology. Our comparison of mtDNA and nDNA datasets revealed however some unexpected highly supported phylogenetic discrepancies (Figure 1B,C). We suggest three explanations for these incongruences. The first is wrong taxonomic assignment during field collection. Accordingly, one or more genes for some species may come from another (similar and mistaken for) species creating conflicting phylogenetic signal and wrong tree topology. Another, in our opinion less likely, explanation involves complex evolutionary events, such as past hybridization between species, which have resulted in different inheritance patterns for either the mtDNA or some regions of the nDNA. The final explanation is the stochasticity embedded in small (four genes) datasets, such as the ones we have used. The stochasticity in the mtDNA tree may have been exacerbated by systematic errors related to the fast-evolving nature of the mtDNA [49,50] and to evident high level of apomorphies, as revealed by the longer terminal branches in the mtDNA tree compared with the nDNA one. The different phylogenetic signal, however, does not seem to relate to the amount of missing data as the mitochondrial alignment is more complete than the nuclear one (38% of missing data in mtDNA vs. 45% in nDNA). Whatever the source of the topological discrepancies between datasets, our result point toward the limitation of a PCR-scaled approach for Aedini phylogeny and point toward future studies based on whole mtDNA and genome-scaled nDNA dataset. Indeed, undetected stochastic and systematic type of errors may also affect the concatenated dataset, as we have shown that the phylogenetic signal is unstable at many nodes when employing different replacement models and statistical frameworks (Figure 2B and Figure S4). In particular, the poor support using heterogeneous models may be due to the ability of this model to detect saturated or fast-evolving sites [45]. Under this scenario, the highly supported Clade B when using homogeneous GTR model (Figures 1A and 2A) may be the result of a systematic error. The various phylogenetic incongruences we observe using different replacement models (Figure 2A,B) reinforce what we have found when comparing nuclear and mitochondrial data. They alert us of possible systematic and stochastic errors. We advocate to adopt a cautious, conservative way in interpreting our (but also other available [21,31]) trees of the Aedini based on few genetic makers, as seemingly high supports (as in our Figure 1A) are not consistent over data type (Figure 1B,C and Figure 2C), method of inference (Figure 2A) or replacement model (Figure 2B). In perspective, our data indicate that the phylogeny of Aedini should be resolved with confidence only using a genome-scaled nuclear and a complete mtDNA dataset as done in other dipteran studies [24].

Our divergence estimates using both the concatenated and the mtDNA and nDNA datasets (Figures 3 and 4) are concordant in indicating that mosquito radiated from the mid-Jurassic on and that Aedini radiation started in the mid-Cretaceous, quite concomitant with the origin and the earliest diversifications of mammals first, and later birds during the Cretaceous. We cautiously speculate that there may have been a general history of co-radiation (the available data do not provide enough evidence to advocate co-evolution) between the Aedini and warm-blooded vertebrates. In support of this hypothesis, the Aedini group has a specific preference for mammals and birds [51]. The fact that a relaxed clock better fits our Aedini concatenated dataset is not surprising considering that a large variety of ecologically characterizes mosquitoes and demographic habits [21], which can be responsible for different generation times and therefore different branch rates [24]. The mean posterior rate for the mtDNA dataset is 1.01×10^{-3} msm, higher than the 9.93×10^{-4} msm estimated for the nuclear genes. The higher mutation rate of mitochondrial genes is expected as the mtDNA is well known to evolve faster than the nuclear genome in animals [52]. Our mean mtDNA rate estimates are, however, circa one order of magnitude smaller than the mitochondrial COI rate of coleopterans (1.17×10^{-2} msm) inferred by Papadopoulou et al. [53]. This can be explained by different timespan between the latter and our dataset. Indeed, shallow phylogenetic studies consistently provide with faster evolutionary rate than deep phylogenies [54,55]. Our nuclear rate estimates are in line instead with those inferred over long phylogenetic distances using Ecdysozoa nuclear data (mean 1.01×10^{-3} msm, [56], but lower than those based on mitochondrial *Drosophila* data (7.9×10^{-3}) [24]; this indicates that mosquitoes may have been characterized during their radiation by an overall smaller number of generations per year compared to *Drosophila*. We found that the nDNA data of most lineages within Clade A evolves faster than in the lineages of Clade B; this patter is less marked, but conserved in the mtDNA data (Figure 5). A possible explanation for this pattern is that species of Clade A have in general more generation per year than those of Clade B. The two important invasive *Aedes* species, *A. albopictus*, and *A. koreicus* are characterized by markedly higher replacement rate if compared with their respective sister species, *A. flavopictus* and *A. japonicus*; this pattern can be observed for both mitochondrial and nuclear data. Assuming that the instantaneous mutation rate is conserved within the genus, this result suggests that *A. albopictus* and *A. koreicus* are characterized by a higher number of generations per years compared to other closely related *Aedes*, a hypothesis which may at least partially explain their high invasive potential.

Our analyses revealed a consistent chronological incongruence between the phylogenetic signal of nuclear and mitochondrial genes. mtDNA provides divergence times within Aedini significantly older than nDNA. Previous clock studies in insects have shown poor [24] to moderate [30,57] incongruence between nuclear and mitochondrial data. In these analyses mitochondrial and nuclear estimates, although different, were overlapping for what concerns their 95% HPD. In our phylogenies, the 95% HPD do not overlap, indicating a statistically significant incongruence. We have shown that these incongruences do not depend on rough taxa (compare Figure 4 and Figure S5), nor on-site occupancy and gene sampling (compare with Figure S7), although there is a mitigation of the discrepancies when using a heterogeneous model of replacement (compare with Figure S8). We conclude that the mtDNA–nDNA chronological incongruence in Aedini data does not depend on analytical conditions, although the correct interpretation of saturation in both datasets, particularly the mtDNA one may play a certain role. Based on our results, we cannot exclude that there may have been a long history of multiple hybridization events within *Aedes* species which have affected the mitochondrial genome differently than the nuclear one. Indeed, complex phylogenetic signal due to multiple hybridization events has been recently shown in the *Anopheles* mosquitos [58,59]. The observed discrepancies prevent from drawing a conclusion on the actual timing of diversification of model organisms such as *A. albopictus* whose mean split from sister species *A. flavopictus* may dramatically range from 32 MYA using mitochondrial to just 4 MYA using nuclear data. On the light of these results, we advocate that future research should concentrate on determining the biological (or methodological) reason of this discrepancy by comparing timetrees from whole mtDNA genomes with those from genome-scaled sampling of nuclear genes.

In conclusion, we have provided here a detailed analysis of the phylogenetic and chronological signal in currently available nuclear and mitochondrial genes of the Aedini. Overall, our data point toward the limitation of a multigene PCR-scaled approach for Aedini phylogeny and indicate that future research should be based on genome scaled data. Probably our most interesting finding is the strong chronological incongruence between the nuclear and the mitochondrial data. We could exclude various possible misleading factors such as taxa assignment, missing data, and saturation (Figures S5–S7), but could not ultimately test a stochastic effect related to using only eight genes. This is because at present there is not enough data in databases to build a taxon-rich genome-scaled dataset centred on Aedini. The incongruences we have identified do not currently allow defining the exact timing of evolution of important model organisms, such as *A. aegypti* and *A. albopictus* [60]. We advocate that these chronological incongruences should be investigated in future by comparing whole mitogenomes with genome-scaled nuclear data as we have done for example, in *Drosophila* [30].

Supplementary Materials: The following are available online at https://www.mdpi.com/2075-1729/11/3/181/s1.

Author Contributions: O.R.-S. designed the study. N.Z. and O.R.-S. performed phylogenetic and molecular clock analyses, O.R.-S., N.Z. and A.R. interpreted the results, O.R.-S. and N.Z. wrote the article. All authors have read and agreed to the published version of the manuscript.

Funding: This research received no external funding.

Institutional Review Board Statement: Not applicable.

Informed Consent Statement: Not applicable.

Data Availability Statement: All data generated for this study are included in this article and its supplementary information file.

Conflicts of Interest: The authors declare no conflict of interest.

References

1. Reinert, J.F.; Harbach, R.E.; Kitching, I.J. Phylogeny and Classification of Aedini (Diptera: Culicidae), Based on Morphological Characters of All Life Stages. *Zool. J. Linn. Soc.* **2004**, *142*, 289–368. [CrossRef]
2. Pombi, M.; Montarsi, F. Mosquitoes (Culicidae). In *Reference Module in Biomedical Sciences*; Elsevier: Amsterdam, The Netherlands, 2020.
3. Wilkerson, R.C.; Linton, Y.M.; Fonseca, D.M.; Schultz, T.R.; Price, D.C.; Strickman, D.A. Making Mosquito Taxonomy Useful: A Stable Classification of Tribe Aedini That Balances Utility with Current Knowledge of Evolutionary Relationships. *PLoS ONE* **2015**, *10*, e0133602. [CrossRef] [PubMed]
4. Taylor, M.J.; Hoerauf, A.; Bockarie, M. Lymphatic Filariasis and Onchocerciasis. *Lancet* **2010**, *376*, 1175–1185. [CrossRef]
5. Pfeffer, M.; Dobler, G. Emergence of Zoonotic Arboviruses by Animal Trade and Migration. *Parasites Vectors* **2010**, *3*, 1–15. [CrossRef] [PubMed]
6. Silverj, A.; Rota-Stabelli, O. On the correct interpretation of similarity index in codon usage studies: Comparison with four other metrics and implications for Zika and West Nile virus. *Virus Res.* **2020**, *286*, 198097. [CrossRef]
7. Kraemer, M.U.G.; Sinka, M.E.; Duda, K.A.; Mylne, A.Q.N.; Shearer, F.M.; Barker, C.M.; Moore, C.G.; Carvalho, R.G.; Coelho, G.E.; Van Bortel, W.; et al. The Global Distribution of the Arbovirus Vectors Aedes Aegypti and Ae. Albopictus. *eLife* **2015**, *4*, e08347. [CrossRef]
8. Medlock, J.M.; Hansford, K.M.; Schaffner, F.; Versteirt, V.; Hendrickx, G.; Zeller, H.; Van Bortel, W. A Review of the Invasive Mosquitoes in Europe: Ecology, Public Health Risks, and Control Options. *Vector-Borne Zoonotic Dis.* **2012**, *12*, 435–447. [CrossRef]
9. Cameron, E.C.; Wilkerson, R.C.; Mogi, M.; Miyagi, I.; Toma, T.; Kim, H.C.; Fonseca, D.M. Molecular Phylogenetics of Aedes Japonicus, a Disease Vector That Recently Invaded Western Europe, North America, and the Hawaiian Islands. *J. Med. Entomol.* **2010**, *47*, 527–535. [CrossRef] [PubMed]
10. Medlock, J.M.; Hansford, K.M.; Versteirt, V.; Cull, B.; Kampen, H.; Fontenille, D.; Hendrickx, G.; Zeller, H.; Van Bortel, W.; Schaffner, F. An Entomological Review of Invasive Mosquitoes in Europe. *Bull. Entomol. Res.* **2015**, *105*, 637–663. [CrossRef] [PubMed]
11. Grard, G.; Moureau, G.; Charrel, R.N.; Holmes, E.C.; Gould, E.A.; de Lamballerie, X. Genomics and Evolution of Aedes-Borne Flaviviruses. *J. Gen. Virol.* **2010**, *91*, 87–94. [CrossRef] [PubMed]
12. Schaffner, F.; Chouin, S.; Guilloteau, J. First Record of Ochlerotatus (Finlaya) Japonicus Japonicus (Theobald, 1901) in Metropolitan France. *J. Am. Mosq. Control Assoc.* **2003**, *19*, 1–5. [PubMed]
13. Faria, N.R.; Quick, J.; Claro, I.M.; Thézé, J.; De Jesus, J.G.; Giovanetti, M.; Kraemer, M.U.G.; Hill, S.C.; Black, A.; Da Costa, A.C.; et al. Establishment and Cryptic Transmission of Zika Virus in Brazil and the Americas. *Nature* **2017**, *546*, 406–410. [CrossRef] [PubMed]
14. Chen, X.-G.; Jiang, X.; Gu, J.; Xu, M.; Wu, Y.; Deng, Y.; Zhang, C.; Bonizzoni, M.; Dermauw, W.; Vontas, J.; et al. Genome Sequence of the Asian Tiger Mosquito, Aedes Albopictus, Reveals Insights into Its Biology, Genetics, and Evolution. *Proc. Natl. Acad. Sci. USA* **2015**, *112*, e5907–e5915. [CrossRef] [PubMed]
15. Matthews, B.J.; Dudchenko, O.; Kingan, S.B.; Koren, S.; Antoshechkin, I.; Crawford, J.E.; Glassford, W.J.; Herre, M.; Redmond, S.N.; Rose, N.H.; et al. Improved Reference Genome of Aedes Aegypti Informs Arbovirus Vector Control. *Nature* **2018**, *563*, 501–507. [CrossRef]
16. Dritsou, V.; Topalis, P.; Windbichler, N.; Simoni, A.; Hall, A.; Lawson, D.; Hinsley, M.; Hughes, D.; Napolioni, V.; Crucianelli, F.; et al. A Draft Genome Sequence of an Invasive Mosquito: An Italian Aedes Albopictus. *Pathog. Glob. Health* **2015**, *109*, 207–220. [CrossRef] [PubMed]
17. Seidel, B.; Montarsi, F.; Huemer, H.P.; Indra, A.; Capelli, G.; Allerberger, F.; Nowotny, N. First Record of the Asian Bush Mosquito, Aedes Japonicus Japonicus, in Italy: Invasion from an Established Austrian Population. *Parasites Vectors* **2016**, *9*, 284. [CrossRef]

18. Montarsi, F.; Drago, A.; Martini, S.; Calzolari, M.; De Filippo, F.; Bianchi, A.; Mazzucato, M.; Ciocchetta, S.; Arnoldi, D.; Baldacchino, F.; et al. Current Distribution of the Invasive Mosquito Species, Aedes Koreicus [Hulecoeteomyia Koreica] in Northern Italy. *Parasites Vectors* **2015**, *8*, 1–5. [CrossRef]
19. Capelli, G.; Drago, A.; Martini, S.; Montarsi, F.; Soppelsa, M.; Delai, N.; Ravagnan, S.; Mazzon, L.; Schaffner, F.; Mathis, A.; et al. First Report in Italy of the Exotic Mosquito Species Aedes (Finlaya) Koreicus, a Potential Vector of Arboviruses and Filariae. *Parasites Vectors* **2011**, *4*, 188. [CrossRef]
20. Huber, K.; Jansen, S.; Leggewie, M.; Badusche, M.; Schmidt-Chanasit, J.; Becker, N.; Tannich, E.; Becker, S.C. Aedes Japonicus Japonicus (Diptera: Culicidae) from Germany Have Vector Competence for Japan Encephalitis Virus but Are Refractory to Infection with West Nile Virus. *Parasitol. Res.* **2014**, *113*, 3195–3199. [CrossRef]
21. Soghigian, J.; Andreadis, T.G.; Livdahl, T.P. From Ground Pools to Treeholes: Convergent Evolution of Habitat and Phenotype in Aedes Mosquitoes. *BMC Evol. Biol.* **2017**, *17*, 262. [CrossRef]
22. Ramasamy, S.; Ometto, L.; Crava, C.M.; Revadi, S.; Kaur, R.; Horner, D.S.; Pisani, D.; Dekker, T.; Anfora, G.; Rota-Stabelli, O. The Evolution of Olfactory Gene Families in Drosophila and the Genomic Basis of Chemical-Ecological Adaptation in Drosophila Suzukii. *Genome Biol. Evol.* **2016**, *8*, 2297–2311. [CrossRef]
23. Crava, C.M.; Brütting, C.; Baldwin, I.T. Transcriptome Profiling Reveals Differential Gene Expression of Detoxification Enzymes in a Hemimetabolous Tobacco Pest after Feeding on Jasmonate-Silenced Nicotiana Attenuata Plants. *BMC Genom.* **2016**, *17*, 1005. [CrossRef]
24. Ometto, L.; Cestaro, A.; Ramasamy, S.; Grassi, A.; Revadi, S.; Siozios, S.; Moretto, M.; Fontana, P.; Varotto, C.; Pisani, D.; et al. Linking Genomics and Ecology to Investigate the Complex Evolution of an Invasive Drosophila Pest. *Genome Biol. Evol.* **2013**, *5*, 745–757. [CrossRef]
25. Rota-Stabelli, O.; Ometto, L.; Tait, G.; Ghirotto, S.; Kaur, R.; Drago, F.; González, J.; Walton, V.M.; Anfora, G.; Rossi-Stacconi, M.V. Distinct Genotypes and Phenotypes in European and American Strains of Drosophila Suzukii: Implications for Biology and Management of an Invasive Organism. *J. Pest Sci.* **2020**, *93*, 77–89. [CrossRef]
26. Feuda, R.; Dohrmann, M.; Pett, W.; Philippe, H.; Rota-Stabelli, O.; Lartillot, N.; Wörheide, G.; Pisani, D. Improved Modeling of Compositional Heterogeneity Supports Sponges as Sister to All Other Animals. *Curr. Biol.* **2017**, *27*, 3864–3870.e4. [CrossRef]
27. Hirano, T.; Saito, T.; Tsunamoto, Y.; Koseki, J.; Ye, B.; Do, V.T.; Miura, O.; Suyama, Y.; Chiba, S. Enigmatic Incongruence between MtDNA and NDNA Revealed by Multi-Locus Phylogenomic Analyses in Freshwater Snails. *Sci. Rep.* **2019**, *9*, 6223. [CrossRef]
28. Near, T.J.; Eytan, R.I.; Dornburg, A.; Kuhn, K.L.; Moore, J.A.; Davis, M.P.; Wainwright, P.C.; Friedman, M.; Smith, W.L. Resolution of Ray-Finned Fish Phylogeny and Timing of Diversification. *Proc. Natl. Acad. Sci. USA* **2012**, *109*, 13698–13703. [CrossRef] [PubMed]
29. Zheng, Y.; Peng, R.; Kuro-O, M.; Zeng, X. Exploring Patterns and Extent of Bias in Estimating Divergence Time from Mitochondrial DNA Sequence Data in a Particular Lineage: A Case Study of Salamanders (Order Caudata). *Mol. Biol. Evol.* **2011**, *28*, 2521–2535. [CrossRef]
30. Wahlberg, N.; Weingartner, E.; Warren, A.D.; Nylin, S. Timing Major Conflict between Mitochondrial and Nuclear Genes in Species Relationships of Polygonia Butterflies (Nymphalidae: Nymphalini). *BMC Evol. Biol.* **2009**, *9*, 92. [CrossRef] [PubMed]
31. Reidenbach, K.R.; Cook, S.; Bertone, M.A.; Harbach, R.E.; Wiegmann, B.M.; Besansky, N.J. Phylogenetic Analysis and Temporal Diversification of Mosquitoes (Diptera: Culicidae) Based on Nuclear Genes and Morphology. *BMC Evol. Biol.* **2009**, *9*, 298. [CrossRef] [PubMed]
32. Da Silva, A.F.; Machado, L.C.; de Paula, M.B.; da Silva Pessoa Vieira, C.J.; de Morais Bronzoni, R.V.; de Melo Santos, M.A.V.; Wallau, G.L. Culicidae Evolutionary History Focusing on the Culicinae Subfamily Based on Mitochondrial Phylogenomics. *Sci. Rep.* **2020**, *10*, 18823. [CrossRef]
33. Reisen, W.K. Update on Journal Policy of Aedine Mosquito Genera and Subgenera. *J. Med. Entomol.* **2016**, *53*, 249. [CrossRef] [PubMed]
34. Abascal, F.; Zardoya, R.; Telford, M.J. TranslatorX: Multiple Alignment of Nucleotide Sequences Guided by Amino Acid Translations. *Nucleic Acids Res.* **2010**, *38*, 1–31. [CrossRef]
35. Katoh, K.; Rozewicki, J.; Yamada, K.D. MAFFT Online Service: Multiple Sequence Alignment, Interactive Sequence Choice and Visualization. *Brief. Bioinform.* **2018**, *20*, 1160–1166. [CrossRef] [PubMed]
36. Kück, P.; Meusemann, K. FASconCAT: Convenient Handling of Data Matrices. *Mol. Phylogenet. Evol.* **2010**, *56*, 1115–1118. [CrossRef] [PubMed]
37. Stamatakis, A. RAxML Version 8: A Tool for Phylogenetic Analysis and Post-Analysis of Large Phylogenies. *Bioinformatics* **2014**, *30*, 1312–1313. [CrossRef] [PubMed]
38. Lartillot, N.; Lepage, T.; Blanquart, S. PhyloBayes 3: A Bayesian Software Package for Phylogenetic Reconstruction and Molecular Dating. *Bioinformatics* **2009**, *25*, 2286–2288. [CrossRef] [PubMed]
39. Bouckaert, R.; Vaughan, T.G.; Barido-Sottani, J.; Duchêne, S.; Fourment, M.; Gavryushkina, A.; Heled, J.; Jones, G.; Kühnert, D.; De Maio, N.; et al. BEAST 2.5: An Advanced Software Platform for Bayesian Evolutionary Analysis. *PLoS Comput. Biol.* **2019**, *15*, 1–28. [CrossRef]
40. Benton, M.J.; Donoghue, P.C.J. Paleontological Evidence to Date the Tree of Life. *Mol. Biol. Evol.* **2007**, *24*, 26–53. [CrossRef]
41. Misof, B.; Liu, S.; Meusemann, K.; Peters, R.S.; Donath, A.; Mayer, C.; Frandsen, P.B.; Ware, J.; Flouri, T.; Beutel, R.G.; et al. Phylogenomics Resolves the Timing and Pattern of Insect Evolution. *Science* **2014**, *346*, 763–767. [CrossRef] [PubMed]

42. Borkent, A.; Grimaldi, D.A. The Earliest Fossil Mosquito (Diptera: Culicidae), in Mid-Cretaceous Burmese Amber. *Ann. Entomol. Soc. Am.* **2004**, *97*, 882–888. [CrossRef]
43. Bouckaert, R.; Heled, J.; Kühnert, D.; Vaughan, T.; Wu, C.H.; Xie, D.; Suchard, M.A.; Rambaut, A.; Drummond, A.J. BEAST 2: A Software Platform for Bayesian Evolutionary Analysis. *PLoS Comput. Biol.* **2014**, *10*, 1–6. [CrossRef] [PubMed]
44. Lartillot, N.; Philippe, H. A Bayesian Mixture Model for Across-Site Heterogeneities in the Amino-Acid Replacement Process. *Mol. Biol. Evol.* **2004**, *21*, 1095–1109. [CrossRef] [PubMed]
45. Rota-Stabelli, O.; Daley, A.C.; Pisani, D. Molecular Timetrees Reveal a Cambrian Colonization of Land and a New Scenario for Ecdysozoan Evolution. *Curr. Biol.* **2013**, *23*, 392–398. [CrossRef] [PubMed]
46. Drummond, A.J.; Ho, S.Y.W.; Phillips, M.J.; Rambaut, A. Relaxed Phylogenetics and Dating with Confidence. *PLoS Biol.* **2006**, *4*, 699–710. [CrossRef]
47. Richards, S.; Murali, S.C. Best Practices in Insect Genome Sequencing: What Works and What Doesn't. *Curr. Opin. Insect Sci.* **2015**, *7*, 1–7. [CrossRef] [PubMed]
48. Van der Kooi, C.J.; Ollerton, J. The Origins of Flowering Plants and Pollinators. *Science* **2020**, *368*, 1306–1308. [CrossRef]
49. Rota-Stabelli, O.; Telford, M.J. A Multi Criterion Approach for the Selection of Optimal Outgroups in Phylogeny: Recovering Some Support for Mandibulata over Myriochelata Using Mitogenomics. *Mol. Phylogenet. Evol.* **2008**, *48*, 103–111. [CrossRef] [PubMed]
50. Bernt, M.; Braband, A.; Middendorf, M.; Misof, B.; Rota-Stabelli, O.; Stadler, P.F. Bioinformatics Methods for the Comparative Analysis of Metazoan Mitochondrial Genome Sequences. *Mol. Phylogenet. Evol.* **2013**, *69*, 320–327. [CrossRef]
51. Reeves, L.E.; Holderman, C.J.; Blosser, E.M.; Gillett-Kaufman, J.L.; Kawahara, A.Y.; Kaufman, P.E.; Burkett-Cadena, N.D. Identification of Uranotaenia Sapphirina as a Specialist of Annelids Broadens Known Mosquito Host Use Patterns. *Commun. Biol.* **2018**, *1*, 1–8. [CrossRef]
52. Rota-Stabelli, O.; Kayal, E.; Gleeson, D.; Daub, J.; Boore, J.L.; Telford, M.J.; Pisani, D.; Blaxter, M.; Lavrov, D.V. Ecdysozoan Mitogenomics: Evidence for a Common Origin of the Legged Invertebrates, the Panarthropoda. *Genome Biol. Evol.* **2010**, *2*, 425–440. [CrossRef] [PubMed]
53. Papadopoulou, A.; Anastasiou, I.; Vogler, A.P. Revisiting the Insect Mitochondrial Molecular Clock: The Mid-Aegean Trench Calibration. *Mol. Biol. Evol.* **2010**, *27*, 1659–1672. [CrossRef]
54. Ho, S.Y.W.; Lanfear, R.; Bromham, L.; Phillips, M.J.; Soubrier, J.; Rodrigo, A.G.; Cooper, A. Time-Dependent Rates of Molecular Evolution. *Mol. Ecol.* **2011**, *20*, 3087–3101. [CrossRef] [PubMed]
55. Ho, S.Y.W.; Lo, N. The Insect Molecular Clock. *Aust. J. Entomol.* **2013**, *52*, 101–105. [CrossRef]
56. Guidetti, R.; McInnes, S.J.; Cesari, M.; Rebecchi, L.; Rota-Stabelli, O. Evolutionary Scenarios for the Origin of an Antarctic Tardigrade Species Based on Molecular Clock Analyses and Biogeographic Data. *Contrib. Zool.* **2017**, *86*, 97–110. [CrossRef]
57. Andújar, C.; Serrano, J.; Gómez-Zurita, J. Winding up the Molecular Clock in the Genus Carabus (Coleoptera: Carabidae): Assessment of Methodological Decisions on Rate and Node Age Estimation. *BMC Evol. Biol.* **2012**, *12*, 40. [CrossRef] [PubMed]
58. Thawornwattana, Y.; Dalquen, D.; Yang, Z. Coalescent Analysis of Phylogenomic Data Confidently Resolves the Species Relationships in the Anopheles Gambiae Species Complex. *Mol. Biol. Evol.* **2018**, *35*, 2512–2527. [CrossRef]
59. Foster, P.G.; de Oliveira, T.M.P.; Bergo, E.S.; Conn, J.E.; Sant'Ana, D.C.; Nagaki, S.S.; Nihei, S.; Lamas, C.E.; González, C.; Moreira, C.C.; et al. Phylogeny of Anophelinae Using Mitochondrial Protein Coding Genes. *R. Soc. Open Sci.* **2017**, *4*, 170758. [CrossRef]
60. Palatini, U.; Masri, R.A.; Cosme, L.V.; Koren, S.; Thibaud-Nissen, F.; Biedler, J.K.; Krsticevic, F.; Johnston, J.S.; Halbach, R.; Crawford, J.E.; et al. Improved Reference Genome of the Arboviral Vector Aedes Albopictus. *Genome Biol.* **2020**, *21*, 215. [CrossRef] [PubMed]

Article

Improving Phylogenetic Signals of Mitochondrial Genes Using a New Method of Codon Degeneration

Xuhua Xia [1,2]

1. Department of Biology, University of Ottawa, 30 Marie Curie, Ottawa, ON K1N 6N5, Canada; xxia@uottawa.ca
2. Ottawa Institute of Systems Biology, 451 Smyth Road, Ottawa, ON K1H 8M5, Canada

Received: 8 July 2020; Accepted: 27 August 2020; Published: 30 August 2020

Abstract: Recovering deep phylogeny is challenging with animal mitochondrial genes because of their rapid evolution. Codon degeneration decreases the phylogenetic noise and bias by aiming to achieve two objectives: (1) alleviate the bias associated with nucleotide composition, which may lead to homoplasy and long-branch attraction, and (2) reduce differences in the phylogenetic results between nucleotide-based and amino acid (AA)-based analyses. The discrepancy between nucleotide-based analysis and AA-based analysis is partially caused by some synonymous codons that differ more from each other at the nucleotide level than from some nonsynonymous codons, e.g., Leu codon TTR in the standard genetic code is more similar to Phe codon TTY than to synonymous CTN codons. Thus, nucleotide similarity conflicts with AA similarity. There are many such examples involving other codon families in various mitochondrial genetic codes. Proper codon degeneration will make synonymous codons more similar to each other at the nucleotide level than they are to nonsynonymous codons. Here, I illustrate a "principled" codon degeneration method that achieves these objectives. The method was applied to resolving the mammalian basal lineage and phylogenetic position of rheas among ratites. The codon degeneration method was implemented in the user-friendly and freely available DAMBE software for all known genetic codes (genetic codes 1 to 33).

Keywords: codon degeneration; phylogenetic conflict; mtDNA; deep phylogeny; ratite

1. Introduction

In a multiple sequence alignment, there are historical signals, such as the number of nucleotide substitutions, that are typically proportional to the divergence time, and non-historical signals that are typically not proportional to the divergence time [1]. Non-historical signals include compositional bias [2–4] and conflicting signals between codons and amino acids (AAs) [2,3,5,6]. Codon degeneration, when properly implemented, can minimize or eliminate these non-historical signals in aligned sequences [7]. I will outline these two sources of undesirable signals, detail a codon degeneration method, and apply the method to mammalian and avian mitochondrial sequences to resolve (1) the phylogeny of basal eutherian lineages and (2) the phylogenetic position of rheas among ratites.

1.1. Nucleotide Composition Bias

Nucleotide composition bias refers to the phenomenon in which distantly related taxa share similar nucleotide frequencies, leading to a spurious similarity between such taxa [8,9]. The problem is particularly serious when one aims to construct a universal tree [10].

Nucleotide composition bias can arise via either shared selection or shared mutation. For example, to stabilize the stem-loop secondary structure in their rRNAs, thermophilic bacteria tend to have not only GC-rich stems but also longer stems, regardless of their phylogenetic affinity [11]. Such convergent evolution due to shared high ambient temperature has long been identified as a potential source of homoplasy [12]. In particular, mesophiles, such as *Deinococcus* and *Bacillus* species, are relatively AU-rich in their 16S rRNA, in contrast to thermophiles, such as *Aquifex, Thermotoga,* and *Thermus* species (Figure 1A). Foster [13] used this example to illustrate the importance of accommodating such composition heterogeneity in phylogenetics. Conventional phylogenetic methods that do not accommodate such compositional heterogeneity consistently group the two mesophiles together (Figure 1B). However, when additional parameters are allowed to model lineage-specific nucleotide frequencies, a correct topology (Figure 1C) was recovered. For this reason, much effort has been spent in the search for efficient methods to model a nonstationary substitution process [13–18].

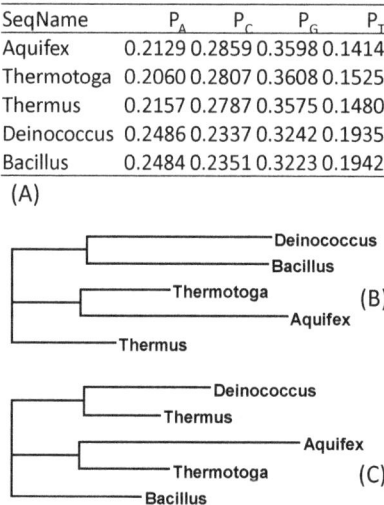

SeqName	P_A	P_C	P_G	P_T
Aquifex	0.2129	0.2859	0.3598	0.1414
Thermotoga	0.2060	0.2807	0.3608	0.1525
Thermus	0.2157	0.2787	0.3575	0.1480
Deinococcus	0.2486	0.2337	0.3242	0.1935
Bacillus	0.2484	0.2351	0.3223	0.1942

(A)

Figure 1. Similarity in nucleotide frequencies between *Deinococcus* and *Bacillus* favors them being clustered together. (**A**) Nucleotide frequencies of 16S rRNA from five prokaryotes [13]. (**B**) The same phylogenetic tree produced from software PhyML [19] with GTR with or without using a gamma distribution to accommodate the rate heterogeneity. (**C**) After degenerating the sequences to purines and pyrimidines, the correct phylogenetic tree was recovered from PhyML with the same options.

Nucleotide and codon degeneration [7,20,21] offers a simple alternative to modeling a nonstationary substitution process. After degenerating the sequences into purine and pyrimidine, the correct tree was recovered by using either likelihood methods or a distance-based method using GTR or simpler substitution models (Figure 1C). Note that using LogDet [8] and paralinear [9] distances, as implemented in DAMBE [22], invariably led to the wrong tree in Figure 1B instead of that in Figure 1C when the sequences were not degenerated. Thus, LogDet and paralinear distances do not accommodate compositional heterogeneity as claimed.

Another selection-mediated source of composition bias is tRNA-mediated selection on codon usage bias. Bacteriophages in *Escherichia coli* exhibit similar codon usage to its host genes (especially highly expressed ones), regardless of their phylogenetic affinity, presumably to take advantage of differential tRNA availability in the host tRNA pool [23–25]. Such tRNA-mediated composition bias also occurs in mitochondrial sequences. For example, some bivalve and chordate species have two Met tRNAs, tRNA$^{Met/CAU}$ and tRNA$^{Met/UAU}$, where CAU and UAU are anticodons, to translate Met AUG and AUA codons. In contrast, most other species only have a single tRNA$^{Met/CAU}$ to translate both Met codons AUG and AUA, where the nucleotide C in the first anticodon site is modified to pair with both A and G [26]. The independent gain of tRNA$^{Met/UAU}$ has resulted in the convergent increase of AUA codon usage in bivalve and chordate species [27,28].

In addition to responding to the shared selection, composition bias can also arise through shared mutation bias [29]. Diverse parasitic bacterial lineages are almost invariably AT-rich [30,31], presumably because spontaneous mutations tend to be AT-biased [32,33]. This also occurs in ancient DNA with differential nucleotide decay [34]. Mitochondrial [35] and bacterial nuclear genomes [36] both exhibit strand bias, where a gene that has switched strand during evolution experiences dramatically different mutation spectra and accumulates substitutions rapidly, leading to an extraordinarily long branch involving the strand-switched gene [27]. In particular, composition bias often changes direction rapidly [37,38], and is therefore not proportional to time.

1.2. Conflicting Signal between Nucleotide and Amino Acid Sequences

There are often phylogenetic conflicts between nucleotide-based and AA-based analyses [2,3,5]. Several codon families contribute to this discrepancy, and I will illustrate this with five examples. First, nearly all genetic codes (except for genetic codes 3 and 23) have TTR and CTN encoding amino acid Leu, and TTY encoding amino acid Phe (Figure 2A), where R stands for purine, N for any nucleotide, and Y for pyrimidine. Leu codons TTA and TTG are more similar to Phe codons TTC and TTT than to synonymous Leu codons CTC and CTT. If we use the match/mismatch score matrix in Figure 2F, the alignment score between nonsynonymous Leu codon TTR and Phe codon TTY is 15 but the alignment score between the synonymous TTR and CTY is only 0 (Figure 2A). The alignment score is an index of sequence similarity. Two aligned sequences with a large alignment score are more similar to each other than two with a small alignment score. Two sequences with a high alignment score are also expected to have a smaller evolutionary distance between them than two sequences with a low alignment score. Thus, at the AA level, sequences L1 to L6 (Figure 2A) are identical but differ from sequences F1 and F2. However, at the nucleotide level, L5 and L6 are more similar to F1 and F2 than to L2 and L4 (Figure 2A).

Second, in most genetic codes (except for genetic codes 2, 5, 9, 13, 14, 21, 24, and 33), AGR and CGN encode Arg, and AGY encodes Ser. Arg codon AGR is more similar to Ser codons AGY (with an alignment score of 30; Figure 2B) than to synonymous Arg codons CGC and CGT (with an alignment score of −30; Figure 2B). Again, at the AA level, sequences R1 to R6 are identical to each other but differ from S1 and S2 (Figure 2B). In contrast, at the nucleotide level, sequences R5 and R6 are more similar to S1 and S2 than they are to R2 and R4 (Figure 2B), leading to conflicts between the AA signals and nucleotide signals.

Examples 3 to 5 are from specialized genetic codes. In genetic code 25 (Candidate Division SR1 and Gracilibacteria Code), TGA is a Gly codon instead of a stop codon, as in the standard code. This Gly codon TGA is more similar to Trp codon TGG (with an alignment score of 60; Figure 2C) than to other synonymous Gly codons, with an alignment score varying from −30 to 30 (Figure 2C). In genetic codes 24 (Rhabdopleuridae Mitochondrial Code) and 33 (Cephalodiscidae Mitochondrial UAA-Tyr Code), AGG is a Lys codon, which is more similar to the Ser codon AGA (alignment score = 60) than to the synonymous codon AAA (alignment score = 30; Figure 2D). Finally, in genetic code 13, Gly codons AGA and AGG are more similar to Ser codon AGC and AGT, with an alignment score of 30, than to synonymous Gly codons GGC and GGT (with an alignment score of 0; Figure 2E). We need to find a codon degeneration method that will ideally achieve the objective of finding synonymous codons in Figure 2 that are more similar to each other than they are to nonsynonymous codons. If we designate a minimum similarity between synonymous codons as $S_{min.S}$ (which equals 0 for Leu codons in Figure 2A) and a maximum similarity between nonsynonymous codons as $S_{max.NS}$ (which equals 60 between Leu codons and Phe codons in Figure 2A), we wish to have a codon degeneration method that ideally yields $S_{min.S} > S_{max.NS}$.

The codon degeneration method I present here was previously implemented by myself for the standard genetic code [7]. I extended the implementation to support all known genetic codes summarized in Xia [39], plus two more recent codes, including 14 genetic codes that are specific for mitochondrial genomes. I applied the codon degeneration method to the analysis of mitochondrial sequences to (1) resolve basal eutherian lineages and (2) elucidate phylogenetic placement of rheas among ratites.

```
L1 CTA              (A)         R1 CGA                    (B)
L2 CTC  30                      R2 CGC  30
L3 CTG  60  30                  R3 CGG  60  30
L4 CTT  30  60  30              R4 CGT  30  60  30
L5 TTA  60   0  30   0          R5 AGA  30 -30   0 -30
L6 TTG  30   0  60   0  60      R6 AGG   0 -30  30 -30  60
F1 TTC   0  60   0  30  30  30  S1 AGC -30  30 -30   0  30  30
F2 TTT   0  30   0  60  30  60  S2 AGT -30   0 -30  30  30  60

G1 GGA              (C)         K1 AAA                    (D)
G2 GGC  30                      K2 AAG  60
G3 GGG  60  30                  K3 AGG  30  60
G4 GGT  30  60  30              S1 AGA  60  30  60
G5 TGA  30 -30   0 -30
W1 TGG   0 -30  30 -30  60                                (F)
                                   A   C   G   U   R   Y   M   K   ?
                                A  30 -30   0 -30  15 -30   0 -15  -8
G1 GGA                          C -30  30 -30   0 -30  15   0 -15  -8
G2 GGC  30          (E)         G   0 -30  30 -30  15 -30 -15   0  -8
G3 GGG  60  30                  U -30   0 -30  30 -30  15 -15   0  -8
G4 GGT  30  60  30              R  15 -30  15 -30  15 -30  -8  -8  -8
G5 AGA  60   0  30   0          Y -30  15 -30  15 -30  15  -8  -8  -8
G6 AGG  30   0  60   0  60      M   0   0 -15 -15  -8  -8   0 -15  -8
S1 AGC   0  60   0  30  30  30  K -15 -15   0   0  -8  -8 -15   0  -8
S1 AGT   0  30   0  60  30  60  ?  -8  -8  -8  -8  -8  -8  -8  -8  -8
```

Figure 2. Synonymous codon families in which some synonymous codons are less similar to each other than they are to some nonsynonymous codons. (**A**) Leu codons TTA and TTG are more similar to Phe codons TTC and TTT than to synonymous Leu codons CTC and CTT. (**B**) Arg codons AGA and AGG are more similar to Ser codons AGC and AGT than to synonymous Arg codons CGC and CGT. (**C**) In genetic code 25, the Gly codon TGA is more similar to the Trp codon TGG than to other synonymous Gly codons. (**D**) In genetic codes 24 and 33, Lys codon AGG is more similar to Ser codon AGA than to the synonymous Lys codon AAA. (**E**) In genetic code 13, Gly codons AGA and AGG are more similar to Ser codons AGC and AGT than to synonymous Gly codons GGC and GGT. (**F**) The match/mismatch matrix for producing the alignment scores in (**A**–**E**). The scores involving ambiguous codes are averages, e.g., the score for A/R is the average of A/A and A/G = (30 + 0)/2 = 15.

2. Materials and Methods

2.1. The "Principled" Codon Degeneration and Two Alternatives

One could perform three different types of codon degeneration to alleviate the problems caused by composition bias and conflict signals between nucleotide and AA sequences. First, we may just degenerate the third codon position (Figure 3), which has been used to alleviate the phylogenetic bias caused by divergent sequences with similar nucleotide compositions [40–42]. For example, the sequences from arthropod taxa [20] differ significantly in the nucleotide frequencies, with GC content at the third codon position ($GC_3\%$) varying from 37.88% to 80.42% in the three ostracods and from 24.10% to 64.40% in arachnids [7]. The degeneration also reduced conflicting signals between nucleotide similarity and AA similarity. Take the Leu and Phe codons Figure 3A for example. $S_{min.S} = 22$ and $S_{max.NS} = 30$. Although this falls short of achieving $S_{min.S} > S_{max.NS}$, it is much better than the nondegenerated case where $S_{min.S} = 0$ and $S_{max.NS} = 60$ (Figure 2A). The $S_{min.S}$ and $S_{max.NS}$ values for all five illustrated cases in Figures 2 and 3 are listed in columns 3 and 4 in Table 1. The difference between $S_{min.S}$ and $S_{max.NS}$ have all changed in the right direction, i.e., $S_{min.S}$ has increased and $S_{max.NS}$ has mostly decreased (Table 1).

The second type of codon degeneration, which we previously named "principled" [7], degenerates the two Leu codons TTA and TTG to YTR and Leu codons CTA, CTG, CTC, and CTT to CTN (Figure 4A). The conceptual principle is that any codon degeneration should not lead to a degenerated codon losing its AA identity. In other words, a degenerated codon should never include nonsynonymous codons.

The operational principle of this codon degeneration is that for two synonymous codon subfamilies of different sizes (e.g., one with four codons and the other with two, as in the case of Leu codons in Figure 2A), codon positions 1 or 2 are degenerated only in the smaller codon subfamily. For example, CTN includes four codons and TTR includes two codons (i.e., the smaller of the two); therefore, the first codon position of only the smaller TTR family is degenerated to YTR. Note that degenerating the first codon position of CTN further to YTN, as in Regier et al. [20], would violate the conceptual principle because YTN encompasses both Leu and Phe codons. If two codon subfamilies differ in both the first and second codon positions, then they are treated as separate codon families and degenerated independent of each other. For example, Ser codons in the standard code is degenerated into AGY and UCN because any further degeneration would violate the conceptual principle.

```
   L1 CT?                  (A)          R1 CG?                              (B)
   L2 CT?  52                           R2 CG?  52
   L3 CT?  52 52                        R3 CG?  52 52
   L4 CT?  52 52 52                     R4 CG?  52 52 52
   L5 TTR  22 22 22 22                  R5 AGR  -8 -8 -8 -8
   L6 TTR  22 22 22 22 75               R6 AGR  -8 -8 -8 -8 75
   F1 TTY  22 22 22 22 30 30            S1 AGY  -8 -8 -8 -8 30 30
   F2 TTY  22 22 22 22 30 30 75         S2 AGY  -8 -8 -8 -8 30 30 75

   G1 GG?                  (C)          K1 AAR                              (D)
   G2 GG?  52                           K2 AAR  75
   G3 GG?  52 52                        K3 AGG  45 45
   G4 GG?  52 52 52                     S1 AGA  45 45 60
   G5 TGA  -8 -8 -8 -8                                                      (F)
   W1 TGG  -8 -8 -8 -8 60          | A   C   G   U   R   Y   M   K   ?
                                 A | 30 -30  0 -30  15 -30  0 -15 -8
                                 C |-30  30 -30  0 -30  15  0 -15 -8
   G1 GG?                               G |  0 -30  30 -30  15 -30 -15  0  -8
   G2 GG?  52                  (E)      U |-30  0 -30  30 -30  15 -15  0  -8
   G3 GG?  52 52                        R | 15 -30  15 -30  15 -30  -8 -8  -8
   G4 GG?  52 52 52                     Y |-30  15 -30  15 -30  15  -8 -8  -8
   G5 AGR  22 22 22 22                  M |  0   0 -15 -15  -8  -8   0 -15 -8
   G6 AGR  22 22 22 22 75               K |-15 -15  0   0  -8  -8 -15  0  -8
   S1 AGY  22 22 22 22 30 30            ? | -8  -8  -8  -8  -8  -8  -8  -8 -8
   S1 AGY  22 22 22 22 30 30 75
```

Figure 3. Degenerating only the third codon position. (**A–F**) The same as in Figure 2, but with the third codon site degenerated and the alignment scores recalculated.

Table 1. $S_{min.S}$ (minimum of the sequence similarity between synonymous codons) and $S_{max.NS}$ (maximum of the sequence similarity between nonsynonymous codons) for the illustrated codon families in Figure 2.

Case [1]	Similarity [2]	No [3]	Third Only [4]	Principled [5]
(A)	$S_{min.S}$	0	22	37
	$S_{max.NS}$	60	30	22
(B)	$S_{min.S}$	−30	−8	22
	$S_{max.NS}$	30	30	0
(C)	$S_{min.S}$	−30	−8	22
	$S_{max.NS}$	60	60	30
(D)	$S_{min.S}$	30	45	60
	$S_{max.NS}$	60	60	45
(E)	$S_{min.S}$	0	22	37
	$S_{max.NS}$	60	30	22

[1] (A–E) refers to the five illustrated cases in Figure 2. [2] Nucleotide similarity as measured using the alignment score computed with the score matrix in Figure 3F. [3] No codon degeneration. [4] Degenerated third codon positions only. [5] "Principled" degeneration.

```
L1 CT?                    (A)         R1 CG?                            (B)
L2 CT? 52                             R2 CG? 52
L3 CT? 52 52                          R3 CG? 52 52
L4 CT? 52 52 52                       R4 CG? 52 52 52
L5 YTR 37 37 37 37                    R5 MGR 22 22 22 22
L6 YTR 37 37 37 37 60                 R6 MGR 22 22 22 22 45
F1 TTY 22 22 22 22 15 15              S1 AGY -8 -8 -8 -8  0  0
F2 TTY 22 22 22 22 15 15 75           S2 AGY -8 -8 -8 -8  0  0 75

G1 GG?                    (C)         K1 AAR                            (D)
G2 GG? 52                             K2 AAR 75
G3 GG? 52 52                          K3 ARG 60 60
G4 GG? 52 52 52                       S1 AGA 45 45 45
G5 KGA 22 22 22 22
W1 TGG -8 -8 -8 -8 30                                                   (F)
```

	A	C	G	U	R	Y	M	K	?
A	30	-30	0	-30	15	-30	0	-15	-8
C	-30	30	-30	0	-30	15	0	-15	-8
G	0	-30	30	-30	15	-30	-15	0	-8
U	-30	0	-30	30	-30	15	-15	0	-8
R	15	-30	15	-30	15	-30	-8	-8	-8
Y	-30	15	-30	15	-30	15	-8	-8	-8
M	0	0	-15	-15	-8	-8	0	-15	-8
K	-15	-15	0	0	-8	-8	-15	0	-8
?	-8	-8	-8	-8	-8	-8	-8	-8	-8

```
G1 GG?
G2 GG? 52                 (E)
G3 GG? 52 52
G4 GG? 52 52 52
G5 RGR 37 37 37 37
G6 RGR 37 37 37 37 60
S1 AGY 22 22 22 22 15 15
S1 AGY 22 22 22 22 15 15 75
```

Figure 4. "Principled" codon degeneration. (**A–F**) The same as in Figure 2, but with "principled" degeneration and recalculated alignment scores.

The Leu codons YTR and CTN, as well as the Phe codon family TTY, were degenerated according to this "principled" codon degeneration (Figure 4A). Similarly, the two Arg subfamilies (Figure 2B) were degenerated to CGN and MGR (Figure 4B). Note that only the smaller subfamily has the first codon position degenerated to M (standing for either A or C). The two Gly subfamilies (Figure 4C) were degenerated to GGN and KGA (where K stands for either G or T). Again, only the smaller codon family has its first codon position degenerated. The two Lys subfamilies, one with two codons and one with a single codon AGG (Figure 2D), were degenerated to AAR and ARG (Figure 4D), with only the smaller codon subfamily (i.e., AGG) having its second codon position degenerated. Degenerating the larger subfamily AAR further to ARR would have violated the conceptual principle because ARR encompasses not only the three Lys codons but also the Ser codon AGA (Figure 4D). The two Gly subfamilies (Figure 4E) were degenerated to GGN and RGR. In short, the conceptual principle is maintained by sticking to the operational principle of degenerating the first or second codon position of the smaller codon subfamily. This codon degeneration function can be accessed in DAMBE by clicking "Sequences|Sequence manipulation|Degenerate synonymous codons".

The benefit of the "principled" codon degeneration is clearly visible in $S_{min.S}$ and $S_{max.NS}$ (last column in Table 1) or by contrasting Figures 3 and 4. The nucleotide similarities among synonymous codons have consistently increased and similarities between nonsynonymous codons have consistently increased. For example, before the codon degeneration, Leu codons TTA and TTG had a nucleotide similarity of 0 to synonymous Leu codons CTC and CTT, which is much smaller than that to the two nonsynonymous Phe codons TTC and TTT ($S_{max.NS} = 60$; Table 1). After the "principled" codon degeneration, $S_{min.S}$ increased to 37 and $S_{max.NS}$ decreased to 22. Thus, the nucleotide similarity and AA similarity are no longer conflicting.

The third type of codon degeneration is used in Regier et al. [20] and violates the conceptual principles above. For example, it degenerates all six Leu codons in Figure 2A to YTN. This obscures the differences between the six Leu codons and the two Phe codons (TTY) because YTN encompasses both. With the "principled" codon degeneration, synonymous Leu codons are all more similar to each other than they are to the two nonsynonymous Phe codons (Figure 4A), with $S_{min.S} = 37$ and $S_{max.NS} = 22$ (Table 1). This third type of codon degeneration results in all six Leu codons and the two Phe codons having the same nucleotide similarity with $S_{min.S} = S_{max.NS} = 37$. The same is true for the Arg codon family in Figure 2B. Regier et al. [20] degenerated all six Arg codons to MGN, which again violates the conceptual principle because MGN encompasses both the six Arg codons and the two Ser codons in Figure 2B. With the "principled" codon degeneration, the six synonymous Arg codons are more similar to each other than to the two Ser codons (Figure 4B), with $S_{min.S} = 22$ and $S_{max.NS} = 0$. The third type of codon degeneration renders $S_{min.S} = S_{max.NS} = 22$. Furthermore, note that the codon sequences can no longer be translated back into amino acid sequences after this third type of codon degeneration. Losing codon identity represents a significant loss of evolutionary information since we can no longer perform codon-based analysis. As pointed out before [7], Miyata's distance is 2.73 between Arg and Ser and 0.63 between Phe and Leu [43], and empirical data suggests that replacement with synonymous codons is more likely than between Arg and Ser or between Phe and Leu codons, according to Figure 13.1 in Xia [44]. Therefore, it is not a good idea to treat nonsynonymous codons as equivalent to synonymous codons.

2.2. Mitochondrial Data and Phylogenetic Analysis

I used two sets of mitochondrial sequences to evaluate the phylogenetic performance of the "principled" codon degeneration, addressing two phylogenetic problems. The first involved the resolution of basal eutherian lineages. I downloaded mitochondrial genomes from 11 mammalian species representing the basal eutherian lineages: *Sus scrofa* (mtDNA accession NC_000845), *Loxodonta africana* (NC_000934), *Equus caballus* (NC_001640), *Dasypus novemcinctus* (NC_001821), *Oryctolagus cuniculus* (NC_001913), *Artibeus jamaicensis* (NC_002009), *Orycteropus afer* (NC_002078), *Galeopterus variegatus* (NC_004031), *Mus musculus* (NC_005089), *Delphinus capensis* (NC_012061), and *Homo sapiens* (NC_012920). The 13 protein-coding genes were extracted using DAMBE [45]. Codon sequences were aligned against aligned AA sequences as an automated process in DAMBE using MAFFT [46,47] with the most accurate but slower LINSI option ("–localpair" and "–maxiterate = 1000"). Individually aligned sequences were then concatenated into the Supplemental File mammal_MAFFT_SuperMatrix.FAS. Individuals and their aligned lengths in the same order as in concatenated supermatrix were ATP6_ATP8 (882), COX1 (1551), COX2 (687), COX3 (783), CYTB (1143), ND1 (954), ND2 (1047), ND3 (351), ND4 (1377), ND4L (294), ND5 (1833), and ND6 (537).

The second phylogenetic problem was about the phylogenetic position of rhea in ratites. I used concatenated mitochondrial coding sequences from 11 mitochondrial genomes [48], including seven paleognathes: *Struthio camelus* (ostrich, GenBank ACCN: NC 002785), *Dromaius novaehollandiae* (emu, NC 002784), *Casuarius casuarius* (cassowary, NC 002778), *Apteryx haastii* (kiwi, NC 002782), *Dinornis giganteus* (extinct moa, NC 002672), *Rhea pennata* (rhea, NC 002783), *Eudromia elegans* (tinamou, NC 002772), and four neognathes: *Gallus gallus* (chicken, NC 001323), *Branta canadensis* (Canada goose, NC 007011), *Phoenicopterus roseus* (flamingo, NC 010089), and *Rhynochetos jubatus* (kagu, NC 010091). Coding sequences were extracted using DAMBE, individually aligned, and then concatenated. The supermatrix is included as the Supplemental File Bird_MAFFT_SuperMatrix.FAS.

PhyML [19] was used for phylogenetic reconstruction using the GTR+Γ substitution model (the best model based on likelihood ratio tests or information-theoretic indices). The tree improvement option "-s" was set to "BEST" (best of NNI and SPR searches). The "-o" option was set to "tlr", which optimized the topology, the branch lengths, and the rate parameters.

3. Results

3.1. Codon Degeneration Increased the Phylogenetic Resolution Power in Early Mammalian Lineages

The aligned mitochondrial sequences representing basal eutherian lineages were analyzed without (Figure 5A) and with the "principled" codon degeneration (Figure 5B). The two resulting topologies were identical. The topology was well corroborated using diverse data, and in particular, validated by the sharing of retroelements [49]. The two trees (Figure 5) differ in support values for internal nodes.

The tree generated using the "principled" codon degeneration (Figure 5B) had substantially higher support values for some nodes than the tree generated without codon degeneration (Figure 5A).

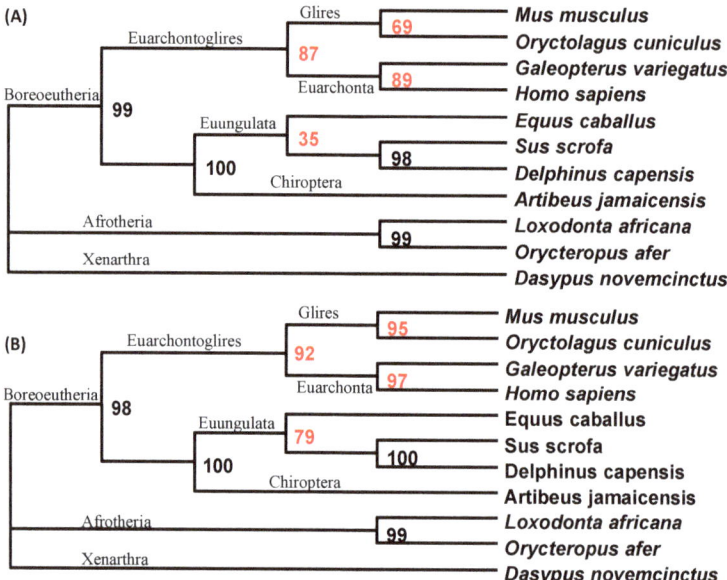

Figure 5. Codon degeneration improved the phylogenetic resolution of mammals. (**A**) PhyML results from sequences without codon degeneration. (**B**) PhyML results from sequences after the "principled" codon degeneration. The corresponding support values differing by ≥5% between (**A**,**B**) are highlighted in red.

3.2. Codon Degeneration Challenged the Conventional Phylogenetic Placement of Rhea

The phylogenetic position of flightless ratites and tinamous have recently been elucidated by using ancient DNA from museum specimens [50–52]. However, the phylogenetic position of rheas is inconsistent, being placed at the root or close to the root of ratites and tinamous in an analysis of mitochondrial genes [50,51], but closer to the emu–cassowary–kiwi clade in an analysis when nuclear genes are used [52]. Here, I show that the placement of rhea close to the root of the ratites and tinamous was due to composition bias that could be corrected using codon degeneration.

The phylogenetic analysis was again performed without (Figure 6A) and with the "principled" codon degeneration (Figure 6B). One may get an intuitive sense of the compositional bias by examining the proportion of nucleotide C (P_C), which is the most abundant nucleotide in these avian mitochondrial genes in the sequences. P_C, shown after each species name in Figure 6A, is higher in the four neognathes than the P_C in most paleognathes. *Rhea pennata* happens to have the highest P_C, which spuriously increases its sequence similarity to the four neognathes and pull it toward the root. Furthermore, the four neognathes encode Leu mostly using CUN instead of UUR, with $P_{CUN} = 0.8779$, which is higher than that for the species (excluding *Rhea pennata*) in paleognathes ($P_{CUN} = 0.8381$). *Rhea pennata*, because of its C-richness, has $P_{CUN} = 0.8841$. This increases the chance of Leu at a site being encoded by CUN in *Rhea pennata* + neognathes, but by UUR in other paleognathes, further increasing the spurious sequence similarity between *Rhea pennata* and the four neognathes.

The phylogeny from the codon-degenerated sequences (Figure 6B) has *Rhea pennata* clustered together with emu (*Dromaius novaehollandiae*), cassowary (*Casuarius casuarius*), and kiwi (*Apteryx haastii*). The phylogeny based on AA sequences translated from the coding sequences also had these four species forming a monophyletic cluster. Furthermore, the phylogenetic relationship from nucleotide sequences without compositional bias also suggested a closer relationship between rhea and the kiwi+cassowary+emu clade [52,53]. These multiple lines of evidence suggest that the placement of rhea close to the root of paleognaths [50,51] was due to compositional bias that could be corrected by codon degeneration. Note that the phylogenetic placement of *Rhea pennata* Figure 6B differs from recent publications [52,53] that have the phylogenetic positions of rhea and kiwi swapped. However, these two studies, albeit with an extensive data compilation and comprehensive data analysis, did not pay particular attention to composition bias, and simply asserted that the noncoding sequences they used were less subject to composition bias than coding sequences. Even if the assertion is true, it does not mean that noncoding sequences are immune to composition bias. There is strand-specific nucleotide bias in both nuclear and mitochondrial genomes [27,35,54] such that an inversion event leading to a sequence switching strands typically results in very different substitution patterns.

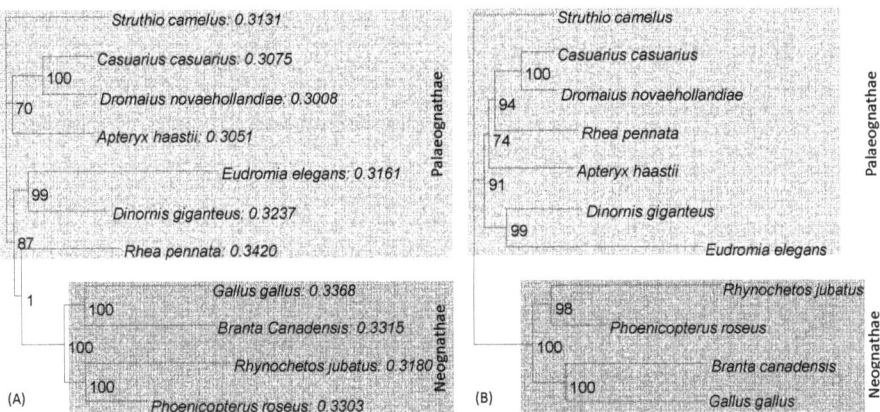

Figure 6. Codon degeneration removed the compositional bias. (**A**) PhyML results from sequences without the codon degeneration, leading to the wrong placement of *Rhea pennata*. The proportion of nucleotide C follows the species name. (**B**) PhyML results from sequences after the "principled" codon degeneration, which recovered the correct phylogeny.

The phylogeny in Figure 6B is consistent with continental vicariance, as illustrated in Figure 7. The geophylogeny (mapping of a phylogeny onto geographic locations) was drawn using PGT software [55]. In the late Cretaceous period (Figure 7, inset A), Africa was separated from South American + Antarctic + Australasia, isolating the ostrich from the rest of the paleognaths (Figure 7). The small and nocturnal ancestor of kiwi should have diverged from the ancestor of the large and diurnal cassowary+emu+rhea clade. The subsequent separation of South American from Antarctica + Australasia resulted in (1) isolation of the rhea lineage from the cassowary + emu lineage and (2) isolation of the tinamou lineage from the moa lineage (Figure 7).

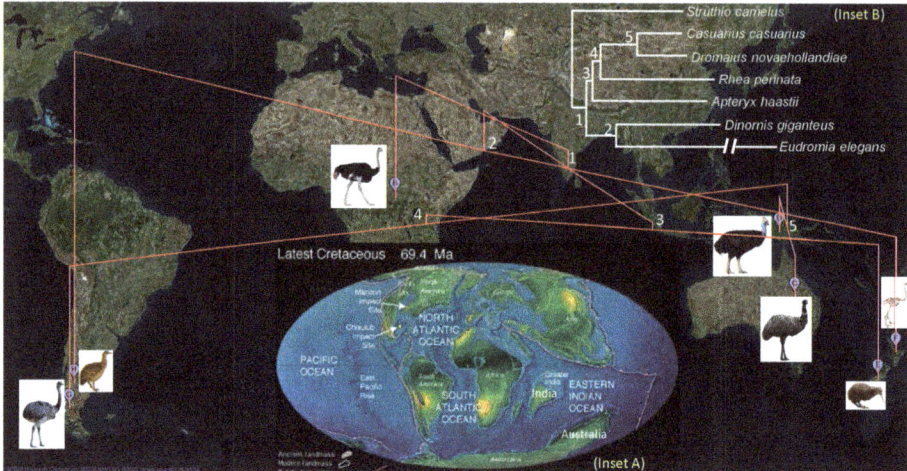

Figure 7. Geophylogeny of seven paleognathes, drawn using PGT software [55]. The geographic positions are approximate, with a single point representing a spatial distribution. The species images are from Wikipedia. Inset **A**: Cretaceous landmasses 69.4 million years ago (credit: US Geological Society, www.usgs.gov). Inset **B**: The phylogeny of the seven paleognathes, with node numbering identical to those on the geophylogeny.

4. Discussion

4.1. When to Use Codon Degeneration?

The codon degeneration method can alleviate compositional bias and reduce differences in phylogenetic analysis from nucleotide and AA sequences when used properly in two specific scenarios. The first scenario is when the composition bias is such that remotely related taxa share similar nucleotide frequencies than closely related species. For example, a GC-rich gene may encode amino acid Leu using CUG, but this codon may change into UUG in a closely related but AT-rich gene. Biased mutation can change directions quite rapidly [11,37,56,57]. Degenerating the third position would remove the difference caused by mutation bias between two closely related species. The second scenario is when the same AA site in a set of aligned sequences is encoded by two blocks of codons, as in the case of Leu codons and Arg codons in the standard genetic code. As is illustrated in Figure 4 and Table 1, the principled codon degeneration reduces the difference between synonymous codons such that the difference in phylogenetic results between the nucleotide-based and AA-based analysis is reduced.

Codon degeneration would not be appropriate when reconstructing the phylogeny of closely related species. For closely related species, sister taxa typically share similar nucleotide frequencies, which are consequently also phylogenetically informative. Furthermore, if Leu at each site is encoded by either UUR or CUN, but never both (and if Arg at each site is encoded by either AGR or CGN, but never both), then the benefit of codon degeneration would be minimal and may not offset the cost of lost information. Typically, only highly diverged sequences may benefit from codon degeneration.

4.2. "Principled" Degeneration versus Degenerating the Third Codon Site Only

The "principled" codon degeneration aims to achieve two objectives: (1) minimize the composition bias and (2) remove conflicting signals between the nucleotide and AA sequences. Degenerating the third codon position should achieve the first objective; therefore, it is interesting to compare the two degeneration methods. I have added phylogenetic results (Figure 8) from sequences degenerated at the third codon only (which should remove most of the composition heterogeneity but does not remove the conflicting signals between the nucleotide and AA sequences). The phylogeny in Figure 8A (with degeneration at the third codon sites only) was comparable to that in Figure 5B (with "principled" codon degeneration). The tree topologies were the same, and the only major difference was the support value of 71 (red in Figure 8A) versus the corresponding value of 92 in Figure 5B.

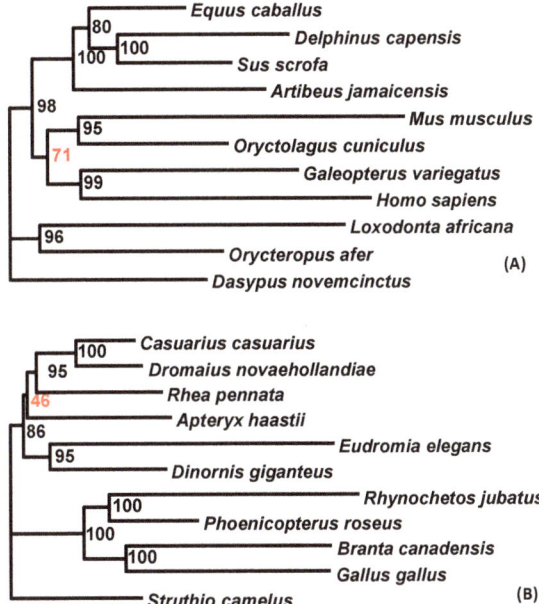

Figure 8. Phylogenetic reconstruction from sequences with only the third codon site degenerated. (**A**) The phylogeny from 11 mammalian species should be compared with the phylogeny in Figure 5B. (**B**) The phylogeny of 11 avian species should be compared with the phylogeny in Figure 6B.

The phylogeny in Figure 8B (with degeneration at the third codon sites only) was comparable to that in Figure 6B (with the "principled" codon degeneration). The topologies were again the same, with the only notable support value of 46 (red in Figure 8B) being substantially lower than the corresponding value of 74 in Figure 6B. The comparisons suggest that the "principled" codon degeneration was more preferable over degenerating the third codon sites only, although more empirical substantiation is needed.

4.3. The Serine Codon Family

The codon degeneration method cannot help with synonymous codons encoded by disjoint blocks of codons, such as Ser codons. Ser is encoded by TCN and AGY for most genetic codes. Sites with Ser codons may distort the phylogenetic signals at the nucleotide level if two closely related taxa happen to have TCN and AGY, respectively, at the same homologous codon site. No codon degeneration method makes two synonymous codons, such as TCN and AGY, more similar to each other than between two nonsynonymous codons, such as Ser codon AGY and Arg codon AGR.

The presence of TCN and AGY codons at the same codon site may cause conflict between nucleotide-based and AA-based analyses [2,3,5,6]. One way to avoid this problem is simply to remove such codon sites. DAMBE offers the option of simply removing sites containing both TCN and AGY codons before nucleotide-based analysis. The function can be accessed by clicking on "Sequence|Sequence manipulation|Remove sites with both UCN and AGY serine codons". The function also provides an option to keep only those codon sites containing UCN and AGY such that it can be checked whether they contribute significant phylogenetic signals. The concatenated mammalian coding sequences (Supplemental File mammal_MAFFT_Supermatrix.FAS) contain 40 codon sites with both UCN and AGY serine codons. I constructed a tree from these 40 codons. The tree shared only a single bipartition with the tree in Figure 5 out of eight bipartitions. This was similar to randomly generated sequences with the same nucleotide frequencies, indicating little information in those codon sites containing both UCN and AGY codons. However, removing these codon sites did not consistently increase the support values for the internal nodes (Figure 9). The phylogeny in Figure 9A was comparable to that in Figure 5A, with the former from sequences after removing the 40 codon sites featuring both UCN and AGY Ser codons, and the latter without removing them. No codon degeneration was done in both cases. The phylogeny in Figure 9B was comparable to that in Figure 5B, both with the "principled" degeneration but they differed in that the former removed the 40 codon sites and the latter did not. The topologies were all the same, and there was no consistent improvement in the support values in both comparisons.

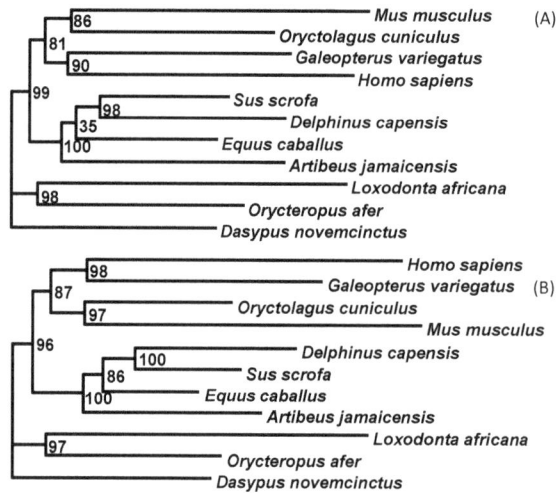

Figure 9. Phylogenetic reconstruction of 11 mammalian species after removing 40 codon sites featuring both UCN and AGY Ser codons: (**A**) without codon degeneration and (**B**) with "principled" codon degeneration.

The avian mitochondrial sequences (Supplemental File bird_MAFFT_Supermatrix.FAS) contained only 14 codon sites featuring both UCN and AGY codons. Removing them did not improve the phylogenetic resolution. Furthermore, there was variation in the Ser encoding in genetic codes 5, 9, 12, 14, 21, 24, and 33, i.e., the Ser codons were not limited to TCN and AGY, such that caution should be exercised when removing codon sites with both UCN and AGY codons because they may not be Ser codons.

5. Conclusions

Codon degeneration methods can improve the phylogenetic signals of highly divergent sequences. It should help to solve the difficult problem of resolving deep phylogeny.

Supplementary Materials: The following are available online at http://www.mdpi.com/2075-1729/10/9/171/s1, mammal_MAFFT_Supermatrix.FAS and bird_MAFFT_Supermatrix.FAS.

Funding: This research was funded by a Discovery Grant from the Natural Science and Engineering Research Council (NSERC, RGPIN/2018-03878) of Canada.

Acknowledgments: I thank L. Jermiin and A. Zwick for their comments and suggestions. Two anonymous reviewers provided excellent comments leading to significant improvement of the manuscript.

Conflicts of Interest: The author declares no conflict of interest. The funders had no role in the design of the study; in the collection, analyses, or interpretation of data; in the writing of the manuscript, or in the decision to publish the results.

References

1. Grundy, W.N.; Naylor, G.J. Phylogenetic inference from conserved sites alignments. *J. Exp. Zool.* **1999**, *285*, 128–139. [CrossRef]
2. Cox, C.J.; Li, B.; Foster, P.G.; Embley, T.M.; Civán, P. Conflicting phylogenies for early land plants are caused by composition biases among synonymous substitutions. *Syst. Biol.* **2014**, *63*, 272–279. [CrossRef] [PubMed]
3. Li, B.; Lopes, J.S.; Foster, P.G.; Embley, T.M.; Cox, C.J. Compositional Biases among Synonymous Substitutions Cause Conflict between Gene and Protein Trees for Plastid Origins. *Mol. Biol. Evol.* **2014**, *31*, 1697–1709. [CrossRef] [PubMed]
4. Criscuolo, A.; Gribaldo, S. BMGE (Block Mapping and Gathering with Entropy): A new software for selection of phylogenetic informative regions from multiple sequence alignments. *BMC Evol. Biol.* **2010**, *10*, 210. [CrossRef]
5. Rota-Stabelli, O.; Lartillot, N.; Philippe, H.; Pisani, D. Serine codon-usage bias in deep phylogenomics: Pancrustacean relationships as a case study. *Syst. Biol.* **2013**, *62*, 121–133. [CrossRef]
6. Zwick, A.; Regier, J.C.; Zwickl, D.J. Resolving Discrepancy between Nucleotides and Amino Acids in Deep-Level Arthropod Phylogenomics: Differentiating Serine Codons in 21-Amino-Acid Models. *PLoS ONE* **2012**, *7*, e47450. [CrossRef]
7. Noah, K.E.; Hao, J.; Li, L.; Sun, X.; Foley, B.; Yang, Q.; Xia, X. Major Revisions in Arthropod Phylogeny Through Improved Supermatrix, with Support for Two Possible Waves of Land Invasion by Chelicerates. *Evol. Bioinform.* **2020**, *16*, 1–12. [CrossRef]
8. Lockhart, P.J.; Steel, M.A.; Hendy, M.D.; Penny, D. Recovering evolutionary trees under a more realistic model of sequence evolution. *Mol. Biol. Evol.* **1994**, *11*, 605–612. [CrossRef]
9. Lake, J.A. Reconstructing evolutionary trees from DNA and protein sequences: Paralinear distances. In Proceedings of the Proceedings of the National Academy of Sciences. *Proc. Natl. Acad. Sci. USA* **1994**, *91*, 1455–1459. [CrossRef]
10. Forterre, P.; Benachenhou-Lafha, N.; Labedan, B. Universal tree of life. *Nature* **1993**, *362*, 795. [CrossRef]
11. Wang, H.C.; Xia, X.; Hickey, D.A. Thermal adaptation of ribosomal RNA genes: A comparative study. *J. Mol. Evol.* **2006**, *63*, 120–126. [CrossRef] [PubMed]
12. Weisburg, W.; Giovannoni, S.; Woese, C. The Deinococcus-Thermus Phylum and the Effect of rRNA Composition on Phylogenetic Tree Construction. *Syst. Appl. Microbiol.* **1989**, *11*, 128–134. [CrossRef]
13. Foster, P.G. Modeling Compositional Heterogeneity. *Syst. Biol.* **2004**, *53*, 485–495. [CrossRef]
14. Galtier, N.; Gouy, M. Inferring pattern and process: Maximum-likelihood implementation of a nonhomogeneous model of DNA sequence evolution for phylogenetic analysis. *Mol. Biol. Evol.* **1998**, *15*, 871–879. [CrossRef] [PubMed]
15. Galtier, N. A Nonhyperthermophilic Common Ancestor to Extant Life Forms. *Science* **1999**, *283*, 220–221. [CrossRef] [PubMed]
16. Blanquart, S.; Lartillot, N. A Bayesian Compound Stochastic Process for Modeling Nonstationary and Nonhomogeneous Sequence Evolution. *Mol. Biol. Evol.* **2006**, *23*, 2058–2071. [CrossRef] [PubMed]
17. Blanquart, S.; Lartillot, N. A Site- and Time-Heterogeneous Model of Amino Acid Replacement. *Mol. Biol. Evol.* **2008**, *25*, 842–858. [CrossRef]
18. Williams, B.A.P.; Cox, C.J.; Foster, P.G.; Szöllősi, G.J.; Embley, T.M. Phylogenomics provides robust support for a two-domains tree of life. *Nat. Ecol. Evol.* **2019**, *4*, 138–147. [CrossRef]
19. Guindon, S.; Gascuel, O. A simple, fast, and accurate algorithm to estimate large phylogenies by maximum likelihood. *Syst. Biol.* **2003**, *52*, 696–704. [CrossRef]
20. Regier, J.C.; Shultz, J.W.; Zwick, A.; Hussey, A.; Ball, B.; Wetzer, R.; Martin, J.W.; Cunningham, C.W. Arthropod relationships revealed by phylogenomic analysis of nuclear protein-coding sequences. *Nature* **2010**, *463*, 1079–1083. [CrossRef]

21. Ishikawa, S.A.; Inagaki, Y.; Hashimoto, T. RY-Coding and Non-Homogeneous Models Can Ameliorate the Maximum-Likelihood Inferences from Nucleotide Sequence Data with Parallel Compositional Heterogeneity. *Evol. Bioinform.* **2012**, *8*, 357–371. [CrossRef] [PubMed]
22. Xia, X. DAMBE6: New Tools for Microbial Genomics, Phylogenetics, and Molecular Evolution. *J. Hered.* **2017**, *108*, 431–437. [CrossRef] [PubMed]
23. Chithambaram, S.; Prabhakaran, R.; Xia, X. Differential Codon Adaptation between dsDNA and ssDNA Phages in Escherichia coli. *Mol. Biol. Evol.* **2014**, *31*, 1606–1617. [CrossRef] [PubMed]
24. Chithambaram, S.; Prabhakaran, R.; Xia, X. The effect of mutation and selection on codon adaptation in Escherichia coli bacteriophage. *Genetics* **2014**, *197*, 301–315. [CrossRef] [PubMed]
25. Prabhakaran, R.; Chithambaram, S.; Xia, X. Aeromonas phages encode tRNAs for their overused codons. *Int. J. Comput. Biol. Drug Des.* **2014**, *7*, 168–182. [CrossRef]
26. Grosjean, H.; De Crécy-Lagard, V.; Marck, C. Deciphering synonymous codons in the three domains of life: Co-evolution with specific tRNA modification enzymes. *FEBS Lett.* **2009**, *584*, 252–264. [CrossRef]
27. Xia, X. Rapid evolution of animal mitochondria. In *Evolution in the Fast Lane: Rapidly Evolving Genes and Genetic Systems*; Singh, R.S., Xu, J., Kulathinal, R.J., Eds.; Oxford University Press: Oxford, UK, 2012; pp. 73–82.
28. Xia, X.; Huang, H.; Carullo, M.; Betrán, E.; Moriyama, E.N. Conflict between Translation Initiation and Elongation in Vertebrate Mitochondrial Genomes. *PLoS ONE* **2007**, *2*, e227. [CrossRef]
29. Muto, A.; Osawa, S. The guanine and cytosine content of genomic DNA and bacterial evolution. In Proceedings of the Proceedings of the National Academy of Sciences. *Proc. Natl. Acad. Sci. USA* **1987**, *84*, 166–169. [CrossRef]
30. Clark, M.A.; Moran, N.A.; Baumann, P. Sequence evolution in bacterial endosymbionts having extreme base compositions. *Mol. Biol. Evol.* **1999**, *16*, 1586–1598. [CrossRef]
31. Xia, X.; Palidwor, G. Palidwor Genomic Adaptation to Acidic Environment: Evidence from Helicobacter pylori. *Am. Nat.* **2005**, *166*, 776. [CrossRef]
32. Li, W.-H.; Gojobori, T.; Nei, M. Pseudogenes as a paradigm of neutral evolution. *Nature* **1981**, *292*, 237–239. [CrossRef] [PubMed]
33. Li, W.-H.; Wu, C.-I.; Luo, C.-C. Nonrandomness of point mutation as reflected in nucleotide substitutions in pseudogenes and its evolutionary implications. *J. Mol. Evol.* **1984**, *21*, 58–71. [CrossRef] [PubMed]
34. LaLonde, M.M.; Marcus, J.M. How old can we go? Evaluating the age limit for effective DNA recovery from historical insect specimens. *Syst. Èntomol.* **2019**, *45*, 505–515. [CrossRef]
35. Xia, X. DNA Replication and Strand Asymmetry in Prokaryotic and Mitochondrial Genomes. *Curr. Genom.* **2012**, *13*, 16–27. [CrossRef] [PubMed]
36. Marín, A.; Xia, X. GC skew in protein-coding genes between the leading and lagging strands in bacterial genomes: New substitution models incorporating strand bias. *J. Theor. Biol.* **2008**, *253*, 508–513. [CrossRef]
37. Nikbakht, H.; Xia, X.; Hickey, D. The evolution of genomic GC content undergoes a rapid reversal within the genus Plasmodium. *Genome* **2014**, *57*, 507–511. [CrossRef]
38. Förstner, K.U.; Von Mering, C.; Hooper, S.D.; Bork, P. Environments shape the nucleotide composition of genomes. *EMBO Rep.* **2005**, *6*, 1208–1213. [CrossRef]
39. Xia, X. Bioinformatics and Translation Elongation. In *Bioinformatics and the Cell: Modern Computational Approaches in Genomics, Proteomics and Transcriptomics*; Springer: Cham, Switzerland, 2018; pp. 197–238.
40. Foster, P.G.; Hickey, D.A. Compositional Bias May Affect both DNA-Based and Protein-Based Phylogenetic Reconstructions. *J. Mol. Evol.* **1999**, *48*, 284–290. [CrossRef]
41. Tarrio, R.; Rodríguez-Trelles, F.; Ayala, F.J. Shared nucleotide composition biases among species and their impact on phylogenetic reconstructions of the Drosophilidae. *Mol. Biol. Evol.* **2001**, *18*, 1464–1473. [CrossRef]
42. Johannsson, S.; Neumann, P.; Wulf, A.; Welp, L.M.; Gerber, H.-D.; Krull, M.; Diederichsen, U.; Urlaub, H.; Ficner, R. Structural insights into the stimulation of S. pombe Dnmt2 catalytic efficiency by the tRNA nucleoside queuosine. *Sci. Rep.* **2018**, *8*, 8880. [CrossRef]
43. Miyata, T.; Miyazawa, S.; Yasunaga, T. Two types of amino acid substitutions in protein evolution. *J. Mol. Evol.* **1979**, *12*, 219–236. [CrossRef] [PubMed]
44. Xia, X. Protein Substitution Model and Evolutionary Distance. In *Bioinformatics and the Cell: Modern Computational Approaches in Genomics, Proteomics and Transcriptomics*; Springer: Cham, Switzerland, 2018; pp. 315–326.
45. Xia, X. DAMBE7: New and Improved Tools for Data Analysis in Molecular Biology and Evolution. *Mol. Biol. Evol.* **2018**, *35*, 1550–1552. [CrossRef] [PubMed]

46. Katoh, K.; Asimenos, G.; Toh, H. Multiple alignment of DNA sequences with MAFFT. *Methods Mol. Biol.* **2009**, *537*, 39–64. [PubMed]
47. Katoh, K.; Kuma, K.-I.; Toh, H.; Miyata, T. MAFFT version 5: Improvement in accuracy of multiple sequence alignment. *Nucleic Acids Res.* **2005**, *33*, 511–518. [CrossRef]
48. Xia, X. Is there a mutation gradient along vertebrate mitochondrial genome mediated by genome replication? *Mitochondrion* **2019**, *46*, 30–40. [CrossRef]
49. Dev, R.R.; Ganji, R.; Singh, S.P.; Mahalingam, S.; Banerjee, S.; Khosla, S. Cytosine methylation by DNMT2 facilitates stability and survival of HIV-1 RNA in the host cell during infection. *Biochem. J.* **2017**, *474*, 2009–2026. [CrossRef]
50. Cooper, A.; Lalueza-Fox, C.; Anderson, S.G.; Rambaut, A.; Austin, J.J.; Ward, R. Complete mitochondrial genome sequences of two extinct moas clarify ratite evolution. *Nature* **2001**, *409*, 704–707. [CrossRef]
51. Mitchell, K.J.; Llamas, B.; Soubrier, J.; Rawlence, N.J.; Worthy, T.H.; Wood, J.R.; Lee, M.S.Y.; Cooper, A. Ancient DNA reveals elephant birds and kiwi are sister taxa and clarifies ratite bird evolution. *Science* **2014**, *344*, 898–900. [CrossRef]
52. Baker, A.J.; Haddrath, O.; McPherson, J.D.; Cloutier, A. Genomic Support for a Moa–Tinamou Clade and Adaptive Morphological Convergence in Flightless Ratites. *Mol. Biol. Evol.* **2014**, *31*, 1686–1696. [CrossRef]
53. Cloutier, A.; Sackton, T.B.; Grayson, P.; Clamp, M.; Baker, A.J.; Edwards, S.V. Whole-Genome Analyses Resolve the Phylogeny of Flightless Birds (Palaeognathae) in the Presence of an Empirical Anomaly Zone. *Syst. Biol.* **2019**, *68*, 937–955. [CrossRef]
54. Xia, X. Mutation and selection on the anticodon of tRNA genes in vertebrate mitochondrial genomes. *Gene* **2005**, *345*, 13–20. [CrossRef] [PubMed]
55. Xia, X. PGT: Phylogeographic Tree for Mapping a Phylogeny onto Geographic Regions. *Glob. Ecol. Biogeogr.* **2019**, *28*, 1195–1199.
56. Xia, X. Extreme Genomic CpG Deficiency in SARS-CoV-2 and Evasion of Host Antiviral Defense. *Mol. Biol. Evol.* **2020**. [CrossRef] [PubMed]
57. Xia, X. DNA Methylation and Mycoplasma Genomes. *J. Mol. Evol.* **2003**, *57*, S21–S28. [CrossRef]

 © 2020 by the author. Licensee MDPI, Basel, Switzerland. This article is an open access article distributed under the terms and conditions of the Creative Commons Attribution (CC BY) license (http://creativecommons.org/licenses/by/4.0/).

life

Article

The Mitochondrial Genome of a Plant Fungal Pathogen *Pseudocercospora fijiensis* (Mycosphaerellaceae), Comparative Analysis and Diversification Times of the Sigatoka Disease Complex Using Fossil Calibrated Phylogenies

Juliana E. Arcila-Galvis [1], Rafael E. Arango [2,3], Javier M. Torres-Bonilla [2,3,4] and Tatiana Arias [1,*,†]

1. Corporación para Investigaciones Biológicas, Comparative Biology Laboratory, Cra 72A Medellín, Antioquia, Colombia; juearcilaga@unal.edu.co
2. Escuela de Biociencias, Universidad Nacional de Colombia-Sede Medellín, Cl 59A Medellín, Antioquia, Colombia; rarango@cib.org.co (R.E.A.); javier.torres@colmayor.edu.co (J.M.T.-B.)
3. Corporación para Investigaciones Biológicas, Plant Biotechnology Unit, Cra 72A Medellín, Antioquia, Colombia
4. Colegio Mayor de Antioquia, Grupo Biociencias, Cra 78 Medellín, Antioquia, Colombia
* Correspondence: tatiana.arias48@tdea.edu.co; Tel.: +57-300-845-6292
† Current address: Tecnológico de Antioquia, Cl 78B Medellín, Antioquia, Colombia.

Citation: Arcila-Galvis, J.E.; Arango, R.E.; Torres-Bonilla, J.M.; Arias, T. The Mitochondrial Genome of a Plant Fungal Pathogen *Pseudocercospora fijiensis* (Mycosphaerellaceae), Comparative Analysis and Diversification Times of the Sigatoka Disease Complex Using Fossil Calibrated Phylogenies. *Life* **2021**, *11*, 215. https://doi.org/10.3390/life11030215

Academic Editor: Andrea Luchetti

Received: 20 January 2021
Accepted: 8 February 2021
Published: 9 March 2021

Publisher's Note: MDPI stays neutral with regard to jurisdictional claims in published maps and institutional affiliations.

Copyright: © 2021 by the authors. Licensee MDPI, Basel, Switzerland. This article is an open access article distributed under the terms and conditions of the Creative Commons Attribution (CC BY) license (https://creativecommons.org/licenses/by/4.0/).

Abstract: Mycosphaerellaceae is a highly diverse fungal family containing a variety of pathogens affecting many economically important crops. Mitochondria play a crucial role in fungal metabolism and in the study of fungal evolution. This study aims to: (i) describe the mitochondrial genome of *Pseudocercospora fijiensis*, and (ii) compare it with closely related species (*Sphaerulina musiva*, *S. populicola*, *P. musae* and *P. eumusae*) available online, paying particular attention to the Sigatoka disease's complex causal agents. The mitochondrial genome of *P. fijiensis* is a circular molecule of 74,089 bp containing typical genes coding for the 14 proteins related to oxidative phosphorylation, 2 rRNA genes and a set of 38 tRNAs. *P. fijiensis* mitogenome has two truncated *cox1* copies, and bicistronic transcription of *nad2-nad3* and *atp6-atp8* confirmed experimentally. Comparative analysis revealed high variability in size and gene order among selected Mycosphaerellaceae mitogenomes likely to be due to rearrangements caused by mobile intron invasion. Using fossil calibrated Bayesian phylogenies, we found later diversification times for Mycosphaerellaceae (66.6 MYA) and the Sigatoka disease complex causal agents, compared to previous strict molecular clock studies. An early divergent *Pseudocercospora fijiensis* split from the sister species *P. musae* + *P. eumusae* 13.31 MYA while their sister group, the sister species *P. eumusae* and *P. musae*, split from their shared common ancestor in the late Miocene 8.22 MYA. This newly dated phylogeny suggests that species belonging to the Sigatoka disease complex originated after wild relatives of domesticated bananas (section Eumusae; 27.9 MYA). During this time frame, mitochondrial genomes expanded significantly, possibly due to invasions of introns into different electron transport chain genes.

Keywords: banana; diversification times; mitochondrial genome; Mycosphaerellaceae; plant pathogens; *Pseudocercospora*; sigatoka disease

1. Introduction

Mycosphaerellaceae is a highly diverse fungal family containing endophytes, saprobes, epiphytes, fungicolous and phytopathogenic species in more than 56 genera [1,2]. Family members can cause significant economic losses to a large number of important plants including ornamentals, food crops and commercially propagated trees [3–8]. Three Mycosphaerellaceae members, *Pseudocercospora eumusae*, *P. fijiensis*, and *P. musae*, [1] are major pathogens of bananas and plantains. They comprise the so-called Sigatoka disease complex which is responsible for one of the most economically destructive diseases for banana growers [9,10]. Diseases caused by these three pathogens induce plant physiological alterations including a reduction in photosynthetic capacity, crop yield, and fruit quality [9]. The Sigatoka disease complex causal agents form a robust clade, with *P. fijiensis* diverging earlier (39.9–30.6 MYA) than sister species *P. eumusae* and *P. musae* (22.6–17.4 MYA) [10–12]. Among them, *Pseudocercospora fijiensis* (teleomorph *Mycosphaerella fijiensis*) is the causal agent of black leaf streak disease (BLSD; aka Black Leaf Spot Disease), one the most damaging and costly diseases for banana and plantain worldwide [13].

Fungal mitochondrial genomes (mitogenomes) are circular or linear, usually AT enriched and range in size from 1.1 kb (*Spizellomyces punctatus*) [14] to 272 kb (*Morchella importuna*) [15]. Their size variation is mostly due to the presence or absence of accessory genes including RNA and DNA polymerases, reverse transcriptases and transposases, mobile introns, and size variation in intergenic regions [16,17]. In spite of the variation in size, their core gene content is largely conserved, even though their relative gene order is highly variable, both between and within the major fungal phyla [18–20]. Mitogenomes have introns and intronic open reading frames (ORFs) classified as group I and group II introns, which differ in their sequence, structure and splicing mechanisms [16,21–25]. Typically, group-II introns contain ORFs that code for reverse-transcriptase-like proteins. In contrast, group-I introns encode proteins with maturase and/or endonuclease activity [16]. Because of the limited comparative analysis of complete fungal mitogenome sequences, it has been difficult to estimate the timeframes and molecular evolution associated with mitochondrial genes or genomes [26].

Mitochondria have proven to be useful in evolutionary biology and systematics because they contain their own genome capable of independent replication, uniparental inheritance [27], near absence of genetic recombination, and uniform genetic backgrounds for some species [28]. Attempts to determine a time frame for fungal evolution are hampered by the lack of reliable fossil records. Hence, so far studies have focused on relating rates of DNA base substitutions and molecular clocks [29], based on the assumption that mutation rates of nuclear genes are similar to their counterparts in organisms with datable fossils [19]. Mitochondria plays a major role in fungal metabolism and fungicide resistance but until now only two annotated mitogenomes have been published in Mycosphaerellaceae (*Zasmidium cellare* and *Zymoseptoria tritici*) [30,31]. Sigatoka disease comparative mitogenome studies will provide answers on the evolution and adaptation of these plant pathogenic fungi.

This study aimed to: (i) sequence and characterize the complete mitogenome of *Pseudocercospora fijiensis*; (ii) compare mitogenomes of *P. fijiensis* with closely related species *P. eumusae*, and *P. musae* (causal agents of Sigatoka), and species with publicly available high throughput data such as *Sphaerulina musiva* and *S. populicola* (causal agents of leaf spot and canker diseases in poplar); and (iii) estimate timeframes and mitochondrial molecular evolution using fossil records to calibrate the Sigatoka disease complex phylogeny. We found that in mitogenomes analyzed herein, there were differences in content of freestanding and intronic Homing Endonuclease Genes (HEGs), genes coding for hypothetical proteins, and accessory genes such as DNA/RNA polymerases, reverse transcriptases and transposases. This work contributes to the understanding of mitogenome organization in Mycospharellaceae. In addition, new fossil calibrations for the Sigatoka's complex species and mitochondrial comparative analysis aid in our understanding of the tempo and mode of evolution of these plant fungal pathogens.

2. Materials and Methods

2.1. Fungal Strain, DNA Extraction, and Library Construction and Sequencing

P. fijiensis isolate 081022 was obtained from naturally infected banana leaves coming from a commercial plantation located in Carepa, Antioquia, Colombia. Taxonomic affiliation has been confirmed based on both morphological criteria and Polymerase Chain Reaction (PCR) [32]. For DNA extraction mycelia from 7-day old culture were transferred to potato dextrose broth and incubated for 5–7 days at room temperature in a rotary shaker. Then, mycelia were harvested from the liquid using vacuum filtration. Total DNA was extracted using a previously described Cetyl Trimethylammonium Bromide (CTAB) method [33]. DNA quality and quantity were measured using a fluorometer (Qubit 3.0, Thermo Fisher Scientific, Waltham, MA, USA). Furthermore, genomic DNA was visualized on 1% agarose gel to check for any break/smear or multiple bands. Library construction was performed using Illumina platform with TruSeq DNA kit (Illumina, San Diego, CA, USA) to acquire as paired-end 2 × 150-bps, with about a 350-bp insert size. Next-generation sequencing was performed by an external service (North Carolina University, Chapel Hill, NC, USA) Hiseq 2500 system®.

2.2. Sequence Sources, Data Filtering and Assemblies

Eleven mitochondrial genome (mitogenomes) sequences were used for this study. Seven belonging to Mycosphaerellaceae: *Pseudocercospora fijiensis* (syn. *Mycosphaerella fijiensis*), *Pseudocercospora eumusae* (syn. *Mycosphaerella eumusae*), *Pseudocercospora musae* (syn. *Mycosphaerella musicola*), *Sphaerulina musiva* (syn. *Septoria musiva*), *Sphaerulina populicola* (syn. *Septoria populicola*), *Zasmidium cellare*, and *Zymoseptoria tritici* (syn. *Mycosphaerella graminicola*). One species is from Capnodiales: *Pseudovirgaria hyperparasitica* and three are from Pleosporales, the sister group of Capnodiales: *Didymella pinodes* (syn. *Mycosphaerella pinodes*), *Parastagonospora nodorum* (syn. *Phaeosphaeria nodorum*), and *Shiraia bambusicola* (Table S1). Mitogenomes were obtained either from our own sequencing data, or sequence data available at GenBank [34], RefSeq [35] or MycoCosm [36]. Authors, seq ID and databases are listed in Table S1.

Read quality was assessed using FastQC v. 0.11.5 [37] for the *P. fijiensis* isolate 081022 raw reads recovered here. Low-quality reads and/or bases were trimmed using Trimmomatic version 0.36 [38]. First, we de novo assembled whole DNA using Spades 3.9.0 (parameter "-careful") [39] at different k-mer sizes (k = 61, 71, 81, and 91). The assembly with the highest N50 and assembly size was scaffolded by SSPACE version 3.0 [40]. Remaining gaps between scaffolds were closed using GapFiller version 1.10 [41] and a final genome assembly was evaluated by REAPR version 1.0.18 [42]. Scaffolds from the whole genome sequencing assembly were mapped to a draft and an unpublished *P. fijiensis* mitogenome available in MycoCosm [36] using Geneious 9.1.5 [43]. Mitogenomes were also filtered from de novo whole-genome assemblies for *Pseudocercospora musae* and *P. eumusae* available online [8,10]. To separate mitochondrial contigs or scaffolds from the nuclear contigs or scaffolds, we used BLASTn [44] and the Electron Transport Chain Conserved Mitochondrial Protein Coding Genes (CMPCGs) compiled from published mitogenomes of *Zasmidium cellare*, *Zymoseptoria tritici*, *Didymela pinodes*, *Phaeosphaeria nodorum* and *Sharaia bambusicola* as queries [30,31,45,46].

Even though the *Sphaerulina musiva* mitogenome was available online [8] we reassembled it using raw reads available at NCBI (SRA: SRR3927043). Our major motivation was a 9322 bp inversion detected around the 10,000 bp position of this mitogenome. This inversion was splitting the gene *nad2* and we wanted to make sure this inversion was present in the *S. musiva* mitogenome. First, raw reads were filtered using BBtools (https://sourceforge.net/projects/bbmap/ (accessed on 10 February 2021)) in Geneious 9.1.5 [43]. Then, MITObim version 1.8 [47] used *cox1* as bait to map all filtered reads that partly or fully overlap with the bait. Eventually, this leads to an extension of the reference sequence and a reduction of gaps until completion of the whole mitogenome [47]. This inversion in the *Sphaerulina musiva* mitogenome was found to be an artifact after reassembling raw reads.

Annotated mitochondrial genomes filtered from whole-genome assembly projects for *Pseudocercospora musae* and *P. eumusae* or reassembled for *Sphaerulina musiva* are available in Figshare (dataset: https://doi.org/10.6084/m9.figshare.12101058.v1 (accessed on 10 February 2021)).

2.3. Annotation

Mitochondrial genomes of *Pseudocercospora fijiensis*, *P. eumusae*, *P. musae*, *Sphaerulina populicola*, *S. musiva* and *Pseudovirgaria hyperparasitica* were annotated in this study using a combination of software. First, predicted ORFs were determined with a translation code for "mold mitochondrial genomes" using Geneious 9.1.5 [43]. Second, genes were identified using BLASTP version 2.4.0 [48] against the non-redundant protein database from NCBI (downloaded August and December 2016); genes were also identified using MITOS [49]. Third, protein domains and sequence patterns were searched with PFAM [50] and PANTHER 11.0 [51]. Additionally, mitogenome annotation was performed using multiple alignments among the fourteen CMPCGs using MUSCLE version 3.8.31 [52] and CLUSTAL W version 2.0 [53]. Inconsistencies regarding length and position of genes was solved paying particular attention to start and stop codons. Identified ORFs larger than 300 bp with start and stop codons that did not show results with the above-mentioned annotation strategies were considered as hypothetical proteins. Circular mitogenome maps were constructed using Geneious 9.1.5. and Geneious prime [43].

2.4. PCR Amplification of cox1 Gene Copies in P. fijiensis

A PCR assay was performed to confirm the presence of two different *cox1* copies: a truncated copy (*cox1_1*) and a complete *cox1* copy with an intron (*cox1_2*). Primers were designed to amplify regions between the first copy (*cox1_1*) and the second (*cox1_2*), including the exons of this last copy. First, a set of primers encompassed *cox1_2 exon1* and *cox1_2 exon 2*. A second pair of primers encompassed *cox1_1* and *cox1_2 exon 2*. PCR amplifications were carried out in a total volume of 10 µL, containing 20 ng genomic DNA, 0.15 µM of each primer, 1× PCR buffer (without $MgCl_2$), 0.75 mM $MgCl_2$, 4 µM of each dNTP and 0.65 U recombinant Taq DNA polymerase (Thermo Fisher Scientific, Waltham, Massachusetts, USA). Cycling parameters were: 3 min at 94 °C, followed by 35 cycles of 30 s at 94 °C, 30 s at a 50 to 60 °C temperature gradient to determine annealing temperature, 1 min at 72 °C, and a final elongation step of 5 min at 72 °C. PCR products were separated by electrophoresis in a 1% (w/v) agarose gel and visualized with GelRed® (Biotium, Fremont, CA, USA) under UV light.

2.5. Transcriptome de novo Assembly

RNA-seq raw reads of *S. musiva* (SRR1652271) and *P. fijiensis* (SRR3593877, SRR3593879) were downloaded from the European Bioinformatics Institute (EMBL EBI) database. Reads were quality filtered and trimmed using BBDuck from BBtools (https://sourceforge.net/projects/bbmap/ (accessed on 10 February 2021)) before carrying out transcriptome de novo assemblies with Trinity version 2.3.1 [54]. The *P. eumusae* (GDIK00000000.1) and *P. musae* assembled transcriptomes (GDIN00000000.1) were also downloaded from GeneBank. RNA-seq Geneious 9.1.5. plugins were used to map the assembled transcripts to mitogenomes of either *S. musiva*, *P. fijiensis*, *P. eumusae* or *P. musae* paying particular attention to gene pairs *atp6-atp8*, *nad2-nad3*.

2.6. RT–PCR Assays for Mitochondrial Gene Pairs of P. fijiensis

Total RNA was extracted from *P. fijiensis* (isolate: 081022) mycelium after fifteen days of culture using TRIzol® (Life Technologies, Carlsbad, CA, USA) according to the manufacturer's instructions. RNA concentrations were measured at 260 nm using a NanoDrop ND-1000 UV-Vis Spectrophotometer (NanoDrop Technologies, Thermo Fisher). DNAse I (Thermo Fisher Scientific, Waltham, MA, USA) was used for cDNA synthesis from RNA as template for amplification using the Maxima First Strand cDNA Synthesis Kit (Thermo Fisher Scientific, Waltham, MA, USA) according to manufacturer's instructions. Primers were designed such that the amplified product encompassed the end of one gene and the beginning of another. We used pairs of NADH-Ubiquinone Oxidoreductase Chain 3 and 2 (*nad3-nad2*) and mitochondrial encoded ATP Synthase Membrane Subunits 6 and 8 (*atp8-atp6*). Both genes and their intergenic sequences were partially amplified. PCR products were run in a 1% agarose gel electrophoresis purified using the GFX PCR DNA and gel band purification kit, according to manufacturer's instructions (GE Healthcare, Chicago, IL, USA). Purified PCR products were sequenced using Sanger Technology at Macrogen Inc. (Seoul, Korea). All sequences are available in GeneBank: *atp6-atp8* cDNA (GenBank: MN171334); *atp6-atp8* DNA (GenBank: MN171335); *nad2-nad3* DNA (GenBank: MN171336); *nad2-nad3* cDNA (GenBank: MN171337); *cob* DNA (GenBank: MN171338); *cob* cDNA (GenBank: MN171339); *nad5* cDNA (GenBank: MN171340).

2.7. Identification of Repetitive Elements

Repetitive sequences in mitogenomes of Mycosphaerellaceae were identified and annotated using Geneious Primer Tandem Repeats Finder and using a minimum repeat length of 100, excluding repeats up to 10 bp longer [43]. Simple sequence repeat (SSR) markers and loci were identified using the MicroSAtellite Identification tool (MISA) [55] (https://doi.org/10.6084/m9.figshare.12101013 (accessed on 10 February 2021)).

2.8. Phylogenetic Analysis and Divergence Times Estimates

Until now, only nuclear markers and strict clock calibration have been used to calculate diversification times for the Sigatoka disease complex species. We aimed to compare these analyses with fossil calibrated Bayesian phylogenies and mitochondrial markers. A phylogenetic tree was reconstructed to calculate diversification times. Since most mitogenomes are uniparentally inherited we used our mitochondrial phylogeny to compare topologies with already published nuclear ones for the Sigatoka disease complex species. *Didymella pinodes* [56], *Pseudovirgaria hyperparasítica* [57], *Phaeosphaeria nodorum* [45] and *Shiraia bambusicola* [46] were used as outgroups. Ten core mitochondrial genes (*cox1, cox3, atp6, cob, nad1, nad2, nad4, nad4L, nad5, nad6*) were aligned one by one for all species using CLUSTAL W version 2.0 [53]. Then, all aligned genes were concatenated in a single alignment for phylogenetic reconstruction. *Cox2, atp8, atp9* and *nad3* were excluded from the alignment either because they could not be fully recovered in *P. musae* (*atp9, nad3*) or because they were missing in outgroups *S. bambusicola* (*atp8 atp9*), *P. nodorum* (*atp8, atp9*) and *D. pinodes* (*atp8, atp9, cox2*).

A Generalized time-reversible (GTR) model was used with an estimated gamma parameter of rate heterogeneity to build maximum likelihood (ML) trees using the Randomized Accelerated Maximum Likelihood RAxML version 8.0 [58] and PhyML version 3.0 [59] programs. One hundred bootstrapped trees were generated and used to assign bootstrap support values to the consensus trees. A Bayesian phylogeny and divergence time analysis was carried out using BEAST2 version 2.5.1 [60]. Separate partitions for each gene were created with BEAUti2 (available in BEAST2). More suitable substitution models for each gene were found using the software package jModelTest2 version 2 [61] according to the Bayesian Information Criterion (BIC) [62]. To accommodate for rate heterogeneity across the branches of the tree we used an uncorrelated relaxed clock model [63] with a lognormal distribution of rates for each gene estimated during the analyses. We also used a strict clock for further comparison of results.

The fossil Metacapnodiaceae [64] was used, assuming this to be a common ancestor of the order Capnodiales with a minimum age of 100 MYA (gamma distribution, offset 100, mean 180, maximum softbound 400). Capnodiales nodes were constrained to monophyly based on the results obtained from ML analysis. A birth/death tree prior was used to model the speciation of nodes in the topology, with gamma priors on the probability of splits and extinctions. We used vague priors on the substitution rates for each gene (gamma distribution with mean 0.2 in units of substitutions per site per time unit). All XML files used to build our Bayesian phylogenies are available at Figshare (https://doi.org/10.6084/m9.figshare.12101055.v1 (accessed on 10 February 2021)). To ensure convergence we ran analyses five times for 50 million generations each, sampling parameters every 5000 generations, assessing convergence and sufficient chain mixing (Effective Sample Sizes > 200) using Tracer version 1.5 [65]. After removal of 20% of each run as burn-in, the remaining trees were combined using LogCombiner (available in BEAST2), summarized as maximum clade credibility (MCC) trees in TreeAnnotator (available in BEAST2), and visualized using FigTree version 1.3.1 [66].

3. Results

3.1. P. fijiensis Mitochondrial Genome

The whole-genome assembly of *P. fijiensis* had an N50 = 39,827 bp and a size of 70.53 Mb in 8040 scaffolds. Recovered scaffolds had an estimated genome coverage of 50X. To separate scaffolds belonging to *P. fijiensis* mitogenome, whole-genome scaffolds were mapped to a draft and unpublished *P. fijiensis* mitogenome (see Table S1). Scaffolds belonging to the mitogenome of *P. fijiensis* were recovered and assembled in a circular sequence deposited in GenBank (accession number: MK754071).

P. fijiensis mitogenome is a circular molecule of 74,089 bp in length containing 14 Electron Transport Chain Conserved Mitochondrial Protein Coding Genes (CMPCG), two ribosomal RNAs, 38 tRNA genes and twelve putative Open Reading Frames (ORFs) of unknown function (Figure 1, Table 1). CMPCGs included ATP Synthase subunits (*atp6*, *atp8*, and *atp9*), Cytochrome Oxidase subunits I, II, and III (*cox1*, *cox2*, and *cox3*), Nicotinamide Adenine Dinucleotide Ubiquinone Oxidoreductase subunits *nad2* and *nad1, 3, 5, 6, nad4L* (Figure 1A, Table 1).

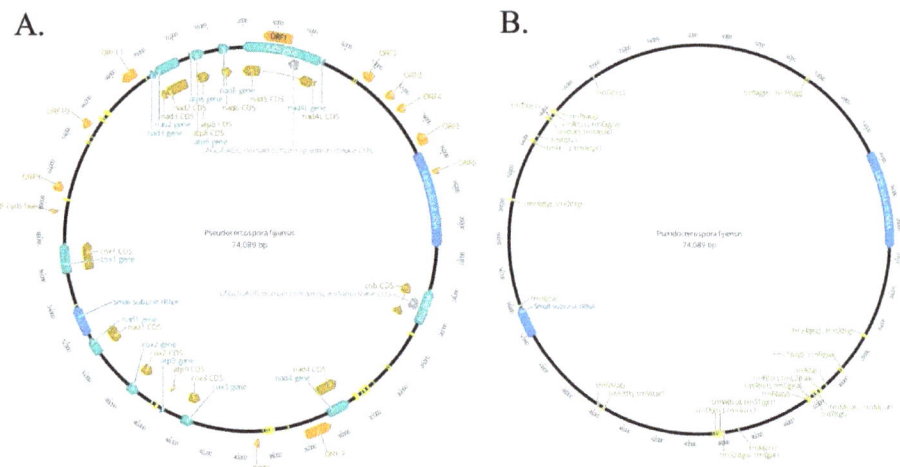

Figure 1. Circular map of the mitogenome of *Pseudocercospora fijiensis*. Genes are visualized by arrows, which are presented in a clockwise direction (forward). (**A**) Annotated introns, hypothetical Open Reading Frames (ORFs), group I mobile introns (Homing Endonuclease Genes (HEGs)) and genic regions; clusters of tRNA and rRNA are also indicated here. (**B**) Annotated tRNA and rRNA regions. Green color arrows: protein-coding genes; yellow color arrows: Coding regions CDS and introns; orange arrows: hypothetical ORFs; blue arrows: genes of large and small ribosomal subunits; bright yellow arrows: tRNAs genes; grey arrow: HEGs. Circular mitogenomes were generated by using the Geneious Primer [43].

Twelve putative ORFs of unknown function were predicted to produce hypothetical proteins containing 77 to 583 amino acids, which were described as ORF1-ORF11 and ORF *Cytb*-like. CMPCGs covered 65.55% of the genome (including 11.77% putative ORFs) (Figure 1A, Table 1). tRNA genes and both rnl and rns corresponded to 8.8% and 10.67% respectively (Figure 1B, Table 1). These values were similar to those reported for other representatives of the Ascomycota phylum (Table 2). Overall, *P. fijiensis* mtDNA has a nucleotide composition of: 36.6% of A, 13.4% of C, 12.6% of G and 37.7% of T. GC-content was 26% with coding and non-coding parts of the genome having, on average, the same GC-percentage.

Table 1. Characteristics and organization of annotated genes in the mitogenome of P. fijiensis.

Gene	Start Position	Stop Position	Length (bp)	Direction	Start Codon	Stop Codon	GC Contents	Product
nad6	435	1109	675	forward	ATA	TAA	23%	subunit 6 of NADH dehydrogenase
nad5	1922	7012	5091	reverse	ATA	TAA	24.8	NADH dehydrogenase subunit 5
ORF1	3253	5001	1749	reverse	ATG	TAA	22.8	hypothetical protein
LAGLIDADG	5042	5722	681	reverse	TTA	TAA	24.4	LAGLIDADG domain containing endonuclease
nad4L	7009	7278	270	reverse	TTA	TAA	23.7	NADH dehydrogenase subunit 4L
trnN(gtt)	9370	9440	71	reverse	-	-	-	tRNA-Asn
trnP(tgg)	9480	9552	73	reverse	-	-	-	
ORF2	9600	10271	672	reverse	ATG	TAA	29.9	hypothetical protein
ORF3	11452	11973	522	forward	TTA	TAG	26.2	hypothetical protein
ORF4	12481	12969	489	forward	ATG	TAA	33.5	hypothetical protein
ORF5	14399	15151	753	reverse	ATA	TAA	-	hypothetical protein
Large subunit rRNA	15433	21413	5981	forward	-	-	-	large subunit ribosomal RNA
ORF6	16699	17037	339	forward	ATT	TAA	14.7	hypothetical protein
cob	23991	26209	2219	reverse	TTA	TAA	29.3	cytochrome b
LAGLIDADG	24749	25492	744	reverse	-	-	-	LAGLIDADG domain containing endonuclease
trnH(gtg)	26876	26948	73	reverse	-	-	-	tRNA-His
trnQ(ttg)	26974	27046	73	reverse	-	-	-	tRNA-Gln
trnL1(tag)	29371	29454	84	reverse	-	-	-	tRNA-Leu
trnF(gaa)	29531	29603	73	reverse	-	-	-	tRNA-Phe
trnA(tgc)	30770	30841	72	reverse	-	-	-	tRNA-Ala
trnE(ttc)	31173	31244	72	reverse	-	-	-	tRNA-Glu
trnL2(taa)	31253	31335	83	reverse	-	-	-	tRNA-Leu
trnM(cat)	31478	31550	73	reverse	-	-	-	tRNA-Met
trnM(cat)	31555	31625	71	reverse	-	-	-	tRNA-Met
trnT(tgt)	31714	31784	71	reverse	-	-	-	tRNA-Thr
trnR(tct)	32001	32071	71	reverse	-	-	-	tRNA-Arg
trnC(gca)	32105	32175	71	reverse	-	-	-	tRNA-Cys
trnR(acg)	32212	32290	79	reverse	-	-	-	tRNA-Arg
nad4	32730	34220	1491	reverse	ATA	TAA	25.4	NADH dehydrogenase subunit 4
ORF7	34319	35902	1584	reverse	ATG	TAA	22.9	hypothetical protein
trnR(cct)	36840	36910	71	reverse	-	-	-	tRNA-Arg
trnS2(tga)	37775	37862	88	reverse	-	-	-	tRNA-Ser
trnI(gat)	37911	37982	72	reverse	-	-	-	tRNA-Ile
trnW(tca)	38019	38090	72	reverse	-	-	-	tRNA-Trp
trnS1(gct)	38154	38234	81	reverse	-	-	-	tRNA-Ser
trnD(gtc)	38279	38351	73	reverse	-	-	-	tRNA-Asp
trnG(tcc)	38481	38551	71	reverse	-	-	-	tRNA-Gly
ORF8	38798	39094	297	forward	ATG	TAG	19.5	hypothetical protein
cox3	43007	43816	810	reverse	ATG	TAA	31.7	Cytochrome c oxidase subunit III
atp9	44999	45223	225	reverse	ATG	ATG	40.9	ATP Synthase Membrane Subunit 9
trnK(ttt)	45506	45578	73	reverse	-	-	-	tRNA-Lys
trnV(tac)	45607	45679	73	reverse	-	-	-	tRNA-Val
trnM(cat)	45915	45987	73	reverse	-	-	-	tRNA-Met
cox2	46989	47738	750	reverse	TAA	TAA	29.9	cytochrome c oxidase subunit II

Table 1. Cont.

Gene	Start Position	Stop Position	Length (bp)	Direction	Start Codon	Stop Codon	GC Contents	Product
nad1	50440	51600	1161	forward	ATG	TAA	30	NADH dehydrogenase subunit 1
Small subunit rRNA	51849	53773	1925	forward	-	-	-	small subunit ribosomal RNA
trnY(gta)	53797	53881	85	forward	-	-	-	tRNA-Tyr
cox1	55837	57708	1872	forward	TTA	TAG	32.2	cytochrome c oxidase subunit I
ORF cytb-like	59442	59672	231	reverse	TTA	TAA	22.5	cytb-like ORF
trnH(gtg)	60291	60363	73	reverse	-	-	-	tRNA-His
trnQ(ttg)	60389	60461	73	reverse	-	-	-	tRNA-Gln
ORF9	60802	61359	558	forward	ATG	TAA	35.1	hypothetical protein
trnF(—)	64038	64111	74	reverse	-	-	-	tRNA-Phe
trnA(tgc)	64155	64226	72	reverse	-	-	-	tRNA-Ala
trnE(ttc)	64498	64569	72	reverse	-	-	-	tRNA-Glu
ORF10	64714	65247	534	reverse	ATA	TAG	27	hypothetical protein
trnI(tat)	65672	65744	73	reverse	-	-	-	tRNA-Ile
trnM(cat)	65749	65819	71	reverse	-	-	-	tRNA-Met
trnT(tgt)	65908	65978	71	reverse	-	-	-	tRNA-Thr
trnR(tct)	66182	66252	71	reverse	-	-	-	tRNA-Arg
trnC(gca)	66286	66356	71	reverse	-	-	-	tRNA-Cys
trnR(acg)	66393	66465	73	reverse	-	-	-	tRNA-Arg
ORF11	68278	69276	999	reverse	ATG	TAA	21.5	hypothetical protein
trnG(tcc)	69523	69593	71	forward	-	-	-	tRNA-Gly
nad3	69768	70139	372	reverse	ATG	TAA	27.2	NADH dehydrogenase subunit 3
nad2	70140	71819	1680	reverse	ATG	TAA	24.5	NADH dehydrogenase subunit 2
atp8	72609	72758	150	forward	ATT	TAA	24	ATP Synthase Subunit 8
atp6	72801	73589	789	forward	ATA	TAA	27.8	ATP Synthase Subunit 6

Table 2. Comparison of mitogenomes of *Pseudocercospora fijiensis* and closely related species (*Sphaerulina musiva*, *S. populicola*, *P. musae*, *P. eumusae* and *Zasmidium cellare*, *Zymoseptoria tritici*). Intragenic (genic) regions include regions of standard CMPCGs (Conserved Mitochondrial Protein Coding Genes), open reading frame (ORFs), rRNAs, and tRNAs. Intergenic regions include regions among standard CMPCGs, ORFs, rRNAs, and tRNAs.

Item	P. musae	P. eumusae	P. fijiensis	S. musiva	S. populicola	Z. cellare	Z. tritici
Genome size (bp)	51,645	59,236	74,089	53,234	139,606	23,743	43,964
GC content (%)	27.80	27	19.2	24.5	31.70	27.80	31.90
No. of introns	1	1	2	6	28	0	0
No. of standard Protein Coding Genes (CMPCGs)	14	14	14	14	14	14	14
No. of rRNAs	2	2	2	2	2	2	2
No. of tRNAs	24	25	38	29	29	25	27
Genic regions (%)	63.25	40.73	65.55	65.06	98.25	79.39	67.79
Intergenic regions (%)	36.75	59.27	46.45	34.94	1.75	20.51	32.21
Number of GIY-YIG intragenic regions	0	0	0	1	5	0	0
Number of GIY-YIG intergenic regions	1	1	0	0	2	0	0
Number of LAGLIDADG intragenic regions	0	0	0	3	18	0	0
Number of LAGLIDADG intergenic regions	0	0	2	1	3	0	0
Number of Repetitive Sequences	32	43	29	8	20	2	0

The three most frequent codons were TTA (708 counts), ATA (498 counts) and TTT (498 counts) encoded amino acids leucine, isoleucine, and phenylalanine, respectively. These amino acids have hydrophobic side chains commonly found in transmembrane helices. These three codons accounted for 6.9% of all codons in the mitogenome. Eight codons (ATA, ATT, TTA, and TTG encoding methionine; CGA, CGC, CGG encoding arginine and TAG encoding a stop codon) were under-represented, being used from one to five times each. CMPCGs started with ATA, ATG, ATT or TTA encoding methionine translation initiation codon. The preferred stop codon was TAA, present in 21 protein-coding genes; the alternative stop codon was TAG. Codon usage of the ORFs was similar to that of the protein-coding loci. *Atp9* also contained the highest GC contents (40.9%) among all the CMPCGs (Table 1).

Thirty-eight tRNAs encoded by the mitogenome of *P. fijiensis* carry all 20 amino acids (Figure 1B, Table 1). Two tRNA iso-acceptors were identified for serine and leucine, and four for arginine and methionine. Among the 38 tRNAs, tRNA-Val, tRNA-Tyr, tRNA-Asp, tRNA-Lys, tRNA-Asn, tRNA-Pro occurred singly (Table 1). tRNA genes were grouped into nine clusters. (Figure 1B).

Twenty-nine repeat regions ranging from 611 bp to 128 bp were located in non-coding regions of the mtDNA, but only one was present three times (length 193 bp) (Table 2). A search within the *P. fijiensis* mitogenome detected 33 SSR markers. The most common SSR were mononucleotide (30 SSR), and only three dinucleotide SSR were found (https://doi.org/10.6084/m9.figshare.12101058 (accessed on 10 February 2021)). A difference of seven nucleotides between *P. fijiensis* isolate 081022 and the unpublished mitochondrial genome available in MycoCosm was found, being 99% identical (Table S1).

3.1.1. Presence of Truncated Conserved Mitochondrial Protein-Coding Genes (CMPCGs)

P. fijiensis mitogenome had truncated copies of four CMPCGs. They were considered truncated copies because they have PFAM and PANTHER domains and high local sequence similarity (>90%) with a complete CMPCG copy despite being an incomplete sequence without start and/or stop codons. Truncated copies included *atp6* (two truncated copies and one complete gene sequence), and *cox1, cob, nad2* and *atp9* (one truncated copy and one complete gene sequence). PCR amplifications were performed to confirm the presence of CMPCG copies in the *P. fijiensis* mitochondrial genome (Figure 2). The annotation showed that *cox 1* had one complete copy with an intron and a second truncated copy (see Figure 2A). Primers were designed to amplify three intra and intergenic regions in these two copies (Figure 2A–D). The amplified PCR bands were of the expected size showing that the assembly on this region was correct (Figure 2B). The presence of an intron in the complete cytochrome b (*cob*) gene of *P. fijiensis* was also experimentally confirmed by PCR amplification (Figure S1).

3.1.2. Homing Endonucleases and Introns in the *P. fijiensis* Mitogenome

A total of two different mobile introns were annotated in the mitogenome of *P. fijiensis* (Figure 1A). All encoding introns were characterized as group I intron type, which encodes homing endonucleases (HE) [67]. They belonged to the LAGLIDADG family and were 680 bp for an intronic-ORF of 2968 bp length in *nad5* and 743 bp long for an intronic-ORF of 1056 bp length in *cob* (Figure 1, Tables 2 and 3).

Table 3. Comparison between divergence time of some clades among different studies.

	Present Study			Chang et al. 2016
Molecular Clock	Relaxed-Log normal		Strict clock	Strict clock
Type of Data	13 mitochondrial protein-coding genes		13 mitochondrial protein-coding genes	46 nuclear single-copy genes
Divergence Time estimation method	Bayesian analysis in BEAST2 v2.5.1		Bayesian analysis in BEAST2 v2.5.1	Penalized likelihood analysis in the program r8s v1.7
Fossil calibration	Capnodiales-Metacapnodiaceae Fossil		Capnodiales- Metacapnodiaceae Fossil	Dothideomycetes crown group
Capnodiales	108.81 (100.17–121.35 MYA, 95% HPD)		151.96 (101.04–232.179 MYA, 95% HPD)	234.2–180.2 MYA
Mycosphaerellaceae	66.66 (55.47–78.27 MYA, 95% HPD)		97.24 (64.27–155 MYA, 95% HPD)	186.7–143.6 MYA
Sphaerulina	13.39 (9.39–17.69 MYA, 95% HPD)		30.31 (19.69–48.63 MYA, 95% HPD)	near 10 MYA
Pseudocercospora + Sphaerulina	48.1 (39.34–57.86 MYA, 95% HPD)		84.46 (55.6–134.54 MYA, 95% HPD)	146.6–112.8 MYA
Pseudocercospora	13.31 MYA (9.49–17.28 MYA, 95% HPD)		27.66 MYA (17.8–44.27MYA, 95% HPD)	39.9–30.6 MYA
P. eumusae + P. musae	8.22 MYA (5.6–11.07 MYA, 95% HPD)		17.87 MYA (11.5–28.71 MYA, 95% HPD)	22.6–17.4 MYA

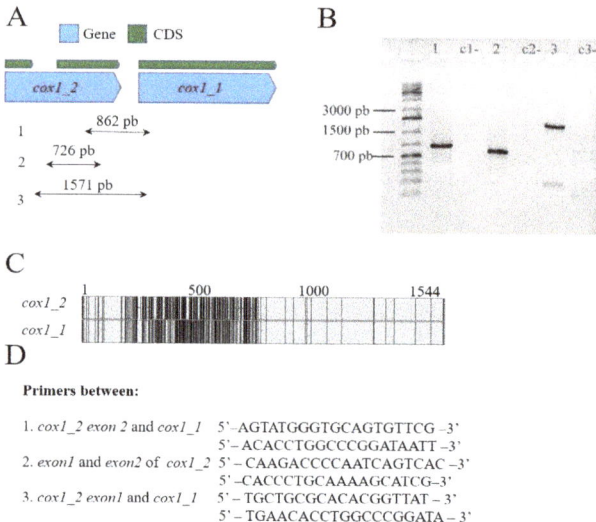

Figure 2. Duplicated Protein-coding genes in the mitogenome of *Pseudocercospora fijiensis*. (**A**) A truncated *cox1* copy and a complete copy with an intron that are localized in tandem. (**B**) PCR amplification including intra and intergenic regions between truncated *cox1* copies. (**C**) Pairwise alignments of truncated *cox1* copies reveal a region of around 600 bp of low similarity between them (non-identical sites appear in black in the alignment). (**D**) Primers used for the amplification of both *cox1* copies.

3.2. Comparative Analysis among Mitochondrial Genomes of Selected Mycosphaerellaceae

The mitochondrial phylogeny of selected species of Mycosphaerellaceae confirms evolutionary relationships found in nuclear phylogenies ((((*Sphaerulina musiva* + *Sphaerulina populicola*) + (*P. fijiensis* + (*P. eumusae* + *P. musae*)) *Zymoseptoria tritici*) *Zasmidium cellare*) [10,12] (Figure 3). We used this phylogeny to compare mitogenomes of *Pseudocercospora fijiensis*, *P. eumusae*, *P. musae*, *Sphaerulina populicola*, and *S. musiva*, and previously published mitogenomes of *Zymoseptoria tritici* and *Zasmidium cellare*. They showed a GC content of between 27% and 32% and a variability in genome size from 23,743 bp in "cellar mold" *Zasmidium cellare* to 136,606 bp in poplar pathogen *Sphaerulina populicola* (Table 2, Figure 3 and Figure S2).

Annotated genes of seven selected Mycosphaerellaceae mitogenomes showed the presence of a set of 14 Conserved Mitochondrial Protein-Coding Genes (CMPCGs), namely the subunits of the electron transport chain complex I (*nad1*, *nad2*, *nad3*, *nad4*, *nad4L*, *nad5* and *nad6*), complex III (*cob*), complex IV (*cox1*, *cox2* and *cox3*), and ATP synthase subunits (*atp6*, *atp8* and *atp9*) (Figure 4, Table S2). There was a small ribosomal subunit RNA (rns), a large ribosomal subunit rRNA (rnl), and a set of 24 to 38 tRNAs (Table 2).

In addition to these core genes, five out of seven mitogenomes also have hypothetical protein coding genes, accessory genes, additional copies of CMPCGs and genetic mobile elements (Figure 3, Table S2). Genome sizes, gene numbers and contents are heterogeneous among species (Table 2, Figure 3). All annotated genomes were found to have truncated gene duplications of some CMPCGs (Table S2). Alignments of truncated gene copies showed their sequences were not identical. However, one of the copies always had a complete coding sequence with a start and a stop codon (https://doi.org/10.6084/m9.figshare.13542191 (accessed on 10 February 2021)). Mitochondrial genomes from *P. musae* and *P. eumusae* also had RNA and DNA polymerases (Table S2).

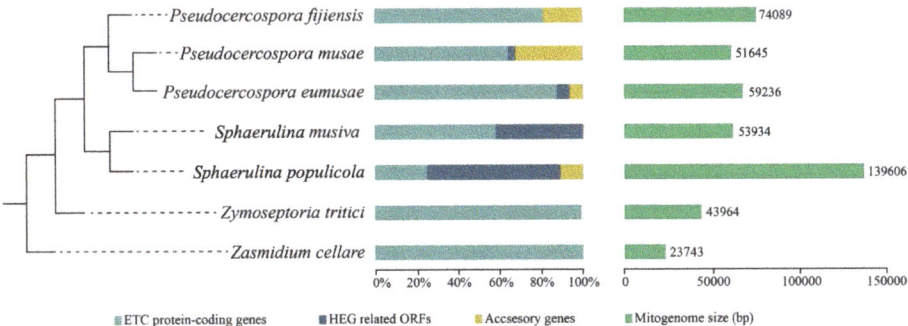

Figure 3. Comparison of mitochondrial genome sizes among Mycosphaerellaceae and contribution of, Conserved Mitochondrial Protein-Coding Genes (CMPCGs), hypothetical proteins, Homing Endonuclease Genes (HEGs) related Open Reading Frames (ORFs) and accessory genes, to genetic content of mitochondrial genomes. Mitochondrial genomes of Mycosphaerellaceae members are different in terms of genome size, and content of accessory genes and HEGs. Some genomes contain only CMPCGs while others exhibit HEG invasion. Phylogenetic relationships were inferred here.

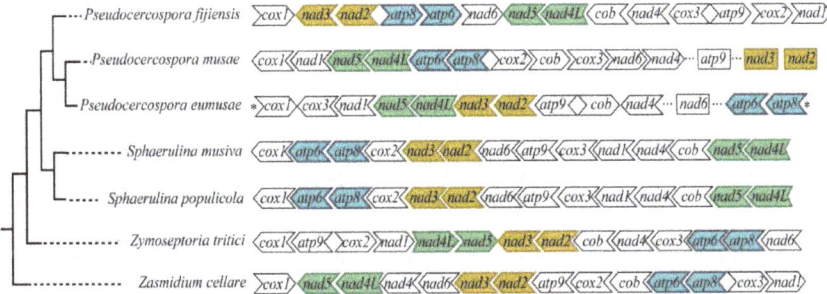

Figure 4. Comparison of Conserved Mitochondrial Protein Coding Genes (CMPCGs) order and orientation among selected Mycosphaerellaceae species. Gene order is not conserved across members even for closely related species except for *Sphaerulina* species. Colored gene pairs were always recovered as neighbors; asterisks indicate *atp8* is neighbor to *cox1* in *P. eumusae*. Ellipsis in *P. musae* and *P. eumusae* mean these genomes were both recovered in different contigs each (two and three respectively); where *nad6* and *atp9* could not be assembled to any contig we not complete a circular mitogenome.

The order of CMPCGs among selected Mycosphaerellaceae mitogenomes was variable, except for sister species *Sphaerulina populicola* and *S. musiva* (Figure 4). Despite this variability, gene pair order was always conserved for gene pairs *nad4L-nad5*, *nad3-nad2* and *atp8-atp6* among all mitogenomes (Figure 4). We mapped RNAseq assembled transcripts from transcriptomes available online for *S. musiva* (SRR1652271), *P. fijiensis* (SRR3593877, SRR3593879), *P. eumusae* (GDIK00000000.1) and *P. musae* (GDIN00000000.1) to *P. fijiensis*, *S. musiva*, *P. musae* and, *P. eumusae* mitochondrial genomes and found neighbor genes were always part of the same transcript for each species. This suggested that they were transcribed as a single mRNA. RT-PCR amplification and subsequent Sanger sequencing confirmed bicistronic expression for *nad3-nad2*, *atp8-atp6* gene pairs in *P. fijiensis* (Figure 5).

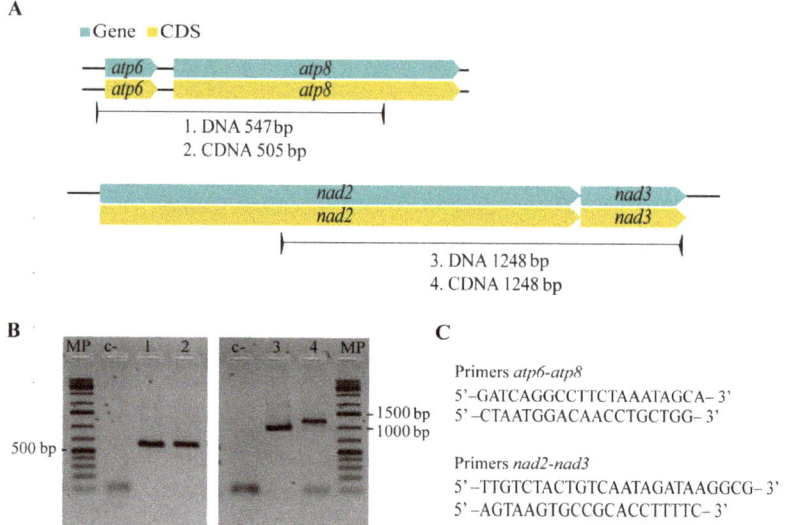

Figure 5. *P. fijiensis* RT-PCR assay to confirm bicistronic expression of genes. (**A**) Primers were designed between neighboring genes with hypothesized bicistronic transcription. (**B**) Gel results of RT-PCR assays for gene pairs *nad2-nad3*, *atp6-atp8* and its intergenic sequences, amplified bands have the expected sequence size. (**C**) Nucleotide sequences of designed primers.

Major differences in genome size among members of Mycosphaerellaceae seemed to be related to the invasive presence of HEG and ORFs in some mitogenomes. *Zasmidium cellare* and *Zymoseptoria tritici* have compact mitogenomes with core mitochondrial genes (CMPCGs) and lack of HEG. While, in other Mycosphaerellaceae, the presence of sequences containing LAGLIDADG or GIY-YIG domains related to Homing Endonuclease Genes (HEGs) was observed (Table 2). These ORFs ranged from one in *Pseudocercospora musae* and *P. eumusae* to 28 in *Sphaerulina populicola* (Table 2). *S. populicola* has the largest mitogenome report in this study (139,606 bp), suggesting HEGs might have caused fragmentation of CMPCGs. In almost all instances, pieces of CMPCGs were collinearly distributed and each fragment was found to be followed by an insertion of a HEG related sequence. An extreme case of HEG invasion to CMPCGs was found in *cox1* of *Sphaerulina populicola* (CDS: 1542 bp). This gene has twelve fragments distributed along 22,070 bp, each of them containing a piece of *cox1* followed by a HEG related domain (Figure S3).

3.3. Inferred Mitochondrial Phylogeny and Diversification Times

A robust phylogeny with posterior probabilities greater than 0.97 was recovered, containing two main lineages (Pleosporales + Capnodiales). Divergence time estimates using fossil calibration are shown in Figure 6, with horizontal bars representing the 95% Highest Posterior Density (HPD) intervals for each node. According to our data, Mycosphaerellaceae diverged from the rest of Capnodiales at the end of the Mesozoic or the early Paleogene, about 66.66 MYA (55.47–78.27 MYA, 95% HPD). The earliest split within Mycosphaerellaceae gave rise to *Zasmidium cellare*. The species *Zymoseptoria tritici* diverged from (*Sphaerulina* + *Pseudocercospora*) also at the end of the Mesozoic or the early Paleogene 59.88 MYA (49.7–70.91 MYA, 95% HPD). The sister genera *Sphaerulina* + *Pseudocercospora* diverged in the Eocene, 48.1 MYA (39.34–57.86 MYA, 95% HPD). The sister clade to the species in the Sigatoka complex includes the species *Sphaerulina musiva* and *S. populicola* sharing their last common ancestor during the Miocene, 13.39 MYA (9.39–17.69 MYA, 95% HPD). The origin of *Pseudocercospora* members of the Sigatoka disease complex in bananas was dated to around 13.31 MYA (9.49–17.28 MYA, 95% HPD) during the Miocene while the sister *species P. eumusae* and *P. musae* split from their shared ancestor in the late Miocene 8.22 MYA (5.6–11.07 MYA, 95% HPD) (Figure 6).

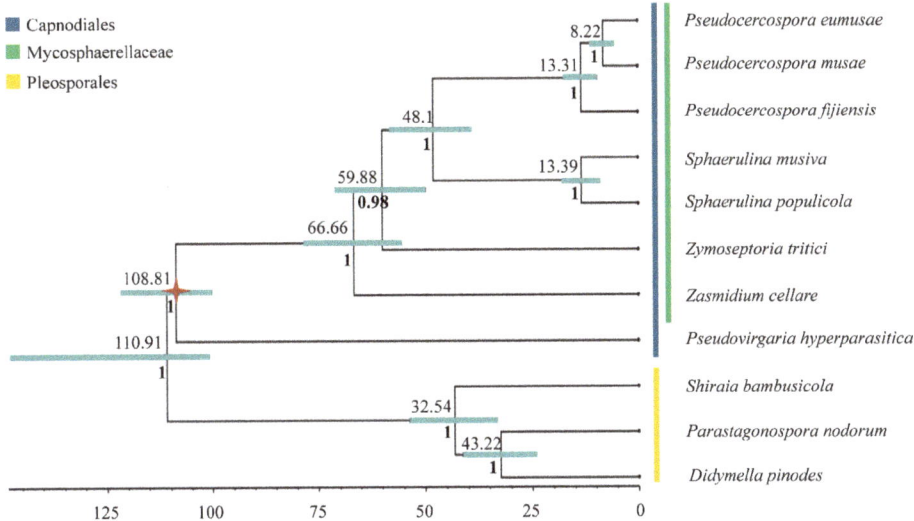

Figure 6. Mycosphaerellaceae Bayesian phylogeny and divergence time estimation. Mycosphaerellaceae Bayesian phylogeny with posterior probabilities in each branch and divergence time estimates calculated using fossil calibrations in BEAST2. Red star represents the fossil calibration placement.

Divergence times were also estimated using strict clock, in order to validate differences between mean node ages using relaxed lognormal clock versus strict clock. For a strict clock, 95% highest posterior density (HPD) intervals were significantly broader (27–131 MY) in comparison to (6–23 MY) and mean node ages were among 9–43 MY older (Table 3) (https://doi.org/10.6084/m9.figshare.13542938.v1 (accessed on 10 February 2021)).

4. Discussion

In this study, the complete mitochondrial genome of a plant pathogenic fungus, *Pseudocercospora fijiensis*, was sequenced and annotated. We also used comparative analysis and fossil calibration phylogenies to further understand the evolution of Mycosphaerellaceae mitogenomes. To date, more than 700 complete fungal mitogenomes are available online, but the mitogenomes of *Pseudocercospora* species have not been reported in the organelle genome database of NCBI (August 2017). The mitogenome of *P. fijiensis* and related species provides a molecular basis for further studies on molecular systematics and evolutionary dynamics of Ascomycota fungi especially belonging to Dothideomycetes.

Ascomycetes mitochondrial genomes like most mitochondrial genomes along the tree of life generally consist of: two ribosomal subunits (*rnl* and *rns*), a distinct set of tRNAs and fourteen genes of the respiratory chain complexes (*cox1*, *cox2*, *cox3*, *cob*, *nad1* to *nad6*, *atp6*, *atp8* and *atp9*) [17]. The mtDNA of *P. fijiensis* contains the 14 mitochondrial inner membrane proteins involved in electron transport and coupled oxidative phosphorylation, as well as rnl and rns (Figure 1). These genes were also found in mitochondrial genomes of selected Mycosphaerellaceae species studied here. Additionally, DNA polymerases, RNA polymerases and Reverse transcriptases were found in *Pseudocercospora musae*, *P. eumusae* and *S. populicola* (Table S1). These polymerases and transcriptases might come from mitochondrial plasmids integrated into their mitochondrial genomes [21,68,69].

A variable number of open reading frames of unknown function and introns related to homing endonuclease genes (HEG), often including GIY-YIG or LAGLIDADG protein domains, were found in several Mycosphaerellaceae genomes including *P. fijiensis* (Table 2;

Figure 1). Despite not being ubiquitous in Ascomycete mitogenomes these mobile elements are fairly common [16,17,22,67,70]. Differences in mitogenome sizes, gene order and gene duplication among Mycosphaerellaceae are attributed to heterogeneous content of accessory genes, and intron mobile sequences (Table 1, Figures 3 and 4).

The comparative study of Mycosphaerellaceae species selected mitogenomes showed that sizes and gene order are not conserved among members of the family (Figures 3 and 4). Divergence times within sister clades *Pseudocercospora* 13.31 MYA (9.49–17.28 MYA, 95% HPD) and *Sphaerulina* 13.39 MYA (9.39–17.69 MYA, 95% HPD) are roughly the same (Table 3; Figure 6). But gene order in *Sphaerulina* species is conserved while in *Pseudocercospora* species it is not (Figures 4 and 6). Nonetheless, some gene pairs were always found together, *atp6-atp8*, *nad2-nad3*, *nad4L-nad5*, in most studied species (Figure 4). Gene order variation among Mycosphaerellaceae could be due to mtDNA rearrangements caused by different processes, such as fusion, fission, recombination, plasmid integration and mobility [21]. Aguileta et al. (2014) compared 38 fungal mitogenomes to understand mitochondrial gene order evolution. They found evidence of gene rearrangements and a relationship with intronic ORFs and repeats. Their results support recombination in all fungal phyla. Despite rearrangements being pervasive in fungal mitogenomes, they found conserved gene pairs *nad2-nad3* and *nad4L-nad5* in most species [20]. Bicistronic transcription of *atp6-atp8*, *nad2-nad3* gene pairs was confirmed experimentally in *P. fijiensis* using RT-PCR (Figure 5). Maintenance of such proximity through evolution is important for mitochondria functions and modifications of such proximity can negatively affect organisms [71].

Even though sister species *Sphaerulina musiva* and *S. populicola* shared core gene content and gene order, their genomes sizes were quite different (53,234 bp and 139,606 bp). This could be due to a widespread occurrence of homing endonucleases HEG related to ORFs and accessory genes in the *S. populicola* mitogenome (Table S1, Figure S2). Mitogenome size variability due to the occurrence of mobile elements and accessory gene invasions was also observed in nine phylogenetically related species belonging to the genera *Aspergillus* and *Penicilium* [72]. In a phytopathogenic fungus *Sclerotinia borealis* mitogenome size expansion was also shown to be due to plasmid-like sequences and HEGs related to ORFs [73].

Mitochondrial gene duplication is seldom described in fungi. HEGs invasion has been previously described in fungal mtDNA, where truncated genes were ubiquitous [74]. *P. fijiensis* and other Mycosphaerellaceae mitogenomes have 2 to12 truncated copies of Conserved Mitocondrial Protein Coding Genes (CMPCGs) (Table S1). Truncated CMPCGs have partial sequences lacking start and/or stop codons. Truncated gene copies did not have a particular distribution pattern; they were either close to each other or dispersed in the genomes. Mardanov et al. (2014) found duplications of truncated extra copies of *atp9* and *atp6* in the phytopathogenic fungus *Sclerotinia borealis* [73]. Two incomplete copies of *atp6* were also found on different strands of the mtDNA of *Shiraia bambusicola* (Pleosporales) [46]. However, it is not common to find highly fragmented genes such as the *cox1* of *Sphaerulina populicola* (Figure S3). A similar case was reported in the mitogenome of *Sclerotinia borealis*, where thirteen introns of *cox1* and truncated copies of CMPCGs were found [73].

Fossil calibrated phylogenies for the Mycospharellaceae had later diversification times 66.66 MYA (55.47–78.27 MYA, 95% HPD) compared to previous studies 186.7–143.6 MYA [10]. The Sigatoka disease complex had an early divergent *Pseudocercospora fijiensis*, that splits from sister species *P. musae* + *P. eumusae* 13.31 MYA (9.49–17.28 MYA, 95% HPD); while sister species *P. eumusae* and *P. musae* split from their shared ancestor in the late Miocene 8.22 MYA (5.6–11.07 MYA, 95% HPD) (Figure 6). Chang et al. (2016) estimated the divergence of *P. fijiensis* from their last common ancestor with *P. musae* + *P. eumusae* to be between 39.9–30.6 MYA and the divergence of *P. musae* and *P. eumusae* to be between 22.6–17.4 MYA. Diversification ages estimated here were based on mitochondrial markers and Bayesian analysis using both relaxed and strict clock models. These have placed all diversification times in the Mycosphaerellaceae at later times than those calculated by

Chang et al. (2016) using nuclear markers, penalized maximum likelihood analysis and strict clock, except for that of *Sphaerulina* (Table 3).

These differences in diversification times compared to Chang et al. (2016) could be due to different calibration points and dating methods. Chang et al. (2016) implemented r8s Likelihood methods [75] and a calibration at the origin of the Dothideomycetes crown group (394–285 MYA) using previous Bayesian estimations [76]. For this study Bayesian analysis in BEAST2 [60] using fossil calibration with a Metacapnodiaceae fossil was implemented [64]. We found that Sigatoka disease members (13.31 MYA (9.49–17.28 MYA, 95% HPD)) appeared after the genus *Musa* (27.9 MYA (21.5–34.4, 95% HPD)) [77]. Species member of this genus within the section Eumusa *sensu latto* comprise cultivated banana and are the host of the Sigatoka disease complex. This naturally prompts a further question: did the Sigatoka disease complex originate through host-tracking evolution? This hypothesis explains why a pathogen is likely to be younger than the host until changes related to the genetics of the pathogens or/and exogenous factors have observed alterations in their virulence spectra [70]. We are currently limited in terms of a good taxonomic sampling and biogeographical analysis for *Pseudocercospora* species in arriving at an answer to this question. A host tracking coevolution hypothesis has also been proposed for *Zymoseptoria tritici* (syn. *Mycosphaerella graminicola*) [5].

5. Conclusions

We successfully sequenced and analyzed mitochondrial genomes of *Pseudocercospora fijiensis, P. eumusae, P. musae, Sphaerulina populicola* and *S. musiva*. A robust mitochondrial phylogeny containing two main lineages (Pleosporales + Capnodiales) was obtained and divergence times were estimated using fossil calibration. Fossil calibrated phylogenies are reported for the first time here for fungal plant pathogens that had later diversification times for the origin of all the species involved, compared to previous studies. Genome size variation and organization among Mycosphaerellaceae could be related to the proliferation of type I and II introns, gene duplications and possible plasmid insertions, phenomena known for many fungal mitogenomes. Despite their order variability, some genes were always recovered as neighbors in all mitogenomes analyzed. Bicistronic expression for *nad3-nad2, atp8-atp6* gene pairs in *P. fijiensis* was confirmed experimentally. Further gene editing and virulence assays will be important to shed light on fungal adaptation and more effective disease control strategies. Phylogenomic studies including a good taxonomic sampling and biogeographical analysis for *Pseudocercospora* species will further clarify whether the Sigatoka disease-causing species virulence flared-up after *Musa* domestication.

Supplementary Materials: The following are available online at https://www.mdpi.com/2075-1729/11/3/215/s1: Table S1. Sources of genomic data used in this study. Table S2: Protein-coding genes annotated in Mycosphaerellaceae mitochondrial genomes, Figure S1: Mitochondrial gene duplication in *Pseudocercospora fijiensis*. A. The intron of *cob*. B. PCR amplification of intra and intergenic sequences among exons. C. Primers used for the amplification of *cob*, Figure S2: Circular map of the mitogenome of *Sphaerulina musiva* and *Sphaerulina populicola*. Smaller 53,234 bp and larger 139,606 bp mitogenomes from selected species of Mycosphaerellaceae used in this study, Figure S3: Homing endonuclease invasion in the mitochondrial genome of *Sphaerulina populicola*. Truncated *cox1* copy possibly due to a pervasive HEGs invasion. Numbers one-12 represent different internal Open Reading Frames (ORFs). Domains content in the ORFs are listed in the table along with their position within *cox1*.

Author Contributions: J.E.A.-G.: Co-developed questions and framework, performed lab work, performed bioinformatic analysis, and wrote manuscript. R.E.A.: Co-developed questions and framework, edit manuscript, provided funding and mentored students. J.M.T.-B.: performed lab work and some bioinformatic analysis. T.A.: Co-developed questions and framework, performed some bioinformatic analysis, wrote and edit manuscript, provided funding and mentored student. All authors have read and agreed to the published version of the manuscript.

Funding: This research was funded by Instituto para el desarrollo de la Ciencia y la Tecnología "Francisco José de Caldas (Colciencias)", Colombia, grant nunmber: 221356934854; The Asociación de

Bananeros de Colombia (Cenibanano-AUGURA), the program "Jovenes Investigadores e Innovadores por la Paz convocatoria 755-2017" funded by Ministerio de Ciencia, Tecnología e Innovación of Colombia.

Data Availability Statement: *Pseudocercospora fijiensis* NADH dehydrogenase subunit 5 (*nad5*) mRNA, partial cds; mitochondrial 502 bp linear mRNA MN171340.1 GI:1817958850. *Pseudocercospora fijiensis* cytochrome b (*cob*) mRNA, partial cds; mitochondrial, 159 bp linear mRNA MN171339.1 GI:1817958848. *Pseudocercospora fijiensis* cytochrome b (*cob*) gene, partial cds; mitochondrial, 1121 bp linear DNA MN171338.1 GI:1817958846. *Pseudocercospora fijiensis* NADH dehydrogenase subunit 2 (*nad2*) and NADH dehydrogenase subunit 3 (*nad3*) mRNAs, partial cds; mitochondrial, 2377 bp linear mRNA MN171337.1 GI:1817958843. *Pseudocercospora fijiensis* NADH dehydrogenase subunit 3 (*nad3*) and NADH dehydrogenase subunit 2 (*nad2*) genes, partial cds; mitochondrial, 1188 bp linear DNA, MN171336.1 GI:1817958840. *Pseudocercospora fijiensis* ATP synthase subunit 8 (*atp8*) gene, complete cds; and ATP synthase subunit 6 (*atp6*) gene, partial cds; mitochondrial, 701 bp linear DNA, MN171335.1 GI:1817958837. *Pseudocercospora fijiensis* ATP synthase subunit 8 (*atp8*) and ATP synthase subunit 6 (*atp6*) mRNAs, partial cds; mitochondrial, 676 bp linear mRNA, MN171334.1 GI:1817958834. Arcila Galvis, Juliana Estefanía; Arias, Tatiana (2020): Data non-curated for NCBI *Pseudocercospora fijiensis* and related Mycosphaerellaceae mitochondrial genomes. Figshare Dataset: https://figshare.com/account/home#/projects/78525 (accessed on 10 February 2021). Arcila Galvis, Juliana Estefanía; Arias, Tatiana (2020): Annotated mitochondrial genomes of Mycosphaerellaceae. Figshare Dataset: https://doi.org/10.6084/m9.figshare.12101058.v1 (accessed on 10 February 2021). Arcila Galvis, Juliana Estefanía; Arias, Tatiana (2020): *P.fijiensis* DATA. Figshare Dataset: https://doi.org/10.6084/m9.figshare.12101013 (accessed on 10 February 2021). Arcila Galvis, Juliana Estefanía; Arias, Tatiana (2020): xml files used to build Bayesian phylogenies. figshare. Figshare Dataset. https://doi.org/10.6084/m9.figshare.12101055.v1 (accessed on 10 February 2021).

Acknowledgments: We thank Michael R. McKain, Yesid Cuesta Astroz and Diego Mauricio Riaño Pachón for their valuable advice during the study, Luis Eduardo Mejia for help with illustrations, Alejandro Rodriguez Cabal for his guidance in Linux and bioinformatics tools, Isabel Cristina Calle for experimental assistance doing PCR assays and Juan Santiago Zuluaga for financial help to support Juliana Arcila work at CIB. Sequence data for *Pseudovirgaria hyperparasitica* were produced by the US Department of Energy Joint Genome Institute http://www.jgi.doe.gov/in (accessed on 10 February 2021) collaboration with the user community.

Conflicts of Interest: The authors have declared that no competing interests exist.

References

1. Wijayawardene, N.N.; Crous, P.W.; Kirk, P.M.; Hawksworth, D.L.; Boonmee, S.; Braun, U.; Dai, D.Q.; D'souza, M.J.; Diederich, P.; Dissanayake, A.; et al. Naming and outline of Dothideomycetes-2014 including proposals for the protection or suppression of generic names. *Fungal. Divers.* **2014**, *69*, 1–55. [CrossRef]
2. Videira, S.I.R.; Groenewald, J.Z.; Nakashima, C.; Braun, U.; Barreto, R.W.; de Wit, P.J.; Crous, P.W. Mycosphaerellaceae—Chaos or clarity? *Stud. Mycol.* **2017**, *87*, 257–421. [CrossRef] [PubMed]
3. Carlier, J.; Zapater, M.-F.; Lapeyre, F.; Jones, D.R.; Mourichon, X. Septoria Leaf Spot of Banana: A Newly Discovered Disease Caused by *Mycosphaerella eumusae* (Anamorph *Septoria eumusae*). *Phytopathology* **2000**, *90*, 884–890. [CrossRef] [PubMed]
4. Thomma, B.P.; Van Esse, H.P.; Crous, P.W.; de Wit, P.J. *Cladosporium fulvum* (syn. *Passalora fulva*), a highly specialized plant pathogen as a model for functional studies on plant pathogenic Mycosphaerellaceae. *Mol. Plant Pathol.* **2005**, *6*, 379–393. [CrossRef] [PubMed]
5. Stukenbrock, E.H.; Banke, S.; Javan-Nikkhah, M.; McDonald, B.A. Origin and Domestication of the Fungal Wheat Pathogen *Mycosphaerella graminicola* via Sympatric Speciation. *Mol. Biol. Evol.* **2006**, *24*, 398–411. [CrossRef]
6. De Lapeyre, L.; Bellaire, D.; Fouré, E.; Abadie, C.; Carlier, J. Black Leaf Streak Disease is challenging the banana industry. *Fruits* **2010**, *65*, 327–342. [CrossRef]
7. Drenkhan, R.; Adamson, K.; Jürimaa, K.; Hanso, M. *Dothistroma septosporum* on firs (*Abies* spp.) in the northern Baltics. *For. Pathol.* **2014**, *44*, 250–254. [CrossRef]
8. Dhillon, B.; Feau, N.; Aerts, A.L.; Beauseigle, S.; Bernier, L.; Copeland, A.; Foster, A.; Navdeep, G.; Henrissat, B.; Herath, P.; et al. Horizontal gene transfer and gene dosage drives adaptation to wood colonization in a tree pathogen. *Proc. Natl. Acad. Sci. USA* **2015**, *112*, 3451–3456. [CrossRef]
9. Arzanlou, M.; Groenewald, J.Z.; Fullerton, R.A.; Abeln, E.C.A.; Carlier, J.; Zapater, M.-F.; Buddenhagen, W.; Viljoen, A.; Crous, P.W. Multiple gene genealogies and phenotypic characters differentiate several novel species of Mycosphaerella and related anamorphs on banana. *Pers. Mol. Phylogeny Evol. Fungi* **2008**, *20*, 19–37. [CrossRef]

10. Chang, T.C.; Salvucci, A.; Crous, P.W.; Stergiopoulos, I. Comparative Genomics of the Sigatoka Disease Complex on Banana Suggests a Link between Parallel Evolutionary Changes in *Pseudocercospora fijiensis* and *Pseudocercospora eumusae* and Increased Virulence on the Banana Host. *PLoS Genet.* **2016**, *12*, e1005904. [CrossRef] [PubMed]
11. Ohm, R.A.; Feau, N.; Henrissat, B.; Schoch, C.L.; Horwitz, B.A.; Barry, K.W.; Condon, B.J.; Copeland, A.C.; Dhillon, B.; Glaser, F.; et al. Diverse Lifestyles and Strategies of Plant Pathogenesis Encoded in the Genomes of Eighteen Dothideomycetes Fungi. *PLoS Pathog.* **2012**, *8*, e1003037. [CrossRef]
12. Arango Isaza, R.E.; Diaz-Trujillo, C.; Dhillon, B.; Aerts, A.; Carlier, J.; Crane, C.F.; de Jong, T.V.; de Vries, I.; Dietrich, R.; Farmer, A.D.; et al. Combating a Global Threat to a Clonal Crop: Banana Black Sigatoka Pathogen *Pseudocercospora fijiensis* (Synonym *Mycosphaerella fijiensis*) Genomes Reveal Clues for Disease Control. *PLoS Genet.* **2016**, *12*, e1005876. [CrossRef]
13. Churchill, A.C.L. *Mycosphaerella fijiensis*, the black leaf streak pathogen of banana: Progress towards understanding pathogen biology and detection, disease development, and the challenges of control. *Mol. Plant. Pathol.* **2011**, *12*, 307–328. [CrossRef] [PubMed]
14. Forget, L.; Ustinova, J.; Wang, Z.; Huss, V.A.R.; Franz Lang, B. *Hyaloraphidium curvatum*: A Linear Mitochondrial Genome, tRNA Editing, and an Evolutionary Link to Lower Fungi. *Mol. Biol. Evol.* **2002**, *19*, 310–319. [CrossRef] [PubMed]
15. Liu, W.; Cai, Y.; Zhang, Q.; Chen, L.; Shu, F.; Ma, X.; Bian, Y. The mitochondrial genome of *Morchella importuna* (272.2 kb) is the largest among fungi and contains numerous introns, mitochondrial non-conserved open reading frames and repetitive sequences. *Int. J. Biol. Macromol.* **2020**, *143*, 373–381. [CrossRef]
16. Hausner, G. Fungal mitochondrial genomes. *Fungal Genomics.* **2003**, *3*, 101.
17. Clark-Walker, G.D. Evolution of Mitochondrial Genomes in Fungi. *Int. Rev. Cytol.* **1992**, *141*, 89–127. [CrossRef]
18. Wolf, K.; Giudice, L.D. The Variable Mitochondrial Genome of Ascomycetes: Organization, Mutational Alterations, and Expression. *Adv. Genet.* **1988**, *25*, 185–308.
19. Paquin, B.; Laforest, M.J.; Forget, L.; Roewer, I.; Wang, Z.; Longcore, J.; Lang, B.F. The fungal mitochondrial genome project: Evolution of fungal mitochondrial genomes and their gene expression. *Curr. Genet.* **1997**, *31*, 380–395. [CrossRef]
20. Aguileta, G.; De Vienne, D.M.; Ross, O.N.; Hood, M.E.; Giraud, T.; Petit, E.; Gabaldón, T. High variability of mitochondrial gene order among fungi. *Genome Biol. Evol.* **2014**, *6*, 451–465. [CrossRef] [PubMed]
21. Kawano, S.; Takano, H.; Kuroiwa, T. Sexuality of Mitochondria: Fusion, Recombination, and Plasmids. *Int. Rev. Cytol.* **1995**, *161*, 49–110. [CrossRef]
22. Burger, G.; Gray, M.W.; Lang, B.F. Mitochondrial genomes: Anything goes. *Trends Genet.* **2003**, *19*, 709–716. [CrossRef] [PubMed]
23. Franz Lang, B.; Laforest, M.-J.; Burger, G. Mitochondrial introns: A critical view. *Trends Genet.* **2007**, *23*, 119–125. [CrossRef] [PubMed]
24. Alverson, A.J.; Wei, X.; Rice, D.W.; Stern, D.B.; Barry, K.; Palmer, J.D. Insights into the Evolution of Mitochondrial Genome Size from Complete Sequences of *Citrullus lanatus* and *Cucurbita pepo* (Cucurbitaceae). *Mol. Biol. Evol.* **2010**, *27*, 1436–1448. [CrossRef]
25. Bernt, M.; Braband, A.; Schierwater, B.; Stadler, P.F. Genetic aspects of mitochondrial genome evolution. *Mol. Phylogenet. Evol.* **2013**, *69*, 328–338. [CrossRef] [PubMed]
26. Torriani, S.F.F.; Penselin, D.; Knogge, W.; Felder, M.; Taudien, S.; Platzer, M.; McDonald, B.A.; Brunner, P.C. Comparative analysis of mitochondrial genomes from closely related *Rhynchosporium* species reveals extensive intron invasion. *Fungal. Genet. Biol.* **2014**, *62*, 34–42. [CrossRef]
27. Basse, C.W. Mitochondrial inheritance in fungi. *Curr. Opin. Microbiol.* **2010**, *13*, 712–719. [CrossRef] [PubMed]
28. Griffiths, A.J.F. Mitochondrial inheritance in filamentous fungi. *J. Genet.* **1996**, *75*, 403–414. [CrossRef]
29. Beimforde, C.; Feldberg, K.; Nylinder, S.; Rikkinen, J.; Tuovila, H.; Dörfelt, H.; Gube, M.; Jackson, D.J.; Reitner, J.; Seyfullah, L.J.; et al. Estimating the Phanerozoic history of the Ascomycota lineages: Combining fossil and molecular data. *Mol. Phylogenet. Evol.* **2014**, *78*, 386–398. [CrossRef]
30. Torriani, S.F.F.; Goodwin, S.B.; Kema, G.H.J.; Pangilinan, J.L.; McDonald, B.A. Intraspecific comparison and annotation of two complete mitochondrial genome sequences from the plant pathogenic fungus *Mycosphaerella graminicola*. *Fungal. Genet. Biol.* **2008**, *45*, 628–637. [CrossRef]
31. Goodwin, S.B.; Mccorison, C.B.; Cavaletto, J.R.; Culley, D.E.; Labutti, K.; Baker, S.E.; Grigoriev, I.V. The mitochondrial genome of the ethanol-metabolizing, wine cellar mold *Zasmidium cellare* is the smallest for a filamentous ascomycete. *Fungal. Biol.* **2016**, *120*, 961–974. [CrossRef] [PubMed]
32. Johanson, A.; Jeger, M.J. Use of PCR for detection of *Mycosphaerella fijiensis* and *M. musicola*, the causal agents of Sigatoka leaf spots in banana and plantain. *Mycol. Res.* **1993**, *97*, 670–674. [CrossRef]
33. Cañas-Gutiérrez, G.P.; Angarita-Velásquez, M.J.; Restrepo-Flórez, J.M.; Rodríguez, P.; Moreno, C.X.; Arango, R. Analysis of the CYP51 gene and encoded protein in propiconazole-resistant isolates of *Mycosphaerella fijiensis*. *Pest. Manag. Sci.* **2009**, *65*, 892–899. [CrossRef]
34. Benson, D.A.; Boguski, M.S.; Lipman, D.J.; Ostell, J.; Ouellette, B.F.F.; Rapp, B.A.; Wheeler, D.L. GenBank. *Nucleic. Acids. Res.* **1999**, *27*, 12–17. [CrossRef] [PubMed]
35. O'Leary, N.A.; Wright, M.W.; Brister, J.R.; Ciufo, S.; Haddad, D.; McVeigh, R.; Rajput, B.; Robbertse, B.; Smith-White, B.; Ako-Adjei, D.; et al. Reference sequence (RefSeq) database at NCBI: Current status, taxonomic expansion, and functional annotation. *Nucleic. Acids. Res.* **2016**, *44*, D733–D745. [CrossRef]

36. Grigoriev, I.V.; Nikitin, R.; Haridas, S.; Kuo, A.; Ohm, R.; Otillar, R.; Riley, R.; Salamov, A.; Zhao, X.; Korzeniewski, F.; et al. MycoCosm portal: Gearing up for 1000 fungal genomes. *Nucleic Acids. Res.* **2014**, *42*, D699–D704. [CrossRef] [PubMed]
37. Andrews, S. FastQC: A Quality Control Tool for High Throughput Sequence Data. 2010. Available online: https://www.bioinformatics.babraham.ac.uk/projects/fastqc/ (accessed on 1 March 2021).
38. Bolger, A.M.; Lohse, M.; Usadel, B. Trimmomatic: A flexible trimmer for Illumina sequence data. *Bioinformatics* **2014**, *30*, 2114–2120. [CrossRef]
39. Bankevich, A.; Nurk, S.; Antipov, D.; Gurevich, A.A.; Dvorkin, M.; Kulikov, A.S.; Kulikov, A.S.; Lesin, V.M.; Nikolenko, S.I.; Pham, S.; et al. SPAdes: A New Genome Assembly Algorithm and Its Applications to Single-Cell Sequencing. *J. Comput. Biol.* **2012**, *19*, 455–477. [CrossRef] [PubMed]
40. Boetzer, M.; Henkel, C.V.; Jansen, H.J.; Butler, D.; Pirovano, W. Scaffolding pre-assembled contigs using SSPACE. *Bioinformatics* **2011**, *27*, 578–579. [CrossRef] [PubMed]
41. Nadalin, F.; Vezzi, F.; Policriti, A. GapFiller: A de novo assembly approach to fill the gap within paired reads. *BMC Bioinform.* **2012**, *13* (Suppl. 14), S8. [CrossRef]
42. Hunt, M.; Kikuchi, T.; Sanders, M.; Newbold, C.; Berriman, M.; Otto, T.D. REAPR: A universal tool for genome assembly evaluation. *Genome Biol.* **2013**, *14*, R47. [CrossRef] [PubMed]
43. Kearse, M.; Moir, R.; Wilson, A.; Stones-Havas, S.; Cheung, M.; Sturrock, S.; Buxton, S.; Cooper, A.; Markowitz, S.; Duran, C.; et al. Geneious Basic: An integrated and extendable desktop software platform for the organization and analysis of sequence data. *Bioinformatics* **2012**, *28*, 1647–1649. [CrossRef] [PubMed]
44. Camacho, C.; Coulouris, G.; Avagyan, V.; Ma, N.; Papadopoulos, J.; Bealer, K.; Madden, T.L. BLAST+: Architecture and applications. *BMC Bioinform.* **2009**, *10*, 421. [CrossRef] [PubMed]
45. Hane, J.K.; Lowe, R.G.T.; Solomon, P.S.; Tan, K.-C.; Schoch, C.L.; Spatafora, J.W.; Crous, P.W.; Kodira, C.; Birren, B.W.; Galagan, J.E.; et al. Dothideomycete plant interactions illuminated by genome sequencing and EST analysis of the wheat pathogen Stagonospora nodorum. *Plant Cell* **2007**, *19*, 3347–3368. [CrossRef]
46. Shen, X.-Y.; Li, T.; Chen, S.; Fan, L.; Gao, J.; Hou, C.-L. Characterization and Phylogenetic Analysis of the Mitochondrial Genome of *Shiraia bambusicola* Reveals Special Features in the Order of Pleosporales. *PLoS ONE* **2015**, *10*, e0116466. [CrossRef] [PubMed]
47. Hahn, C.; Bachmann, L.; Chevreux, B. Reconstructing mitochondrial genomes directly from genomic next-generation sequencing reads—A baiting and iterative mapping approach. *Nucleic Acids Res.* **2013**, *41*, e129. [CrossRef]
48. Johnson, M.; Zaretskaya, I.; Raytselis, Y.; Merezhuk, Y.; McGinnis, S.; Madden, T.L. NCBI BLAST: A better web interface. *Nucleic Acids Res.* **2008**, *36*, W5–W9. [CrossRef] [PubMed]
49. Bernt, M.; Donath, A.; Jühling, F.; Externbrink, F.; Florentz, C.; Fritzsch, G.; Pütz, J.; Middendorf, M.; Stadler, P.F. MITOS: Improved de novo metazoan mitochondrial genome annotation. *Mol. Phylogenet. Evol.* **2013**, *69*, 313–319. [CrossRef] [PubMed]
50. Finn, R.D.; Bateman, A.; Clements, J.; Coggill, P.; Eberhardt, R.Y.; Eddy, S.R.; Heger, A.; Hetherington, K.; Holm, L.; Mistry, J.; et al. Pfam: The protein families database. *Nucleic Acids Res.* **2014**, *42*, D222–D230. [CrossRef] [PubMed]
51. Mi, H.; Huang, X.; Muruganujan, A.; Tang, H.; Mills, C.; Kang, D.; Thomas, P.D. PANTHER version 11: Expanded annotation data from Gene Ontology and Reactome pathways, and data analysis tool enhancements. *Nucleic Acids Res.* **2017**, *45*, D183–D189. [CrossRef]
52. Edgar, R.C. MUSCLE: Multiple sequence alignment with high accuracy and high throughput. *Nucleic Acids Res.* **2004**, *32*, 1792–1797. [CrossRef] [PubMed]
53. Larkin, M.A.; Blackshields, G.; Brown, N.P.; Chenna, R.; McGettigan, P.A.; McWilliam, H.; Valentin, F.; Wallace, I.M.; Wilm, A.; Lopez, R.; et al. Clustal W and Clustal X version 2.0. *Bioinformatics* **2007**, *23*, 2947–2948. [CrossRef] [PubMed]
54. Grabherr, M.G.; Haas, B.J.; Yassour, M.; Levin, J.Z.; Thompson, D.A.; Amit, I.; Adiconis, X.; Fan, L.; Raychowdhury, R.; Zeng, Q.; et al. Full-length transcriptome assembly from RNA-Seq data without a reference genome. *Nat. Biotechnol.* **2011**, *29*, 644–652. [CrossRef] [PubMed]
55. Beier, S.; Thiel, T.; Münch, T.; Scholz, U.; Mascher, M. MISA-web: A web server for microsatellite prediction. *Bioinformatics* **2017**, *33*, 2583–2585. [CrossRef]
56. Okorski, A.; Pszczółkowska, A.; Jastrzębski, J.P.; Paukszto, Ł.; Okorska, S. The complete mitogenome of Mycosphaerella pinodes (Ascomycota, Mycosphaerellaceae). *Mitochondrial DNA Part B* **2016**, *1*, 48–49. [CrossRef]
57. Braun, U.; Crous, P.W.; Groenewald, J.Z.; Scheuer, C. *Pseudovirgaria*, a fungicolous hyphomycete genus. *IMA Fungus* **2011**, *2*, 65–69. [CrossRef]
58. Stamatakis, A. RAxML version 8: A tool for phylogenetic analysis and post-analysis of large phylogenies. *Bioinformatics* **2014**, *30*, 1312–1313. [CrossRef]
59. Guindon, S.; Dufayard, J.F.; Lefort, V.; Anisimova, M.; Hordijk, W.; Gascuel, O. New Algorithms and Methods to Estimate Maximum-Likelihood Phylogenies: Assessing the Performance of PhyML 3.0. *Syst. Biol.* **2010**, *59*, 307–321. [CrossRef]
60. Bouckaert, R.; Heled, J.; Kühnert, D.; Vaughan, T.; Wu, C.-H.; Xie, D.; Suchard, M.A.; Rambaut, A.; Drummond, A.J. BEAST 2: A Software Platform for Bayesian Evolutionary Analysis. *PLoS Comput. Biol.* **2014**, *10*, e1003537. [CrossRef] [PubMed]
61. Darriba, D.; Taboada, G.L.; Doallo, R.; Posada, D. jModelTest 2: More models, new heuristics and parallel computing. *Nat. Methods* **2012**, *9*, 772. [CrossRef]
62. Schwarz, G. Estimating the Dimension of a Model. *Ann. Stat.* **1978**, *6*, 461–464. [CrossRef]

63. Drummond, A.J.; Ho, S.Y.W.; Phillips, M.J.; Rambaut, A. Relaxed phylogenetics and dating with confidence. *PLoS Biol.* **2006**, *4*, 88. [CrossRef] [PubMed]
64. Schmidt, A.R.; Beimforde, C.; Seyfullah, L.J.; Wege, S.-E.; Dörfelt, H.; Girard, V.; Grabenhorst, H.; Gube, M.; Heinrichs, J.; Nel, A.; et al. Amber fossils of sooty moulds. *Rev. Palaeobot Palynol.* **2014**, *200*, 53–64. [CrossRef]
65. Rambaut, A.; Drummond, A. *Tracer: MCMC Trace Analysis Tool, Version 1.5.*; University of Oxford: Oxford, UK, 2009; Available online: http://tree.bio.ed.ac.uk/software/tracer/ (accessed on 1 March 2021).
66. Rambaut, A. FigTree: Tree Figure Drawing Tool, Version 1.3.1. 2006. Available online: http://tree.bio.ed.ac.uk/software/figtree/ (accessed on 1 March 2021).
67. Megarioti, A.H.; Kouvelis, V.N. The Coevolution of Fungal Mitochondrial Introns and Their Homing Endonucleases (GIY-YIG and LAGLIDADG). *Genome. Biol. Evol.* **2020**, *12*, 1337–1354. [CrossRef] [PubMed]
68. Griffiths, A.J. Natural plasmids of filamentous fungi. *Microbiol. Rev.* **1995**, *59*, 673–685. [CrossRef] [PubMed]
69. Cahan, P.; Kennell, J.C. Identification and distribution of sequences having similarity to mitochondrial plasmids in mitochondrial genomes of filamentous fungi. *Mol. Genet. Genom.* **2005**, *273*, 462–473. [CrossRef] [PubMed]
70. Jelen, V.; de Jonge, R.; Van de Peer, Y.; Javornik, B.; Jakše, J. Complete mitochondrial genome of the Verticillium-wilt causing plant pathogen *Verticillium nonalfalfae*. *PLoS ONE* **2016**, *11*, e0148525. [CrossRef]
71. Salvatori, R. *Organization and Regulation of Mitochondrial Gene Expression*; Stockholm University: Stockholm, Sweden, 2020. Available online: https://www.diva-portal.org/smash/get/diva2:1416629/FULLTEXT01.pdf?fbclid=IwAR1Ncc4BTQ26Ygmh_aPBdwvK6HDsivbSSp44OIHnBdMdsSYEzCTqT6P0Bwo (accessed on 1 July 2017).
72. Joardar, V.; Abrams, N.F.; Hostetler, J.; Paukstelis, P.J.; Pakala, S.; Pakala, S.B.; Zafar, N.; Abolude, O.O.; Payne, J.; Andrianopoulos, A.; et al. Sequencing of mitochondrial genomes of nine *Aspergillus* and *Penicillium* species identifies mobile introns and accessory genes as main sources of genome size variability. *BMC Genom.* **2012**, *13*, 698. [CrossRef]
73. Mardanov, A.V.; Beletsky, A.V.; Kadnikov, V.V.; Ignatov, A.N.; Ravin, N.V. The 203 kbp mitochondrial genome of the phytopathogenic fungus *Sclerotinia borealis* reveals multiple invasions of introns and genomic duplications. *PLoS ONE* **2014**, *9*, e107536. [CrossRef]
74. Dassa, B.; London, N.; Stoddard, B.L.; Schueler-Furman, O.; Pietrokovski, S. Fractured genes: A novel genomic arrangement involving new split inteins and a new homing endonuclease family. *Nucleic Acids Res.* **2009**, *37*, 2560–2573. [CrossRef]
75. Sanderson, M.J. r8s: Inferring absolute rates of molecular evolution and divergence times in the absence of a molecular clock. *Bioinformatics* **2003**, *19*, 301–302. [CrossRef] [PubMed]
76. Gueidan, C.; Ruibal, C.; de Hoog, G.S.; Schneider, H. Rock-inhabiting fungi originated during periods of dry climate in the late Devonian and middle Triassic. *Fungal. Biol.* **2011**, *115*, 987–996. [CrossRef] [PubMed]
77. Christelová, P.; Valárik, M.; Eva, H.; Langhe, E.D.; Dole, J. A multi gene sequence-based phylogeny of the Musaceae (banana) family. *BMC Evol. Biol.* **2011**, *11*, 103. [CrossRef] [PubMed]

Article

The Mitochondrial Genome of the Sea Anemone *Stichodactyla haddoni* Reveals Catalytic Introns, Insertion-Like Element, and Unexpected Phylogeny

Steinar Daae Johansen [1,2,*], Sylvia I. Chi [3], Arseny Dubin [1] and Tor Erik Jørgensen [1]

1 Faculty of Biosciences and Aquaculture, Nord University, 8049 Bodø, Norway; aaduu@protonmail.com (A.D.); tor.e.jorgensen@nord.no (T.E.J.)
2 Department of Medical Biology, Faculty of Health Sciences, UiT—The Arctic University of Norway, 9037 Tromsø, Norway
3 Centre for Innovation, Canadian Blood Services, Ottawa, ON K1G 4J5, Canada; sylvia.ighemchi@blood.ca
* Correspondence: steinar.d.johansen@nord.no

Abstract: A hallmark of sea anemone mitochondrial genomes (mitogenomes) is the presence of complex catalytic group I introns. Here, we report the complete mitogenome and corresponding transcriptome of the carpet sea anemone *Stichodactyla haddoni* (family Stichodactylidae). The mitogenome is vertebrate-like in size, organization, and gene content. Two mitochondrial genes encoding NADH dehydrogenase subunit 5 (ND5) and cytochrome c oxidase subunit I (COI) are interrupted with complex group I introns, and one of the introns (ND5-717) harbors two conventional mitochondrial genes (ND1 and ND3) within its sequence. All the mitochondrial genes, including the group I introns, are expressed at the RNA level. Nonconventional and optional mitochondrial genes are present in the mitogenome of *S. haddoni*. One of these gene codes for a COI-884 intron homing endonuclease and is organized in-frame with the upstream COI exon. The insertion-like *orfA* is expressed as RNA and translocated in the mitogenome as compared with other sea anemones. Phylogenetic analyses based on complete nucleotide and derived protein sequences indicate that *S. haddoni* is embedded within the family Actiniidae, a finding that challenges current taxonomy.

Keywords: Actiniaria; group I intron; mtDNA; mitogenome; phylogeny; rearrangement; sea anemone

1. Introduction

Hexacoral mitochondria harbor economically organized and vertebrate-like mitochondrial genomes (mitogenomes) ranging in size from 16 to 22 kb [1]. Most mitogenomes consist of a single circular DNA coding for the same set of two ribosomal RNAs (rRNAs) and 13 hydrophobic oxidative phosphorylation (OxPhos) proteins as compared with that of vertebrates [2]. More than 200 complete hexacoral mitogenome sequences are available representing all five extant orders, i.e., Actiniaria (sea anemones), Zoantharia (colonial anemones), Corallimorpharia (mushroom corals), Antipatharia (black corals), and Scleractinia (stony corals) [1]. Sequencing analyses indicate frequent occurrence of nonconventional and optional genes, as well as a highly reduced tRNA gene repertoire [2–7]. Insertion-like *orfA* is a representative of a widespread nonconventional mitochondrial gene in sea anemones, but its RNA transcript or derived protein has, so far, not been linked to a cellular function [5,8]. The most unusual feature, however, is the presence of complex group I introns [1,9].

Group I introns are mobile genetic elements found in a variety of genetic compartments, including mitochondria [10,11]. Mitochondrial introns in metazoans are scarce and restricted to only a few orders within the Cnidaria, Porifera, or Placozoa phyla [2,12]. Group I introns catalyze their own splicing reaction at the RNA level by a ribozyme consisting of conserved RNA paired segments (named P1 to P9) organized into catalytic, substrate, and scaffold helical stack domains [13–15].

The RNA splicing reaction is well studied [16] and initiated by an exogenous guanosine (exoG) cofactor associated with the P7 paired segment, which subsequently attacks and cleaves RNA at the 5′ splice site (SS) within segment P1. The exoG becomes covalently ligated to the 5′ end of the intron RNA. In the second step of splicing, the terminal intron nucleotide (ωG) replaces exoG in P7 and becomes attacked by the free 3′ hydroxyl group of the upstream exon. The result of splicing is intron excision and exon ligation. Some group I introns carry homing endonuclease genes (HEGs) that promote genetic mobility at the DNA level [15,17,18]. Homing endonucleases are sequence specific DNases that cleave an intron-lacking allele of the host gene, resulting in intron spread by homing. There are several distinct families of HEGs, and mobile group I introns in mitochondria encode homing endonucleases of the LAGLIDADG family [17].

Two mitochondrial genes are found interrupted by complex group I introns in hexacorals. The NADH dehydrogenase subunit 5 gene (ND5) contains an intron at position 717 (human ND5 gene numbering [19]). The ND5-717 intron is obligatory, strictly vertically inherited, and has a fungal origin [19]. ND5-717 varies dramatically in size between hexacoral orders due to a large P8 insertion containing 2–15 conventional OxPhos genes. The shortest form, which is represented by most sea anemone species, corresponds to a P8 insertion of the ND1 and ND3 genes. Both group I intron cis-splicing (most sea anemones) and back-splicing (mushroom corals) appear involved in ND5-717 RNA processing [1,7]. The cytochrome c oxidase subunit I (COI) gene contains mobile-type introns at three different genic positions (720, 867, or 884; human COI gene numbering [19]). Different hexacoral orders tend to harbor COI introns representing distinct evolutionary histories [1,20,21], and sea anemones carry COI-884 group I introns containing LAGLIDADG-type HEGs within P8 [5,8,20].

The family Stichodactylidae consists of only of two genera, *Heteractis* and *Stichodactyla*, and about 10 species [22]. Carpet sea anemones of the genus *Stichodactyla* have a distinct morphology with flattened column and very short tentacles that complicates taxonomy based on external features [23]. Phylogenetic analyses based on partial ribosomal RNA gene sequences from nuclei and mitochondria indicate a relationship between *Stichodactyla* and members of the family Actiniidae [22,24]. Here, we report the complete mitogenome and corresponding transcriptome of the tropical sea anemone *Stichodactyla haddoni*. We characterize two complex catalytic group I introns and provide support of a mitogenome rearrangement involving the nonconventional mitochondrial *orfA*. Mitogenome-based phylogeny indicates that *S. haddoni* is embedded within the family Actiniidae.

2. Materials and Methods

2.1. Animal Collection and Nucleic Acid Isolation

A specimen of Haddon's Carpet Anemone (*Stichodactyla haddoni*) of Indo-Pacific origin was retrieved, in 2013, from a pet shop (Tromsø, Norway) and kept alive for several years in our in-house reef tank at the University of Tromsø (UiT) (Tromsø, Norway). Genomic DNA was isolated from approximately 20 mg of fresh tissue containing tentacles, column, and oral disc using the Wizard Genomic DNA Purification kit (Promega, Madison, WI, USA), according to the manufacturer's instructions. Tissue was homogenized in 600 µL nuclei lysis solution for 20 s at 6000 rpm using Precellys 24 homogenizer (Stretton Scientific, Stretton, UK), and polysaccharide contaminants were removed by phenol/chloroform extraction steps. The purified DNA was eluted in water. Total RNA was isolated using the TRIzol reagent (Thermo Fisher Scientific, Waltham, MA, USA), as previously described [6]. Approximately 30 mg fresh tissue containing tentacles, column, and oral disc was crushed directly in liquid nitrogen, and then in Trizol using the same "Precellys" settings as described above for DNA isolation. The RNA was purified from each tissue separately by repeated chloroform extractions, followed by precipitation in isopropanol overnight at 4 °C. The RNA pellet was washed in 70% ethanol and resuspended in nuclease-free water (Thermo Fisher Scientific, Waltham, MA, USA).

2.2. DNA and RNA Sequencing

Genomic DNA and total RNA were subjected to whole genome and transcriptome Ion Personal Genome Machine (Ion PGM) sequencing, essentially as previously described [6]. In short, all library preparations, template reactions, and sequencing steps were performed according to the standard protocols. Ion Xpress ™ Plus gDNA Fragment Library Preparation kit (Thermo Fisher Scientific, Waltham, MA, USA) was used for DNA library preparation. Approximately 1 μg input genomic DNA was physically sheared for 400 bp selection on a Covaris S2 sonicator (Woburn, MA, USA). The whole transcriptome library was constructed using an Ion Total RNA-Seq Kit v2. Total RNA was polyA-selected using the mRNA DIRECT Purification Kit (Thermo Fisher Scientific, Waltham, MA, USA) and subsequently fragmented enzymatically. The fragmented total RNA was subjected to reverse transcription with a reverse transcriptase mix (Invitrogen 10X SuperScript® III Enzyme Mix, Thermo Fisher Scientific, Waltham, MA, USA). Final preparations and sequencing were performed using the Ion PGM™ Sequencing 400 Kit (Thermo Fisher Scientific, Waltham, MA, USA), according to the manufacturer's protocol and on 316 v2 chips. Selected mtDNA regions, including non-coding intergenic regions, were subjected to PCR amplification, plasmid cloning, and Sanger sequencing using specific primers, essentially as previously described [6,8].

2.3. Mitogenome Assembly and Annotation

The mitogenome sequence of S. haddoni was assembled from the whole genome read pools. The initial sequence was built using MIRA assembler v3.4.1.1 [25] with the *Urticina eques* mitochondrial genome (HG423144) as a reference [5]. Remaining reads were subsequently mapped iteratively to the initial sequence using Mitochondrial baiting and iterative mapping (MITObim) script v1.6 [26] with default settings. Mitogenome annotation was performed using MITOS revision 272 [27], supported by manual correction of coding sequences.

2.4. Mitochondrial Transcriptome

Transcriptome data were generated for S. haddoni and S. helianthus. Mitochondrial mRNAs in sea anemones are polyadenylated [5,8]. Transcripts from S. haddoni were unambiguously determined, using the corresponding mitogenome sequence (MW760873) as reference, by analysis of quality-filtered Ion PGM reads on CLC Genomics Workbench v8.5 (CLC-Bio, Aarhud, Denmark). About 15.45 million poly(A) enriched RNA reads were obtained, corresponding to 5.90 million reads from oral disk, 5.60 million reads from tentacles, and 3.95 million reads from column; 8530 reads (0.055%) were unambiguously identified as mitochondrial transcripts. Transcriptome data of S. helianthus, obtained by Illumina paired reads, was retrieved from the NCBI SRA database (SRR7126073) [28]. About 42.68 million poly(A) enriched RNA reads were obtained, and 12,053 reads (0.028%) were unambiguously identified as mitochondrial transcripts when compared to the S. haddoni mitogenome sequence (MW760873). The Illumina paired reads were mapped by "STAR" version 2.7 (https://pubmed.ncbi.nlm.nih.gov/23104886/ (accessed on 1 March 2021)).

2.5. Phylogenetic Analysis

The sequences were aligned with T-COFFEE v11.00 using t_coffee_msa, mafft_msa, muscle_msa parameters. Alignments for each gene were created and trimmed independently, and then concatenated to make the final sequence. All internal gaps were included. A general time reversible substitution model with discrete gamma distribution for rate heterogeneity across sites (GTRGAMMA) was applied to all partitions. Phylogenetic analyses were conducted using MEGA X software [29] and RAxML version 8.2.12 for ML trees. All sequence alignments were model tested prior to tree constructions. Maximum-likelihood (ML), neighbor joining (NJ), and minimal evolution (ME) methods were used for comparison. The topologies of the trees were evaluated by 500 bootstrap replicates.

3. Results

3.1. Characteristic Features of the S. Haddoni Mitogenome

The complete circular mitogenome (mtDNA) sequence of the *S. haddoni* sea anemone (18.999 bp, GenBank accession number MW760873) was determined on both strands using a combined Ion PGM and Sanger sequencing strategy. The 19 annotated mitochondrial genes (Table S1) were all located on the same strand and correspond to two rRNA genes (encoding SSU and LSU rRNAs), two tRNA genes (encoding tRNA^{fMet} and tRNA^{Trp}), and 15 protein coding genes. The latter group consisted of 13 conventional mitochondrial genes coding for OxPhos proteins common among most metazoan mitogenomes, and two nonconventional mitochondrial protein genes located within intron and intergenic regions (IGR).

All conventional mitochondrial genes, including rRNA and tRNA genes, were organized in a similar order as compared with that of most sea anemone mitogenomes investigated [1], and a linear map of the circular *S. haddoni* mtDNA is presented in Figure 1. Some interesting and unusual features were noted. Firstly, the ND5 gene was interrupted by a complex group I intron (ND5-717) harboring the genes of ND1 and ND3 within its structure. Secondly, the COI gene was intervened by a mobile-like group I intron (COI-884) containing a homing endonuclease gene (HEG). Finally, the nonconventional mitochondrial *orfA* found within IGR-6 in other sea anemones (Table S2) was translocated to IGR-12 in *S. haddoni*.

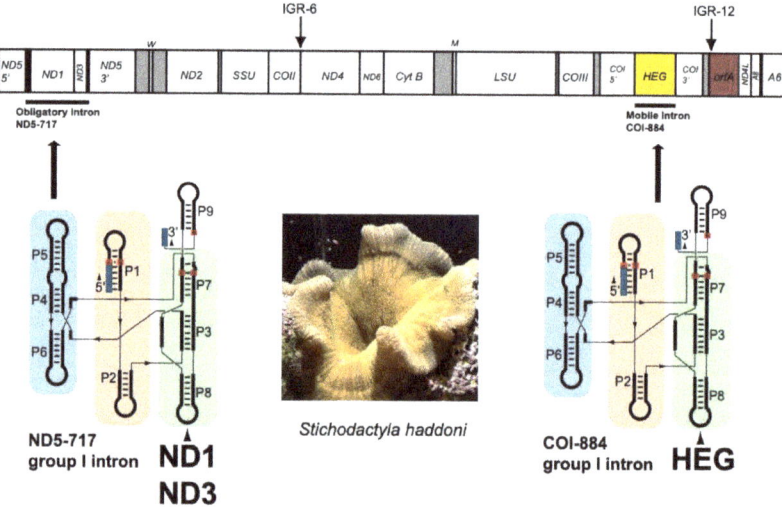

Figure 1. Mitogenome features in *Stichodactyla haddoni*. Gene content and organization of the circular mitogenome presented as a linear map. The mitogenome harbors 15 protein coding genes, two rRNA genes, and 2 tRNA genes. All genes are encoded by the same DNA strand. SSU and LSU, mitochondrial small- and large-subunit rRNA genes. The tRNA genes M and W (tRNA^{fMet} and tRNA^{Trp}) are indicated by the standard one-letter symbols for amino acids. ND1-6, NADH dehydrogenase subunit 1–6 genes; COI-III, cytochrome c oxidase subunit I-III genes; Cyt B, cytochrome b gene; A6 and A8, ATPase subunit 6 and 8 genes; HEG, homing endonuclease gene; *orfA*, open reading frame A gene; IGR-6 and -12, intergenic regions 6 and 12. The ND5-717 and CO-884 introns are schematically indicated below the gene map as paired segment (P1 to P9) catalytic RNA core folds. Segment P8 contains large insertions; ND1 and ND3 genes in ND5-717, and a HEG in COI-884. Photo of *Stichodactyla haddoni* specimen by S.D. Johansen.

3.2. Two Complex Catalytic Group I Introns

Two group I introns were present at conserved sites within the ND5 gene (site 717) and COI gene (site 884). Secondary structure folding of the corresponding RNA confirmed both introns as group I introns (Figure 2). Two interesting structural features were noted in the *S. haddoni* ND5-717 intron (Figure 2a). The terminal nucleotide of the intron, which is catalytical important and universally conserved as a guanosine (ωG) in group I introns, was replaced by ωA. The P8 segment was found to contain a large insertion harboring two OxPhos genes (ND1 and ND3 genes). Splicing of ND5-717 results in a ligated ND5 mRNA and the excised intron RNA is proposed to generate the mRNA precursors of ND1 and ND3 (Figure 3).

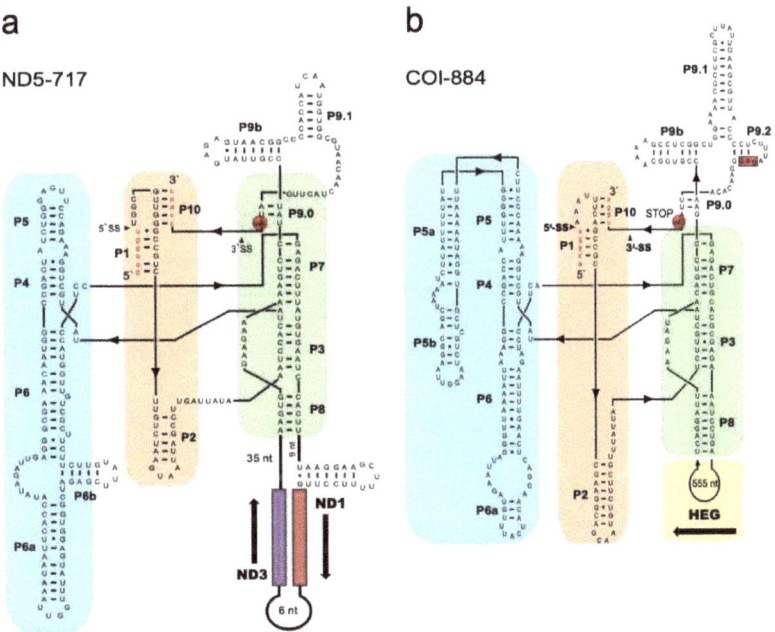

Figure 2. Secondary structure diagram of *Stichodactyla haddoni* mitochondrial group I introns. (**a**) Secondary structure of the ND5-717 group I ribozyme (segments P1 to P10). Flanking ND5 exon sequences shown in red lowercase letters. 5'SS and 3'SS indicate splice sites. The three helical stacks, named Scaffold, Substrate, and Catalytic domains are indicated by blue, yellow, and green boxes, respectively. The last nucleotide of the intron (ω), which is considered to be a universally conserved guanosine (ωG) among group I introns, is ωA in ND5-717 (red circle). The P8 segment harbors the ND1 and ND3 genes. (**b**) Secondary structure of the COI-884 group I ribozyme (segments P1 to P10). Flanking COI exons sequences are in red lowercase letters. The last intron nucleotide (ωG) is indicated (red circle). The three helical stacks are indicated by blue, yellow, and green boxes. The P8 extension contains the HEG insertion. Note that the HEG stop codon (UAG, red box) is located close to the 3' end of the intron sequence.

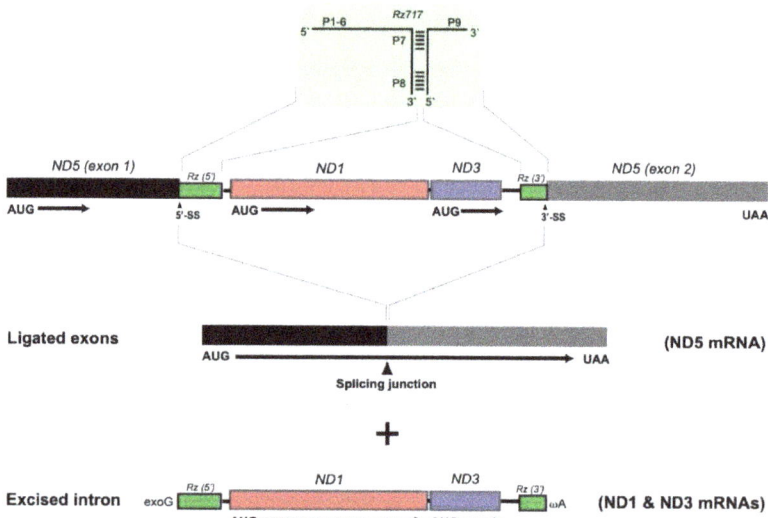

Figure 3. *Stichodactyla haddoni* ND5-717 intron splicing. The group I ribozyme (Rz717, green box) is indicated above the precursor map, and ligated ND5 mRNA is shown below. The ND1 and ND3 mRNAs are proposed generated from the excised intron. Initiation codon (AUG) and stop codon (UAA) of the ND5 mRNA, as well as the splicing junction, are indicated. exoG, exogenous guanosine cofactor; ωA, 3′ terminal intron nucleotide.

The mobile-like COI-884 intron is optional among sea anemones and noted in 81% of inspected species [1]. The secondary structure fold indicates a typical group IC1 intron with a complex back-folded P5 segment (Figure 2b). RNA splicing restores the COI mRNA by exon ligation (Figure 4a). The *S. haddoni* COI-884 intron was found to harbor a homing endonuclease gene (HEG) in segment P8, which expands its 5′ and 3′ ends into the entire intron. Thus, the COI-884 intron sequence has a dual coding potential of a homing endonuclease and a catalytic RNA.

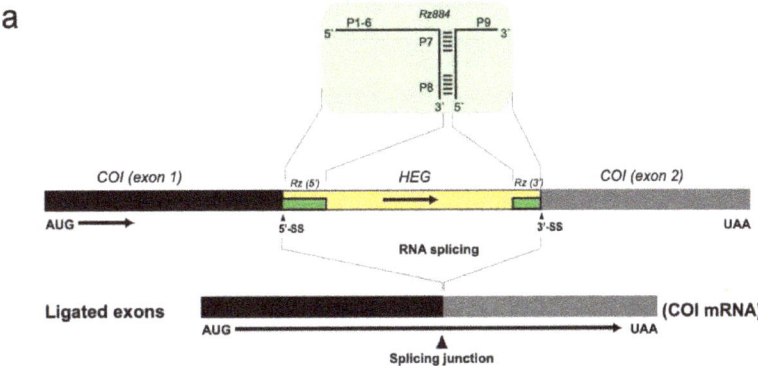
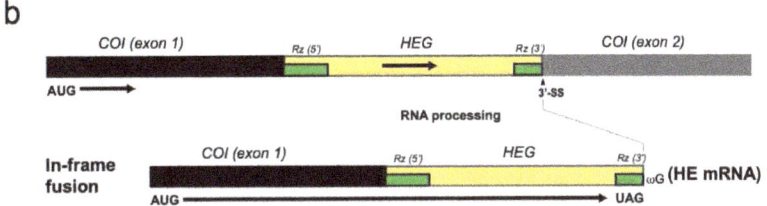
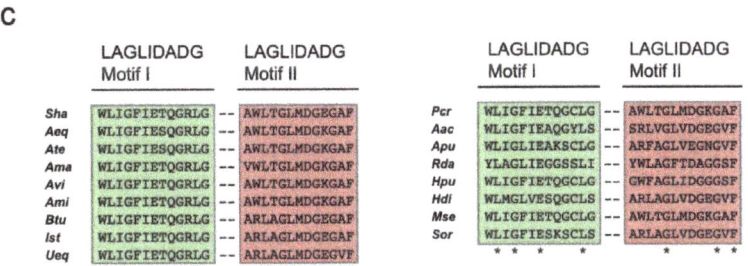

Figure 4. *Stichodactyla haddoni* COI-884 intron splicing and processing. (**a**) The group I ribozyme (Rz884, green box) is indicated above the precursor map, and ligated COI mRNA is shown below. Start (AUG) and stop (UAA) codons are indicated. HEG, homing endonuclease gene. (**b**) Intron RNA processing at the 3′ splice site (SS) generates a COI-HEG in-frame fusion product probably important for HEG expression. HE mRNA, homing endonuclease mRNA. (**c**) LAGLIDADG amino acid sequence motif of the COI-884 encoded homing endonucleases in sea anemones. Each endonuclease contains two copies of the sequence motif (green and red boxes). Left panel, alignment of *Stichodactyla haddoni* (Sha) to eight Actinidae species (Aeq, *Actinia equina*; Ate, *Actinia tenebrosa*; Ama, *Anemonia majano*; Avi, *Anemonia viridis*; Ami, *Anthopleura midori*; Btu, *Bolocera tuediae*; Ist, *Isosicyonis striata*; Ueq, *Urticina eques*). Right panel, alignment of *Phymanthus crucifer* (Pcr) to species representing seven different families (Aac, *Antholoba achates*; Apu, *Aiptasia pulchella*; Rda, *Relicanthus daphneae*; Hpu, *Halcampoides purpurea*; Hdi, *Hormathia digitata*; Mse, *Metridium senile*; Sor, *Sagartia ornata*). *, identical amino acid residues in alignments. See Table S2 for accession numbers and taxonomy.

3.3. Nonconventional Protein Coding Genes

Two nonconventional protein coding genes were noted in *S. haddoni*, i.e., the intron HEG and *orfA*. A closer inspection of the COI-884 intron HEG revealed an in-frame fusion to the upstream COI exon, indicating a COI-HEG fusion strategy in gene expression

(Figure 4b). Furthermore, the *S. haddoni* homing endonuclease belongs to the LAGLIDADG family (Figure 4c), a homing endonuclease family common among mitochondrial and chloroplast mobile introns [17,18]. The gene *orfA* has been previously noted among several sea anemone mitogenomes and suggested to be an insertion-like element [5,8]. The derived protein sequence appeared conserved among sea anemones inspected, and orfA in *S. haddoni* was clearly similar to the C-terminal part of corresponding orfA in *Anthopleura midori* and *U. eques* (Figure S1). Interestingly, *orfA* in *S. haddoni* is located in IGR-12 (between the COI and ND4L genes), which is different from that in other sea anemone mitogenomes (IGR-6, between genes of COII and ND4) (Figure 5 and Table S2).

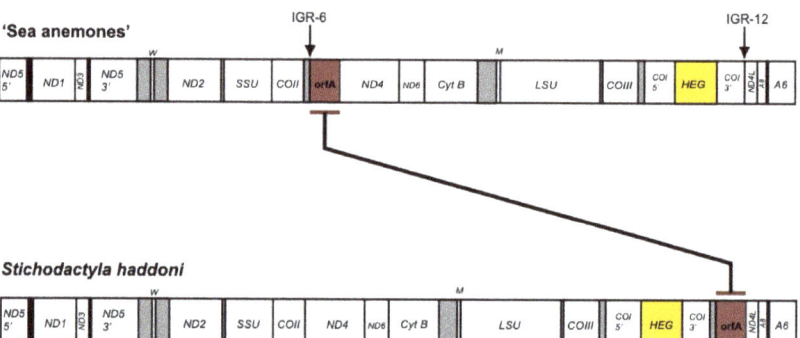

Figure 5. Translocation of mitochondrial *orfA* in the *Stichodactyla haddoni*. The *orfA* is located within IGR-6 in all sea anemones investigated, except IGR-12 in *S. haddoni*. See legend to Figure 1 for mitochondrial gene abbreviations.

3.4. Expression of Mitochondrial Genes

An Ion PGM sequencing approach was used to assess mitochondrial transcripts from *S. haddoni*. The analysis showed that all protein genes were expressed as RNA in all tissue samples assessed (oral disc, tentacles, and column), including the HEG and *orfA*, as well as ligated ND5 and COI mRNAs (Figure 6). The fraction of mitochondrial reads was at 0.055% of total reads, corresponding to 0.074% in oral disc, 0.011% in tentacles, and 0.096 in column. The majority of mitochondrial transcripts, however, was represented by rRNAs (89%). This corresponds well to that observed in other sea anemone species [5,8,30]. To expand the mitochondrial transcriptome assessments, mitochondrial-derived transcripts were mined out from a recently obtained transcriptome dataset in the closely related *S. helianthus* [28]. Illumina paired reads were mapped to the *S. haddoni* mitogenome and normalized against gene sizes (Figure S2a). Similar results were obtained in *S. helianthus* and *S. haddoni*. All conventional and nonconventional mitochondrial genes were clearly expressed as RNA. Both introns were spliced out perfectly, as identified by multiple read sequences representing ligated ND5 mRNA (Figure S2b) and ligated COI mRNA (Figure S2c), and RNA ligation junction consistent with COI-884 full-length circles was observed.

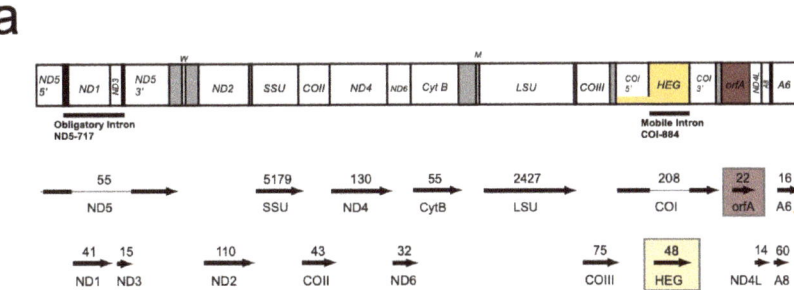

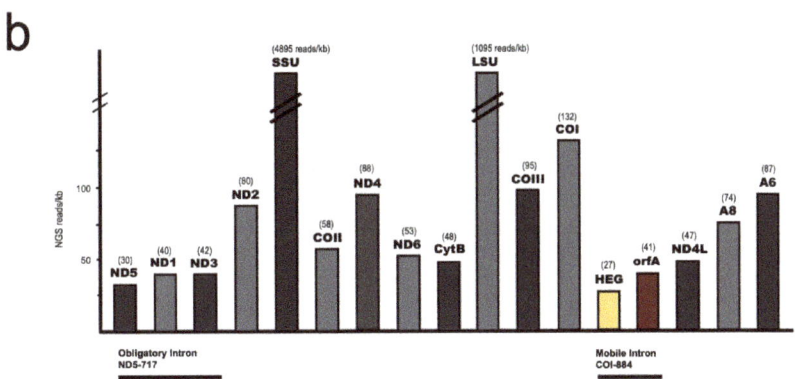

Figure 6. Mitochondrial transcripts from next generation sequencing (NGS) libraries in *Stichodactyla haddoni*. (**a**) Mapping of Ion PGM reads from protein coding and rRNA coding regions. Libraries were based on the poly (A) fraction of total cellular RNA isolated from oral disc, tentacles, and column. Read coverage per gene region is presented below the mitogenome organization map. mRNAs from the nonconventional mitochondrial genes are highlighted in boxes. (**b**) Histograms of estimated normalized read numbers (NGS reads/kb). See legend to Figure 1 for mitochondrial gene abbreviations.

3.5. Mitogenome-Based Phylogeny

To investigate phylogenetic relationships of *S. haddoni* and other sea anemones we performed mitogenome analyses based on concatenated OxPhos genes (13,702 nucleotide positions) and derived protein (3237 amino acid residues) sequences from 23 specimens representing 20 species and 13 Actiniaria families (Table S2). Phylogenetic reconstructions using the ML, NJ, and ME tree-building methods resolved the genera with strong statistical support, and a representative ML tree based on mitochondrial gene sequences is presented in Figure 7. *S. haddoni* (family Stichodactylidae) was found embedded within the family Actiniidae, supported by high bootstrap values by maximum-likelihood (98%), neighbor joining (98%), and minimal evolution (99%) in the nucleotide-based analysis. A similar observation was made for *Phymanthus crucifer* (family Phymanthidae). These unexpected findings were further supported by protein-based phylogenetic reconstructions (Figure S3).

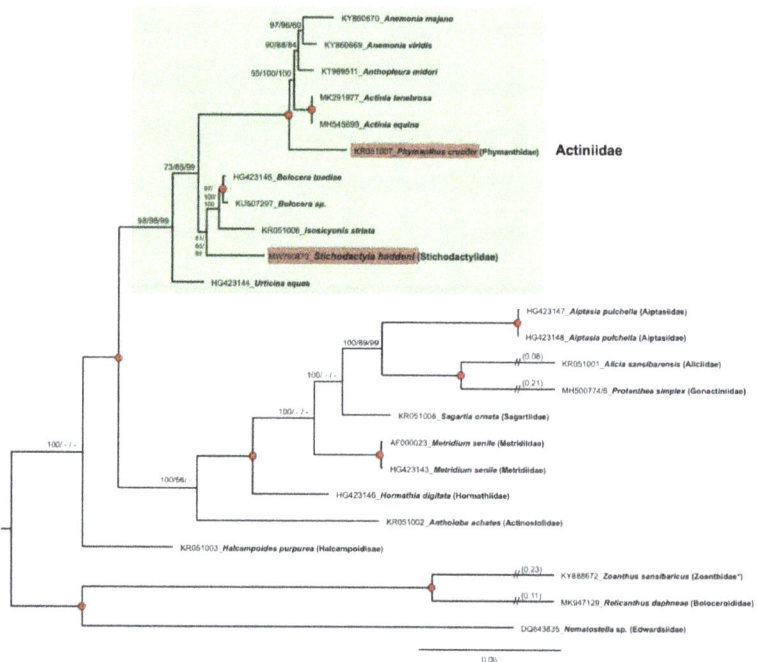

Figure 7. Phylogeny of sea anemones assessed by mitogenome gene sequences. Maximum-likelihood (ML) phylogenetic tree is shown based on alignments of 13,702 nucleotide positions obtained from concatenated genes. Bootstrap values (%) of 500 replications for alternative phylogenetic inference methods are shown at the internal nodes (ML/NJ/ME). Red-filled circles indicate highly significant branch points (bootstrap values of 100%) across the ML, NJ, and ME tree construction methods. Members of the Actiniidae family, including *Stichodactyla haddoni* and *Phymanthus crucifer* are marked by a green box. NJ, neighbor joining; ME, minimal evolution.

4. Discussion

Here, we report the complete mitogenome sequence of the sea anemone *S. haddoni* and corresponding transcriptome and identify several interesting features. These include two complex group I introns, a homing endonuclease gene in fusion with upstream exon, and a translocated *orfA* insertion-like element. Finally, we present a mitogenome-based phylogeny in apparent conflict with current morphology-based taxonomy.

The ND5-717 intron has a sequence organization similar to most other sea anemones [1]. A unique feature, however, is that the last intron nucleotide in ND5-717 is ωA, and not ωG as in most group I introns. A plausible explanation is that ωA prevents 3′ splice site hydrolysis and intron circularization [31], which are RNA processing events that could challenge exon ligation and the generation of ND5 mRNA. An implication of the in-frame COI-HEG fusion strategy is that almost the complete COI-884 intron has protein coding potential, some overlapping with ribozyme coding. A similar strategy has been seen in some introns from fungal mitochondria [32,33]. High-level expression of a homing endonuclease can potentially be hazardous to host genomes due to start activity [17], and downregulation is essential. We observe RNA consistent with COI-884 full-length circles, but a low level of intact fusion transcripts in transcriptome data. This indicates that intron RNA circularization, initiated by 3′ splice site hydrolysis [31], could be involved in downregulation of homing endonuclease expression. Additional experiments on cellular RNAs and appropriate in vitro self-splicing assays are need in order to evaluate this possibility.

The *orfA* sequence represents a widespread nonconventional sea anemone mitochondrial gene, but with unknown cellular function (Table S2). There are several features that link *orfA* to insertion-like elements, such as being located in highly transcribed regions, has a common but sporadic distribution among species, and its expression is induced by environmental stress [8]. Furthermore, *orfA* appears to be evolving under relaxed selection as compared with most OxPhos genes [5]. Here, we report *orfA* in both *S. haddoni* and *S. helianthus* to be expressed as polyadenylated RNA at similar levels to that of most Complex I (NADH dehydrogenase subunit) genes and associated with a gene translocation event from IGR-6 (typical in sea anemones) to IGR-12. Mitochondrial gene rearrangements have been well documented in hexacoral orders [1]. Deep-water species of stony corals (*Madrepora* and *Lophelia*) [19,34] and mushroom corals (*Corallimorphus* and *Corynactis*) [35] show gene order rearrangements as compared with other members of the orders. A dramatic rearrangement is seen in the *Protanthea* deep-water sea anemone [30]. The mitogenome consists of two distinct mitochromosomes and the gene order is heavily scrambled as compared with that of other investigated sea anemones. Interestingly, both strands contain mitochondrial genes, which differ from all other hexacoral mitogenomes.

The sea anemones (Order Actiniaria) represent a large and morphological diverse group of animals divided into approximately 1200 species in 46 families [22]. Several reports have included molecular markers in order to improve the phylogenetic assessments, studies that have applied nuclear and mitochondrial rRNA gene sequences, mitochondrial protein gene sequences, and complete mitogenomes [5,24,30,36–38]. The mitogenome-based phylogeny presented in this study supported a close association between *S. haddoni* (family Stichodactylidae) and members of the family Actiniidae (Figure 7 and Figure S3). Actiniidae constitutes the largest family known among the sea anemones with more than 200 species and 44 genera [22], and our findings corroborate earlier reports that the current Actiniidae taxonomy apparently is polyphyletic [22,24]. A similar finding was noted for *P. crucifer* (family Phymanthidae) [24,39,40], which also appears embedded within the family Actiniidae (Figure 7 and Figure S3). An additional molecular observation in the mitogenomes that link *S. haddoni* and *P. crucifer* to members of Actiniidae is the in-fusion expression mode of intron HEG. This feature is almost exclusively observed among sea anemones of the family Actiniidae (Table S2). Furthermore, a comparison of the conservation of LAGLIDADG sequence motifs of COI-884 intron homing endonucleases also suggests a close relationship of *S. haddoni* and *P. crucifer* to members of the family Actiniidae (Figure 4). Thus, there is an obvious need for revisions in sea anemone taxonomy, especially among Actiniidae and closely related species. One possibility is to add more related taxa (e.g., additional species and genus of the Stichodactylidae) to the mitogenome-based analysis. However, phylogenetic assessments based on rRNA or mitogenome sequence molecular markers are apparently not sufficient to resolve high resolution relationships, and we suggest that a genomic-based phylogeny would probably be more appropriate.

5. Conclusions

The mitogenome of *S. haddoni* harbors two complex catalytic group I introns, i.e., ND5-717 which carries the conventional ND1 and ND3 mitochondrial genes, and COI-884 which contains a nonconventional HEG. The latter was organized in-frame with the upstream COI exon 1, indicating a gene fusion strategy of intron homing endonuclease expression. A second nonconventional mitochondrial gene in *S. haddoni*, i.e., *orfA*, was found translocated and associated with mitogenome rearrangement. Phylogenetic analysis indicates *S. haddoni* to be embedded within the Actiniidae family, suggesting a polyphyletic origin of Actiniidae, and calls for a revision in sea anemone taxonomy.

Supplementary Materials: The following are available online at https://www.mdpi.com/article/10.3390/life11050402/s1, Figure S1: Amino acid alignment of orfA protein in Stichodactyla haddoni (Sha; 174 aa), Anthopleura midori (Ami, 617 aa) and Urticina eques (Ueq, 644 aa). See Table S2 for accession numbers. Identical residues are indicated (*) above aligned sequences, Figure S2: Mitochondrial transcripts from Illumina paired read libraries in Stichodactyla helianthus [28] mapped to the S. haddoni mitogenome sequence (MW760873). (a) Histograms of estimated normalized read numbers (NGS reads/kb). See legends to Figures 1 and 6 for mitochondrial gene abbreviations.

(b) Sashimi plot of ND5 splicing junction, consistent with perfect splicing of the ND5-717 intron. 21 reads mapped unambiguously to the ligated sequence region. (c) Sashimi plot of COI splicing junction, consistent with perfect splicing of the COI-884 intron. 100 reads map unambiguously to the ligated sequence region, Figure S3: Phylogeny of sea anemones assessed by mitogenome protein sequences. Maximum-likelihood (ML) phylogenetic tree is shown based on alignments of 3237 amino acid residues derived from concatenated genes. Bootstrap values (%) of 500 replications for alternative phylogenetic inference methods are shown at the internal nodes (ML/NJ/ME). Red-filled circles indicate highly significant branch points (bootstrap values of 100%) across the ML, NJ, and ME tree construction methods. Members of the Actiniidae family, including S. haddoni and Phymanthus crucifer are marked by a green box. NJ, neighbor joining; ME, minimal evolution, Table S1: Annotation of the complete mitogenome sequence of S. haddoni, Table S2: Key features of sea anemone mitogenomes included in this study.

Author Contributions: S.D.J., S.I.C., A.D. and T.E.J. designed the research; S.I.C. performed Ion PGM sequencing, assembly, and annotation; S.I.C. and T.E.J. performed transcriptome analysis; A.D. performed phylogenetic analysis; S.D.J. analyzed data and wrote the manuscript together with S.I.C., A.D. and T.E.J. All authors have read and agreed to the published version of the manuscript.

Funding: This work was supported by grants from the Tromsø Research Foundation, Research Council of Norway, UiT, The Arctic University of Norway, and Nord University.

Institutional Review Board Statement: Not applicable.

Informed Consent Statement: Not applicable.

Data Availability Statement: Sequencing data is available in GenBank under the accession number MW760873.

Acknowledgments: We thank members of our research teams at Nord University and UiT for interesting discussions. We thank Anita Ursvik and Annica Hedberg at UiT for technical supports.

Conflicts of Interest: The authors declare no conflict of interest.

References

1. Johansen, S.D.; Emblem, Å. Mitochondrial group I introns in hexacorals are regulatory elements. In *Advances in the Studies of the Bentic Zone*; Soto, L.A., Ed.; IntechOpen: London, UK, 2020; Chapter 7; pp. 1–25. [CrossRef]
2. Osigus, H.J.; Eitel, M.; Bernt, M.; Donath, A.; Schierwater, B. Mitogenomics at the base of Metazoa. *Mol. Phylogenet. Evol.* **2013**, *69*, 339–351. [CrossRef]
3. Flot, J.F.; Tillier, S. The mitochondrial genome of *Pocillopora* (Cnidaria: Scleractinia) contains two variable regions: The putative D-loop and a novel ORF of unknown function. *Gene* **2007**, *401*, 80–87. [CrossRef] [PubMed]
4. Beagley, C.T.; Wolstenholme, D.R. Characterization and localization of mitochondrial DNA-encoded tRNA and nuclear DNA-encoded tRNAs in the sea anemone *Metridium senile*. *Curr. Genet.* **2013**, *59*, 139–152. [CrossRef] [PubMed]
5. Emblem, Å.; Okkenhaug, S.; Weiss, E.S.; Denver, D.R.; Karlsen, B.O.; Moum, T.; Johansen, S.D. Sea anemones possess dynamic mitogenome structures. *Mol. Phylogenet. Evol.* **2014**, *75*, 184–193. [CrossRef] [PubMed]
6. Chi, S.I.; Johansen, S.D. Zoantharian mitochondrial genomes contain unique complex group I introns and highly conserved intergenic regions. *Gene* **2017**, *628*, 24–31. [CrossRef]
7. Chi, S.I.; Dahl, M.; Emblem, Å.; Johansen, S.D. Giant group I intron in a mitochondrial genome is removed by RNA back-splicing. *BMC Mol. Biol.* **2019**, *20*, 16. [CrossRef] [PubMed]
8. Chi, S.I.; Urbarova, I.; Johansen, S.D. Expression of homing endonuclease gene and insertion-like element in sea anemone mitochondrial genomes: Lesson learned from *Anemonia viridis*. *Gene* **2018**, *652*, 78–86. [CrossRef] [PubMed]
9. Beagley, C.T.; Okada, N.A.; Wolstenholme, D.R. Two mitochondrial group I introns in a metazoan, the sea anemone *Metridium senile*: One intron contains genes for subunits 1 and 3 of NADH dehydrogenase. *Proc. Natl. Acad. Sci. USA* **1996**, *93*, 5619–5623. [CrossRef]
10. Haugen, P.; Simon, D.M.; Bhattacharya, D. The natural history of group I introns. *Trends Genet.* **2005**, *21*, 111–119. [CrossRef]
11. Nielsen, H.; Johansen, S.D. Group I introns: Moving in new directions. *RNA Biol.* **2009**, *6*, 375–383. [CrossRef] [PubMed]
12. Schuster, A.; Lopez, J.V.; Becking, L.E.; Kelly, M.; Pomponi, S.A.; Worheide, G.; Erpenbeck, D.; Cardenas, P. Evolution of group I introns in Porifera: New evidence for intron mobility and implications for DNA barcoding. *BMC Evol. Biol.* **2017**, *17*, 82. [CrossRef]
13. Cech, T.R.; Damberger, S.H.; Gutell, R.R. Representation of the secondary and tertiary structure of group I introns. *Nat. Struct. Biol.* **1994**, *1*, 273–280. [CrossRef] [PubMed]
14. Vicens, Q.; Cech, T.R. Atomic level architecture of group I introns revealed. *Trends Biochem. Sci.* **2006**, *31*, 41–51. [CrossRef] [PubMed]

15. Hedberg, A.; Johansen, S.D. Nuclear group I introns in self-splicing and beyond. *Mob. DNA* **2013**, *4*, 17. [CrossRef] [PubMed]
16. Cech, T.R. Self-splicing of group I introns. *Annu. Rev. Biochem.* **1990**, *59*, 543–568. [CrossRef]
17. Stoddard, B.L. Homing endonuclease structure and function. *Q. Rev. Biophys.* **2005**, *38*, 49–95. [CrossRef]
18. Hafez, M.; Hausner, G. Homing endonucleases: DNA scissors on a mission. *Genomics* **2012**, *55*, 553–569. [CrossRef]
19. Emblem, Å.; Karlsen, B.O.; Evertsen, J.; Johansen, S.D. Mitogenome rearrangement in the cold-water scleractinian coral *Lophelia pertusa* (Cnidaria, Anthozoa) involves a long-term evolving group I intron. *Mol. Phylogenet. Evol.* **2011**, *61*, 495–503. [CrossRef]
20. Goddard, M.R.; Leigh, J.; Roger, A.J.; Pemberton, A.J. Invasion and persistence of a selfish gene in the Cnidaria. *PLoS ONE* **2006**, *1*, e3. [CrossRef]
21. Celis, J.S.; Edgell, D.R.; Stelbrink, B.; Wibberg, D.; Hauffe, T.; Blom, J.; Kalinowski, J.; Wilke, T. Evolutionary and biogeographical implications of degraded LAGLIDADG endonuclease functionality and group I intron occurrence in stony corals (Scleractinia) and mushroom corals (Corallimorpharia). *PLoS ONE* **2017**, *12*, e0173734. [CrossRef]
22. Daly, M.; Brugler, M.R.; Cartwright, P.; Collins, A.G.; Dawson, M.N.; Fautin, D.G.; France, S.C.; McFadden, C.S.; Opresko, D.M.; Rodriguez, E.; et al. The phylum Cnidaria: A review of phylogenetic patterns and diversity 300 years after Linnnaeus. *Zootaxa* **2007**, *1668*, 127–182. [CrossRef]
23. Attaran-Fariman, G.; Javid, P.; Shakouri, A. Morphology and phylogeny of the sea anemone *Stichodactyla haddoni* (Cnidaria: Anthozoa: Actiniaria) from Chabahar Bay, Iran. *Turk. J. Zool.* **2015**, *39*, 998–1003. [CrossRef]
24. Daly, M.; Chaudhuri, A.; Gusmao, L.; Rodriguez, E. Phylogenetic relationships among sea anemones (Cnidaria: Anthozoa: Actiniaria). *Mol. Phylogenet. Evol.* **2008**, *48*, 292–301. [CrossRef]
25. Chevreux, B.; Wetter, T.; Suhai, S. Genome sequence assembly using trace signals and additional sequence information. *German Conf. Bioinform.* **1999**, *99*, 45–56.
26. Hahn, C.; Bachmann, L.; Chevreux, B. Reconstructing mitochondrial genomes directly from genomic next-generation sequencing reads—A baiting and iterative mapping approach. *Nucleic Acids Res.* **2013**, *41*, e129. [CrossRef]
27. Bernt, M.; Donath, A.; Jühling, F.; Externbrink, F.; Florentz, C.; Fritzsch, G.; Stadler, P.F. MITOS: Improved de novo metazoan mitochondrial genome annotation. *Mol. Phylogenet. Evol.* **2013**, *69*, 313–319. [CrossRef] [PubMed]
28. Rivera-de-Torre, E.; Martinez-del-Pozo, A.; Garb, J.E. *Stichodactyla helianthus*' de novo transcriptome assembly: Discovery of a new actinoporin isoform. *Toxicon* **2018**, *150*, 105–114. [CrossRef]
29. Kumar, S.; Stecher, G.; Tamura, K. MEGA7: Molecular evolutionary genetics analysis version 7.0 for bigger datasets. *Mol. Biol. Evol.* **2016**, *33*, 1870–1874. [CrossRef]
30. Dubin, A.; Chi, S.I.; Emblem, Å.; Moum, T.; Johansen, S.D. Deep-water sea anemone with a two-chromosome mitochondrial genome. *Gene* **2019**, *692*, 195–200. [CrossRef]
31. Nielsen, H.; Fiskaa, T.; Birgisdottir, A.B.; Haugen, P.; Einvik, C.; Johansen, S.D. The ability to form full-length intron RNA circles is a general property of nuclear group I introns. *RNA* **2003**, *9*, 1464–1475. [CrossRef] [PubMed]
32. Lambowitz, A.M.; Belfort, M. Introns as mobile genetic elements. *Annu. Rev. Biochem.* **1993**, *62*, 587–622. [CrossRef]
33. Guo, W.W.; Moran, J.V.; Hoffman, P.W.; Henke, R.M.; Butow, R.A.; Perlman, P.S. The mobile group I intron 3a of the yeast mitochondrial COCI gene codes a 35-kDa processed protein that is an endonuclease but not a maturase. *J. Biol. Chem.* **1995**, *270*, 15563–15570. [CrossRef] [PubMed]
34. Lin, M.F.; Kitahara, M.V.; Tachikawa, H.; Fukami, H.; Miller, D.J.; Chen, C.A. Novel organization of the mitochondrial genome in the deep-sea coral, *Madrepora oculata* (Hexacorallia, Scleractinia, Oculinidae) and its taxonomic implications. *Mol. Phylogenet. Evol.* **2012**, *65*, 323–328. [CrossRef]
35. Lin, M.F.; Kitahara, M.V.; Luo, H.; Tracey, D.; Geller, J.; Fukami, H.; Miller, D.J.; Chen, C.A. Mitochondrial genome rearrangements in the Scleractinia/corallimorpharian complex: Implications for coral phylogeny. *Genome Biol. Evol.* **2014**, *6*, 1086–1095. [CrossRef]
36. Daly, M.; Gusmao, L.C.; Reft, A.J.; Rodriguez, E. Phylogenetic signal in mitochondrial and nuclear markers in sea anemones (Cnidaria, Actiniaria). *Integr. Comp. Biol.* **2010**, *50*, 371–378. [CrossRef]
37. Rodriguez, E.; Barbeitos, M.S.; Brugler, M.R.; Crowley, L.M.; Grajales, A.; Gusmao, L.; Haussermann, V.; Reft, A.; Daly, M. Hidden among sea anemones: The first comprehensive phylogenetic reconstruction of the order Actiniaria (Cnidaria, Anthozoa, Hexacorallia) reveals a novel group of hexacorals. *PLoS ONE* **2014**, *9*, e96998. [CrossRef]
38. Titus, B.M.; Benedict, C.; Laroche, R.; Gusmao, L.C.; Van Deusen, V.; Chiodo, T.; Meyer, C.P.; Berumen, M.L.; Bartholomew, A.; Yanagi, K.; et al. Phylogenetic relationships among the clownfish-hosting sea anemones. *Mol. Phylogenet. Evol.* **2019**, *139*, 106526. [CrossRef] [PubMed]
39. Zhang, L.; Zhu, Q. Complete mitochondrial genome of the sea anemone, *Anthopleura midori* (Actiniaria: Actiniidae). *Mitochondrial DNA* **2017**, *28*, 335–336. [CrossRef] [PubMed]
40. Foox, J.; Brugler, M.; Siddall, M.E.; Rodriguez, E. Multiplexed pyrosequencing of nine sea anemone (Cnidaria: Anthozoa: Hexacorallia: Actiniaria) mitochondrial genomes. *Mitochondrial DNA* **2016**, *27*, 2826–2832. [CrossRef]

Review

mtDNA Heteroplasmy: Origin, Detection, Significance, and Evolutionary Consequences

Maria-Eleni Parakatselaki and Emmanuel D. Ladoukakis *

Department of Biology, University of Crete, 70013 Heraklion, Greece; grad751@edu.biology.uoc.gr
* Correspondence: ladoukakis@uoc.gr

Abstract: Mitochondrial DNA (mtDNA) is predominately uniparentally transmitted. This results in organisms with a single type of mtDNA (homoplasmy), but two or more mtDNA haplotypes have been observed in low frequency in several species (heteroplasmy). In this review, we aim to highlight several aspects of heteroplasmy regarding its origin and its significance on mtDNA function and evolution, which has been progressively recognized in the last several years. Heteroplasmic organisms commonly occur through somatic mutations during an individual's lifetime. They also occur due to leakage of paternal mtDNA, which rarely happens during fertilization. Alternatively, heteroplasmy can be potentially inherited maternally if an egg is already heteroplasmic. Recent advances in sequencing techniques have increased the ability to detect and quantify heteroplasmy and have revealed that mitochondrial DNA copies in the nucleus (NUMTs) can imitate true heteroplasmy. Heteroplasmy can have significant evolutionary consequences on the survival of mtDNA from the accumulation of deleterious mutations and for its coevolution with the nuclear genome. Particularly in humans, heteroplasmy plays an important role in the emergence of mitochondrial diseases and determines the success of the mitochondrial replacement therapy, a recent method that has been developed to cure mitochondrial diseases.

Keywords: mtDNA; heteroplasmy; paternal leakage; NUMTs; selection

Citation: Parakatselaki, M.-E.; Ladoukakis, E.D. mtDNA Heteroplasmy: Origin, Detection, Significance, and Evolutionary Consequences. *Life* **2021**, *11*, 633. https://doi.org/10.3390/life11070633

Academic Editors: Andrea Luchetti and Federico Plazzi

Received: 17 June 2021
Accepted: 24 June 2021
Published: 29 June 2021

Publisher's Note: MDPI stays neutral with regard to jurisdictional claims in published maps and institutional affiliations.

Copyright: © 2021 by the authors. Licensee MDPI, Basel, Switzerland. This article is an open access article distributed under the terms and conditions of the Creative Commons Attribution (CC BY) license (https://creativecommons.org/licenses/by/4.0/).

1. Introduction

The strict maternal transmission of mtDNA results in homoplasmic individuals, who typically have a single mtDNA haplotype, the maternal one. However, heteroplasmy (the simultaneous presence of two or more types of mtDNA in the same individual) has often been reported in several animal species [1–4]. Given that the uniparental transmission of the mtDNA is one of the most general rules in biology and that mtDNA has been extensively used as a genetic marker for phylogenetic studies due to its maternal transmission, the scarce evidence for mtDNA heteroplasmy in the late 1980s and 1990s attracted attention from the scientific community. At that time, heteroplasmy was considered as an interesting exception of the strict maternal mtDNA inheritance.

Recently, heteroplasmy has been extensively studied thanks to the modern sequencing techniques. These studies, most of which were conducted in model organisms, revealed that heteroplasmy was more widespread than it was previously believed, particularly as low frequency variants [5], and that both drift and selection play a role in its dynamics within individuals and among generations.

2. The Generation and the Study of Heteroplasmy

2.1. The Sources of Heteroplasmy

Heteroplasmy can primarily occur through somatic mutagenesis during an individual's lifetime and through leakage of paternal mtDNA in the zygote during fertilization.

Recent studies, using modern sequencing techniques, have revealed that heteroplasmy due to somatic mutations might be prevalent among the individuals of a population [4–7]. The higher mutation rate in animal mtDNA and the huge copy number of mtDNAs, compared to the nuclear DNA, is expected to create variants of mtDNA which will co-exist with the common, maternal haplotype. These variants should differ from the common haplotype in one or few SNPs and are expected to appear in extremely low frequency [5,6]. Indeed, a detailed study in humans revealed that almost all individuals were heteroplasmic for different mitotypes which appeared in very low frequencies (below 1.5%) [5]. For this reason, this source of heteroplasmy should be cautiously identified in experiments because it can be confused with sequencing errors.

Heteroplasmy due to paternal leakage is generated by the presence of sperm's mtDNA in low frequency within some individuals of the population, and it has been observed in several species [8–10]. Paternal leakage can be considered as an inherent inadequacy of the otherwise strict mechanisms that protect maternal mtDNA transmission [11], in a similar way that mutations escape the repairing machinery of the cell. These mechanisms are variable and include the prevention of the paternal mtDNA to enter the egg during fertilization [12], the production of spermatozoa without mtDNA [13], and the destruction of the sperm's mitochondria after their entrance in the egg during fertilization [14]. Sperm's mtDNA can enter the egg and occasionally remain in the zygote in a heteroplasmic condition.

Oocytes that are already heteroplasmic can also potentially produce heteroplasmic individuals (maternal transmission of heteroplasmy), but these oocytes must have become heteroplasmic with one of the two primary ways. Inheritance of heteroplasmy has been observed in several organisms, such as human [15,16], crustaceans [17], and *Drosophila* [18].

A unique case of heteroplasmy has been observed in bivalves. This type of heteroplasmy is not due to paternal leakage, biparental transmission, or somatic mutations, but due to a specific way of mtDNA transmission that is called doubly uniparental inheritance (DUI) [19,20]. In DUI, the sperm is homoplasmic for a mtDNA type, called M, and the eggs are homoplasmic for the maternal type called F [21]. During fertilization and the subsequent development, the F type of mtDNA is spread in the somatic tissues of males and females, as well as in the germline of females. The M type is basically restricted to the male germline, producing males that are mosaics of the F type in their somatic tissues and the M type in their germline. This separation is not very strict because the M type has also been observed in somatic tissues. The peculiar case of heteroplasmy in many bivalves is the byproduct of the co-existence of two distinct uniparental mtDNA transmission routes in the same species: an egg and a sperm transmission route. The presence of paternal mtDNA in male bivalves with DUI is not due to paternal leakage, but due to paternal transmission of mtDNA in male lineage. Paternal leakage in these species can only be observed in females, in which paternal mtDNA has rarely been observed in some individuals [22].

Experimentally, heteroplasmy due to paternal leakage is more easily detected, compared to the mutation generated heteroplasmy, because the maternal and the paternal mtDNA differ in several nucleotide positions. It is more difficult to distinguish between heteroplasmy due to paternal leakage from that due to already heteroplasmic eggs unless the sequence of the paternal mitotype is known beforehand.

2.2. Measuring Heteroplasmy

To study the dynamics of heteroplasmy and for comparative purposes, it is not sufficient to report its presence in the organisms, but we also need to know the relative quantity of the mitotypes. There are two ways to quantify heteroplasmy. The first is to measure the proportion of individuals that are heteroplasmic in a population. This type has been used to quantify the proportion of heteroplasmic individuals in *Drosophila* natural populations [8] or the proportion of heteroplasmic progeny of particular crosses [18,23]. The second is to measure the frequency of the rare haplotype(s) relative to the common one, which is also known as "level of heteroplasmy". This has been used to quantify heteroplasmy within individuals [5–7] or within tissues [24,25].

2.3. The Hierarchical Levels for Studying Heteroplasmy

Heteroplasmy can be studied at hierarchical levels (Table 1). In a broad sense, the first, basic level is the population level. A population can be considered as "heteroplasmic" because it contains individuals with different mtDNA haplotypes. The second hierarchical level of heteroplasmy is the individual level. An individual is heteroplasmic if its tissues are either heteroplasmic or homoplasmic for alternative haplotypes. The latter is not very common, but it is not unknown in nature. For example, in the doubly uniparental inheritance (DUI) of mtDNA described above, the sperm is homoplasmic for the M type and the somatic tissues are homoplasmic for the F [21]. Therefore, males are heteroplasmic as individuals, but most of their tissues are homoplasmic for alternative mtDNA types. The third level of heteroplasmy is the tissue level. A tissue can be heteroplasmic, but its cells can be homoplasmic for alternative haplotypes. A fourth level of heteroplasmy would be when a cell is heteroplasmic, but its mitochondria are homoplasmic for different haplotypes. To our knowledge, the third and the fourth levels of heteroplasmy have not been directly observed, but they can be deduced from indirect observations. For example, the observation that shifts in heteroplasmy levels occurred in different human oocytes, which can lead to oocytes being homoplasmic for alternative mtDNA haplotypes [26], or that heteroplasmic cows producing homoplasmic progeny for alternative haplotypes in a few generations [27] implying heteroplasmy of the third level. Similarly, the observation that heteroplasmic stem cells result in stem cells homoplasmic for the alternative mitotypes [28] implies the heteroplasmy of the fourth level. Finally, the fifth and the lowest level of heteroplasmy would be the mitochondrion level. A single mitochondrion can contain different haplotypes among the several mtDNA molecules that it possesses. This heteroplasmy level has been observed experimentally [29], but it can also be deduced retrospectively because mutation-caused heteroplasmy or mtDNA recombination should have originated in single heteroplasmic mitochondria. Modern methods have been developed to detect directly or indirectly heteroplasmy in single cells. Such methods include an initial step of single cell isolation or propagation which is followed by massive parallel sequencing [30–32]. These advanced methods identify heteroplasmy at the single cell level but they cannot distinguish yet between levels four and five.

Table 1. Hierarchical levels for the study of heteroplasmy. 1. The population level: different individuals might contain different mtDNA haplotypes. 2. The individual level: the individuals are heteroplasmic, but the tissues are homoplasmic for alternative haplotypes. The different organs are an oversimplified representation for the different tissues. 3. The tissue level: The tissue can be heteroplasmic, but its cells can be homoplasmic for different haplotypes. 4. The cell level: the cell can be heteroplasmic, but its mitochondria can be homoplasmic for alternative haplotypes. 5. The mitochondrion level: the mitochondrion can contain different mtDNA haplotypes.

Levels of Heteroplasmy	Description	Experimental Evidence for b
1. Heteroplasmic population	Individuals are (a) heteroplasmic or (b) homoplasmic for alternative haplotypes	All organisms
2. Heteroplasmic individual	Tissues are (a) heteroplasmic or (b) homoplasmic for alterative haplotypes	Bivalves (ref. [22])
3. Heteroplasmic tissue	Cells are (a) heteroplasmic or (b) homoplasmic for alternative haplotypes	Indirect evidence and refs. [26,27]
4. Heteroplasmic cell	Mitochondria are (a) heteroplasmic or (b) homoplasmic for alterative haplotypes	Indirect evidence and (ref. [28])
5. Heteroplasmic mitochondrion	Mitochondria are heteroplasmic.	Direct observation in ref. [29]

The hierarchical study of heteroplasmy is obviously a simplified approach of what happens in nature. For instance, in mussel bivalves, the sperm is homoplasmic for the paternal haplotype (M) [20], but other tissues can be heteroplasmic [33]. In addition, this hierarchy implies that each higher level is heteroplasmic if any of its lower levels is heteroplasmic. An individual (level 2) can be homoplasmic if the population (level 1) is heteroplasmic, but it would necessarily be heteroplasmic if its tissues (level 3) or its cells (level 4) or its mitochondria (level 5) are heteroplasmic.

2.4. Techniques for Detection of Heteroplasmy

The first reported case of mtDNA heteroplasmy was in *Drosophila mauritiana*, where two size variants of mtDNA coexisted in virgin eggs of single females. The two mtDNA variants were detected using restriction enzymes [34]. In general, the most common method for detecting heteroplasmy before the advent of PCR was the isolation of whole mtDNA, its digestion with restriction enzymes, followed by visualization of the restriction pattern either with ethidium bromide or with hybridization with labeled probes [35–38]. The application of PCR boosted the detection of heteroplasmy because it was accompanied by more precise techniques in identification of sequence variation. For example, single stranded conformation polymorphism (SSCP) can detect even a single nucleotide polymorphism between two small amplified DNA segments [39,40]. Other targeted PCR based methods that have been used for detecting heteroplasmy include PCR-RFLP [41,42], cloning of the amplified fragment, and sequencing several clones [43,44], or even direct sequencing of the PCR product and identifying double peaks in the chromatogram [45]. Targeted PCR based techniques still remain popular because they are accurate and easily handled even in a small lab. PCR has not only been used for the detection of heteroplasmy but also for its quantification. Quantifying the level of heteroplasmy can either exclud e [46] or include qPCR [47,48]. Sometimes, qPCR can be modified into more sophisticated methods in order to quantify more accurately or more specifically the level of heteroplasmy. Such sophisticated qPCR methods include TaqMan approach [49], ARMS-qPCR [28,50,51], or droplet digital PCR (ddPCR) [31,52,53].

Targeted PCR methods have been proven powerful in detection of heteroplasmy but have some soft points. Firstly, they are restricted to a small region of mtDNA, and they might miss variation in other mtDNA regions. Secondly, miscalculation of heteroplasmy level is possible, due to non-specific binding of primers to the template. Thirdly, some knowledge of the template sequence is needed for all PCR-based methods. More recently, various next-generation sequencing (NGS) methods have revolutionized the study of heteroplasmy because they could detect and quantify very rare heteroplasmic variants using a great variety of sophisticated experimental techniques and bioinformatic tools [54–58]. The NGS methods revealed the extent of heteroplasmy in tissues and organisms, gave insight on its causes, and identified its role in mitochondrial diseases and aging [6,30,32]. Most of the modern methodological strategies include the generation of mtDNA-enriched libraries either with capture- or with targeted PCR-based methods, followed by NGS. In capture-based methods, biotinylated single strand DNA probes, 300–360 bases long, are used as baits for the mtDNA. After two rounds of hybridization, the library is sufficiently mtDNA-enriched and sequencing analysis takes place [59,60]. In PCR-based methods, the whole mtDNA genome is amplified with primers specifically designed for the mtDNA prior to sequencing [60–62]. Alternatively, a specific mtDNA region is PCR amplified and the amplicons are massively sequenced [5,63]. MtDNA haplotypes retrieved from whole-genome sequences have also been used to detect and quantify new mtDNA mutations that occur in heteroplasmy [64,65].

Recently, researchers took advantage of the properties of the Transposase-Accessible Chromatin Assay combined with sequencing (ATAC-seq). ATAC-seq is targeting the non-chromatinized DNA, so it is routinely used to study the sequence dynamics of the transcriptionally active regions of the nuclear DNA. Xu et al. observed that ATAC-seq libraries are highly enriched with mtDNA, since mtDNA is not chromatinized and therefore it is highly accessible [66]. Using this approach, researchers studied the heteroplasmy shifts that happen in single cells during hematopoietic differentiation.

The time period during which heteroplasmy is studied can be divided into three eras, according to technical innovations that have been recruited in heteroplasmy research. The first is the pre-PCR era. During this period, heteroplasmy was firstly discovered, but the scientific yield was poor. In a PubMed search with the key words "heteroplasmy" and "mtDNA", only 23 papers appeared in the 1980s. The second is the PCR targeting-era, in which the recruitment of PCR methods revolutionized the study of heteroplasmy, as it has been done in other fields. In the same PubMed search, 274 papers appeared in the 1990s and 508 in the decade 2000–2009. The third is the NGS era, in which massive sequencing techniques added to the standard PCR methods and allowed the detailed detection and quantification of heteroplasmy. This yielded 879 papers in the decade 2010–2020.

2.5. Heteroplasmy and NUMTs

Nuclear mitochondrial DNA (NUMT) are fragments of mtDNA which have moved to the nucleus [67] is an important source of error in the study of heteroplasmy. The PCR primers that are used for mtDNA amplification can also bind in the NUMT sequence due to their sequence similarity. Then, both molecules (mtDNA and NUMT) are amplified and the result can be passed for heteroplasmy.

NUMTs were discovered using traditional sequencing methods [67], but, albeit not immune, these methods are less susceptible in detecting false heteroplasmy. The mtDNA copies vastly outnumber the nuclear copies in the cells. This difference is reflected to the PCR template DNA, and, in turn, to the amplicon which will be sequenced. Consequently, the signal of the NUMTs might be faint or absent relative to the signal of mtDNA in the chromatogram, and it will be ignored. False heteroplasmy due to NUMTs is expected to be more commonly found in modern sequencing methods that can detect very rare haplotypes. Indeed, it has been proposed that mtDNA heteroplasmies below a 2% level could be actually NUMTs [68], and one should be alerted to distinguish real from false heteroplasmy.

The first NUMT annotations in the genome of *Drosophila* [69] and human [70] showed that the average length of the NUMTs is small, and, therefore, if heteroplasmy was observed in large mtDNA fragments, this was true heteroplasmy. However, NUMTs in the length of almost the whole mtDNA have been observed recently in bat [71], in *Drosophila* (Parakatselaki, Rand and Ladoukakis, unpublished) and in humans [72,73]. The length of the NUMT, as well as the sequence divergence from the mtDNA, might depend on the time that the translocation occurred, expecting that the older the NUMT, the shorter and more divergent from the mtDNA would be. However, this hypothesis remains to be tested. If this is the case, then younger NUMTs would be more easily confused as alternative mtDNA haplotypes in the same individual.

Despite being crucial for heteroplasmy detection, it turns out that NUMT identification is not an easy task for several reasons. First, it is difficult to annotate the NUMTs in the nuclear genome. Many methods for NUMT annotation rely on identifying the sites that the NUMT has embedded in the nuclear genome. The NGS reads which include the insertion point should be chimeric containing a part of mitochondrial-like and a part of nuclear sequence. However, these NGS reads are difficult to be distinguished from the artificially chimeric sequences, which are inherent errors of modern sequencing methods [74] and are removed during the cleaning process. Second, due to the short length of the NGS reads, mtDNA reads can be easily confused with NUMT reads, and this makes the full recovery of the NUMT sequence difficult. Finally, if the mtDNA translocation is very recent, then it would segregate in the population as polymorphism. If the sequenced genome comes from an individual that does not contain the NUMT, then it will not be discovered despite its presence in the population.

An example, which shows the difficulty to distinguish between real and false heteroplasmy, comes from the discussions which were raised from a recent publication which showed biparental transmission of mtDNA in humans. In this study, several individuals in three families appeared heteroplasmic for both maternal and paternal mtDNA [10]. Some scientists pointed out that this observation would be compatible with the presence of NUMTs [75] or other methodological and analytical issues [76]. Even though a detailed study has shown that large NUMTs in Y chromosome might imitate biparental transmission of mtDNA [72], there is no conclusive data whether the heteroplasmy that was observed at the first place was due to biparental mtDNA transmission or not. We faced the same problem in *Drosophila* (Parakatselaki, Rand and Ladoukakis, unpublished). Perhaps, this distinction is easy to make if the NUMTs are short, highly divergent from the mtDNA, containing nonsense mutations. However, it is more difficult in recent, large NUMTs with low sequence divergence from the mtDNA.

While traditional lab techniques have been employed in the past, and they still have merit for NUMT detection, the modern, massive sequencing data, need sophisticated bioinformatic approaches to identify NUMTs in the genomes. Such methods have been recently developed [54,55,77–80], but they are more efficient in human genome. Even in that case, the bioinformatic tools which have been developed for general use might not be adequate to detect NUMTs that are large and recently embedded in the nuclear genome [72,73]. In that case, more targeted bioinformatical tools are needed in order to identify accurately the NUMTs.

These recent studies have revealed that any observation of heteroplasmy, including the already reported cases, should first exclude the possibility that this is false, NUMT originated heteroplasmy.

3. The Applications of Heteroplasmy

3.1. Heteroplasmy and Diseases

Heteroplasmy is crucial for a group of disorders associated with dysfunctional mitochondria, known as mitochondrial diseases [81]. These malfunctions are caused by mutations located either in the nuclear-encoded genes which function in mitochondria or in the mitochondrial genes themselves. The emergence of the mitochondrial diseases caused by mutations in the mitochondrial genome is hard to predict, due to the non-Mendelian transmission of mtDNA [82,83]. Whether a pathology will emerge or not depends on the proportion of mutant mtDNA molecules relative to the wild-type molecules, in other words the relative heteroplasmy level. In most cases, wild-type molecules are able to compensate for the malfunction caused by mtDNA mutations. However, if the proportion of the mutated variants exceeds a certain threshold, then the wild-type mtDNA is insufficient to mask the deficiency caused by the high mutation load and consequently pathologies arise. This situation is known as the threshold effect in mitochondrial diseases [4,84]. The threshold level at which symptoms arise is different, depending on the mutation type and the tissue, but typically varies between 60% and 80%. This means that the majority of the mutations have a "recessive" phenotype [85,86]. For a more detailed review on the association of heteroplasmy and diseases, see reference [4].

The frequency of the mitochondrial diseases caused by mtDNA mutations is estimated at 1:4300 [87]. Mostly affected are the organs that rely on aerobic metabolism; therefore, the mtDNA mutations are linked with cardiovascular, neurological, and age-related degenerative diseases [88]. Currently, these diseases are not curable, but there are attempts to cure or to prevent them. mtDNA modification are promising techniques for this direction [89–91]. In addition, a set of mitochondrial replacement techniques (MRT) have been recently developed for the prevention of mitochondrial diseases. These techniques result in zygotes with mtDNA from a healthy donor, which rely either on the transfer of pronuclei from a zygote to another [92], or the transfer of the metaphase II spindle from the mother oocyte to the healthy donor oocyte [93], or the transfer of some ooplasm from a healthy donor to the affected oocyte [94]. Regardless of the technique, a small number of mutated mitochondria would be also transferred. As a result, the embryo will mainly contain the wild-type mtDNA of the new egg, but it will also contain a small amount of diseased mtDNA in heteroplasmy [95]. Due to the non-mitotic replication of the mtDNA, the dynamics of heteroplasmy level in the embryo can threaten the success of the MRT [96]. If the heteroplasmy level follows a random trajectory and if the initial proportion of the mutated mtDNA in the zygote is known, then the probability of predominance of the diseased mtDNA and the failure of MRT can be estimated. This probability is expected to be low but not negligible because the initial mutated mtDNA is rare in the zygote [96,97]. However, an older study in human cell lines has shown that the nuclear background can practically select the mtDNA haplotype that they will coexist with [98]. This suggests that the dynamics of heteroplasmy can be non-random and perhaps the coevolution of the mutated mtDNA with its nuclear background might favor this specific combination. However, more studies are needed to investigate this phenomenon.

3.2. Heteroplasmy and mtDNA as Genetic Marker

The asexual transmission of mtDNA is one of its fundamental assets as a molecular genetic marker [99]. Heteroplasmy could potentially affect the validity of mtDNA as a marker with three ways. First, in case alternative haplotypes are sequenced from different individuals, then the comparison between individuals would erroneously increase their genetic difference. To make this clearer, let us imagine that two individuals are heteroplasmic for two mtDNA haplotypes A and B. In addition, during the experimental process, the A haplotype by chance is amplified and sequenced in the one individual and the B in the other. The comparison of the haplotypes would erroneously show that the two individuals are genetically different. Second, heteroplasmy can lead to interlineage recombination, which cannot exist under the strict asexual mtDNA transmission. Recombination is expected to mix the evolutionary histories of the different parts of mtDNA molecule and introduce noise in phylogenies which are based on mtDNA [100]. Finally, false heteroplasmy due to NUMTs, and biased amplification and sequencing can lead to comparisons between nuclear (NUMT) and real mtDNA sequences or between solely NUMT sequences, which obviously will lead to incorrect results.

The first two ways are not expected to significantly affect the use of mtDNA as a marker because heteroplasmy is rare both as percentage of individuals within a population and as levels of heteroplasmy within an individual. Furthermore, mtDNA recombination, albeit observable [101,102] (but see reference [103]), seems to happen at very low rates. The false heteroplasmy due to NUMTs on the other hand can significantly affect the information of mtDNA as a marker because NUMTs are common and easily amplified.

4. Heteroplasmy and mtDNA Evolution

4.1. Selection and Drift on the Heteroplasmy Levels

Even though drift plays an important role in shifts in heteroplasmy levels, selection also seems to be in action in several cases. Selection has been described to act on heteroplasmy levels with four different outcomes: first, it guarantees maternal mtDNA transmission, second, it removes heteroplasmic variants that are malfunctional and cause diseases, third, it can maintain a low heteroplasmy level as the first step for recombination and the subsequent removal of deleterious mutations, and, fourth, it can act to improve the mito-nuclear linkage and reduce the effects of sexual antagonism, which are produced by strict maternal transmission [104]. In the first case, the target of selection is the heteroplasmy levels within an individual, i.e., selection will reduce the heteroplasmy levels, resulting in homoplasmic individuals. In the second case, selection will act against deleterious mtDNA variants. In the third case, the levels of heteroplasmy should be controlled in the population level and both purifying and positive selection should act producing balanced heteroplasmy levels. In the fourth case, heteroplasmy due to paternal leakage should be controlled to increase the linkage between mitochondrial and nuclear DNA [104].

4.2. Selection against Heteroplasmy to Support Maternal Transmission

Uniparental transmission of cytoplasmic genetic elements (chloroplastic and mitochondrial DNA) is one of the most common rules in biology. In most organisms, this transmission occurs through the mother. For reasons that still remain unclear [105] (but see [106] for a relative discussion), mtDNA is inherited exclusively from one parent. Even in several bivalves for which both parents can transmit their mtDNA to their progeny, this happens in a way that, within the same species, two independent transmission routes exist, one maternal and one paternal, and the rule of uniparental transmission of mtDNA is not violated [22]. Therefore, it is expected that heteroplasmy occurring through paternal leakage would be removed by purifying selection. The variety of pre- and post-fertilization mechanisms that ensure the maternal inheritance of mtDNA [12] fall in this category of selection.

4.3. Selection on Heteroplasmy to Control Deleterious mtDNA Mutations

As Stewart and Chinery [4] review in detail, in many cases, mtDNA variants that contain deleterious mutations can coexist in heteroplasmy with wild type variants. In some cases, the deleterious variants can increase their frequency and can cause mitochondrial diseases. In this case, purifying selection can act either in the germline or in the somatic tissues in order to remove mtDNA copies with deleterious mutations. In the first case, selection will prevent the transmission of these mutations in the next generation. In the second case, it will prevent malfunctional mtDNAs to increase their frequency in the somatic tissues and cause disease. However, the overall dynamics of heteroplasmy depend on both selection and drift.

4.3.1. Dynamics of Heteroplasmy in the Germline

Drift seems to be the prevalent process which determines the heteroplasmy in germ line. The genetic bottleneck which occurs during oogenesis and was first observed in heteroplasmic Holstein cows [27] explained sufficiently the shifts that were observed in the levels of heteroplasmy in mothers and progeny. Since then, the bottleneck in the female germline has been confirmed in many vertebrate species, including mice [107], salmon [108], and zebrafish [109]. In humans, the estimated size of the bottleneck was about 30–35 mtDNAs [110]. More recent estimates support an even more severe mtDNA reduction in the germline, with only nine mtDNA genomes being transmitted on average and with a variable-size bottleneck [6].

Despite the important role of drift on heteroplasmy levels in germline, selection has also been observed in several cases. Purifying selection acting in the germline was first reported in mice. Fan and colleagues observed a dramatic decrease of a frameshift mutation in the NAD6 gene, from 47% to 14% in two successive generations and a complete loss of the mutation within few generations [111]. Furthermore, a different group generated mtDNA mutations using a mutator gene of *pol* γ and studied their dynamics for six generations [112]. They observed that the frequency of non-synonymous mutations in mtDNA was decreased compared to the synonymous ones, suggesting a selective mechanism that prevents deleterious mutations from passing from one generation to another. Elimination of detrimental mutations has also been observed during *Drosophila* germline development, both by purifying selection acting against mutated mtDNA genomes [113] and by selective propagation of the wild-type genome during oogenesis [114].

Purifying selection has been also found to act in the human germline. Sequencing of 39 mother-child pairs showed that non-synonymous mutations were less transmitted compared to synonymous ones, suggesting that potential pathogenic variants are eliminated in the germline [110]. These findings are in line with the results from a different study [15], which reports that the female germline is capable of recognizing and removing the detrimental mtDNA haplotypes, preventing their transmission. Furthermore, De Fanti and colleagues provided evidence that mtDNA mutations are counter-selected in human oocytes during the expulsion of the first and second polar bodies [115]. The time point at which selection occurs has been recently identified at the Carnegie stages 12–21 [26].

Mutant haplotypes can be under selective pressure either at the level of cell or organelle [116]. Interestingly, genes related with oxidative metabolism and mtDNA replication and transcription were found to be upregulated in primordial germ cells. It was suggested that the shift from glycolytic to oxidative metabolism at that stage, combined with the bottleneck that results in mitochondria with various levels of heteroplasmy, put the mtDNA under a selective pressure, which results in maintaining only the functional mtDNAs [26]. Indeed, there is evidence that mtDNA segregation and haplotype preference are driven by the effect of the mtDNA haplotypes on ROS signaling and OXPHOS functionality and that the strength of purifying selection acting on the germline is greatly dependent on the nuclear context [117].

Physical separation within a single cell has also been documented to favor the action of purifying selection. More specifically, decreased levels of the pro-fusion protein Mitofusin results in mitochondrial fragmentation during early stages of *Drosophila* oogenesis. This means that the defective function of mutated genomes cannot be complemented by functional genomes, helping the purifying selection to eliminate them [118].

4.3.2. Dynamics of Heteroplasmy in Somatic Tissues

It seems that selection is less effective to remove deleterious mutations in somatic tissues, which are thought to follow more neutral segregation patterns [119,120]. Somatic bottlenecks are generally less severe compared to germline ones [121]; however, there are exceptions, like the extreme bottleneck observed in human hair follicles [122]. Relaxed bottlenecks result in reduced variance in the levels of heteroplasmy among cells. Therefore, in this case, drift rather than selection is thought to be the primary driving force of heteroplasmy dynamics [26].

However, several cases of selection against or for heteroplasmy have been documented in somatic tissues. In dividing cells, heteroplasmy shifts happen due to random segregation of the mtDNA haplotypes to the daughter cells, but, if a mutation has a strong effect on the cellular function, purifying selection will act to eliminate it. This is the case for *MTTL1* m.3243A>G mutation in humans, which is exponentially decreased in blood cells over time [123]. Moreover, patients with the pathogenic mutations m.3243A>G and m.8344A>G were found to carry less mutation load in the mitotic gastrointestinal epithelial cells when compared to smooth muscle cells, suggesting the action of purifying selection [124]. In non-dividing cells, mtDNA is copied under a relaxed replication pattern, ensuring the maintenance of the mtDNA quantity. However, a mutated haplotype can be increased through this process, leading to a heteroplasmy shift within a single cell over time [125] that can be explained by random drift model [126–128].

Surprisingly, signatures of positive selection on heteroplasmy have also been found for specific mtDNA haplotypes at specific tissues [6]. Very recently, a study shed light on key factors that shape the heteroplasmy dynamics on the somatic tissues. Specifically, it was concluded that the differential effect on the OXPHOS system caused by the different mtDNA haplotypes is cell-type specific, leading to different mtDNA preference and different heteroplasmy levels across different cell types. They also suggested that heteroplasmy dynamics are greatly influenced by mitonuclear interactions and other environmental factors [129].

4.4. Evolutionary Significance of Heteroplasmy

The maternal transmission of mtDNA results in homoplasmic individuals, which implies a lack of inter-lineage recombination. Non-recombining genomes are subject to deleterious mutation accumulation, a process known as Muller's ratchet [130]. MtDNA seems to have escaped the ratchet because it has survived for more than two billion years [131] and selection acts efficiently, as suggested by the very low dN/dS ratio, particularly in animals' mtDNA [132]. The survival of the uniparentally transmitted mtDNA for such a long time has been called "the mitochondrial paradox" [133]. Many mechanisms have been proposed to act in order for mtDNA to escape Muller's ratchet [133]. One such mechanism, with experimental support, is the genetic bottleneck that was described above [134]. The genetic bottleneck decreases the variation of the mtDNA within cells producing cells with marginal levels of heteroplasmy, but increases the variation among cells, increasing the efficacy of selection [135], removing cells which contain deleterious mtDNA variants. Given that the bottleneck has been observed in many organisms beyond mammals, it is expected to play a fundamental role in the mitochondrial paradox [4]. The genetic bottleneck hypothesis though has the soft points that many organisms cannot have a bottleneck [136], that it can be active only against relatively strongly deleterious mutations (slightly deleterious mutations can be fixed in each oocyte by genetic drift), and it can be applicable only to mutations that are expressed in the eggs or in the zygotes. An alternative, non-mutually exclusive hypothesis suggests that mtDNA overcomes Muller's ratchet by allowing a low rate of paternal leakage, which leads to recombination with the maternal lineage [105,137]. The level of heteroplasmy needed for this process is low because the recombination rate sufficient for cancelling Muller's ratchet is extremely low [138].

Apart from its contribution in the "mitochondrial paradox", heteroplasmy produced by paternal leakage can also have other evolutionary consequences. A byproduct of strict maternal mtDNA transmission is the accumulation of mutations in mtDNA that can be deleterious for males but beneficial or neutral for females, a hypothesis that has been described as "mother's curse" [139,140]. It has been proposed that such male specific deleterious effects can be dampened by paternal leakage [141]. In addition, the strict maternal transmission of mtDNA increases the linkage between mitochondrial and nuclear genomes, necessary for the smooth mito-nuclear function. This linkage, however, is reduced either by mutations or by recombination that occurs in the male nuclear genome. A recent study proposes that paternal leakage can strengthen the mitonuclear linkage mitigating the sexual antagonism between the genomes [104].

These hypotheses attribute a functional role to heteroplasmy itself and suggest that paternal leakage should not exceed a certain level, for uniparental transmission of mtDNA not to be threatened, but cannot be totally depleted in order to allow a low recombination rate or to mitigate the sexual antagonism and its detrimental results. This implies that heteroplasmy level should be controlled by selection. The compromise between maternal transmission of mtDNA and paternal leakage can be compared with the compromise between faithful DNA replication and the generation of variation with the mutational process, which maintains the mutation rate in low but non-zero levels. Despite being far from proven, this hypothesis is supported by several lines of experimental evidence. First, heteroplasmy has been observed in a variety of organisms as we have mentioned above. Second, homologous mtDNA recombination is pervasive in plants [142] and has also been observed in several animal species such as mice [143], lizards [3], fish [144], scorpions [145], flies [146], humans [147] (but see [148]), and several others [101,149]. Third, a recent study has shown that the rate of heteroplasmy is non-randomly distributed across *Drosophila* families, and that this characteristic can be inherited [18]. Theoretical studies have suggested that paternal leakage itself might be an evolvable trait, due its role in the survival of mtDNA [104]. In addition, an experimental study has shown that artificially heteroplasmic mice for two non-diseased mtDNA haplotypes showed severe behavioral and cognitive malfunctions [25]. Therefore, given that the different haplotypes did not contain deleterious mutations, heteroplasmy was detrimental by itself. It remains to be studied whether heteroplasmy creates problems in compatibility between the two haplotypes or of the two haplotypes with the nucleus.

The hypothesis of functional heteroplasmy is similar to that of J. Maynard Smith on the benefit of sex [150]. Sexual reproduction allows recombination, which in turn facilitates selection to fix beneficial mutations and to remove deleterious ones. Particularly for the mtDNA, the target of selection is the heteroplasmy level and the unit of selection is the mtDNA population because, if recombination happens in some individuals, the whole population of mtDNA can be benefited. If proven, this would be an example of group selection.

5. Conclusions

Heteroplasmy of mtDNA was observed almost forty years ago. The first evidence was scarce and was considered as interesting exceptions in the strict maternal mtDNA transmission. Within a few previous years, there was fundamental reconsideration of several aspects regarding heteroplasmy. Albeit not exhaustively, we tried to review some of these aspects in this paper. First, recent studies, employing traditional and modern techniques, have revealed that heteroplasmy might be more common than previously believed in several animals, including humans. Second, it seems that both drift and selection shape heteroplasmy dynamics in individuals and in populations. Third, the involvement of selection in heteroplasmy dynamics suggests that it plays a substantial role in mitochondrial function which is related with mitochondrial diseases. It is also related with the success of the modern mitochondrial replacement therapy. Fourth, heteroplasmy might be important not only for the mitochondrial function but also for the evolution and the survival of mtDNA itself, as a first step for interlineage recombination, and the escape of mtDNA from Muller's ratchet as well as for the mitigation of sexually antagonistic effects, resulting from strict maternal mtDNA transmission. If this hypothesis will be supported by data, then it will change our view for the strict maternal transmission of the mtDNA. However, researchers that work on mtDNA heteroplasmy should be careful to exclude from their studies two sources of potential errors which mimic heteroplasmy: the technical errors (PCR, sequencing bioinformatic parameters, sample contamination) and the presence of mitochondrial copies in the nucleus (NUMTs). For the near future, we need research in-breadth and in-depth in order to understand the uniparental transmission of mtDNA and the functional and evolutionary role of heteroplasmy. The in-breadth research would extend our knowledge on heteroplasmy beyond model species. The in-depth research would accurately estimate the heteroplasmy level in many individuals and tissues per species, which will reveal the exact contribution of drift and selection in heteroplasmy function and dynamics.

Author Contributions: All authors contribute equally. All authors have read and agreed to the published version of the manuscript.

Funding: The project was supported by Special Account of Research, University of Crete (KA10137). M-E.P. has been supported by a scholarship from the Hellenic Foundation for Research and Innovation (HFRI).

Institutional Review Board Statement: Not applicable.

Informed Consent Statement: Not applicable.

Data Availability Statement: Not applicable.

Conflicts of Interest: The authors declare no conflict of interest.

References

1. Mastrantonio, V.; Latrofa, M.S.; Porretta, D.; Lia, R.P.; Parisi, A.; Iatta, R.; Dantas-torres, F.; Otranto, D.; Urbanelli, S. Paternal Leakage and MtDNA heteroplasmy in *Rhipicephalus* spp. ticks. *Sci. Rep.* **2019**, 1–8. [CrossRef] [PubMed]
2. Wen, M.; Peng, L.; Hu, X.; Zhao, Y.; Liu, S.; Hong, Y. Transcriptional quiescence of paternal MtDNA in cyprinid fish embryos. *Sci. Rep.* **2016**, *6*, 28571. [CrossRef] [PubMed]
3. Ujvari, B.; Dowton, M.; Madsen, T. Mitochondrial DNA recombination in a free-ranging australian lizard. *Biol. Lett.* **2007**, *3*, 189–192. [CrossRef] [PubMed]
4. Stewart, J.B.; Chinnery, P.F. Extreme heterogeneity of human mitochondrial DNA from organelles to populations. *Nat. Rev. Genet.* **2021**, 106–118. [CrossRef]
5. Payne, B.A.I.; Wilson, I.J.; Yu-Wai-Man, P.; Coxhead, J.; Deehan, D.; Horvath, R.; Taylor, R.W.; Samuels, D.C.; Santibanez-Koref, M.; Chinnery, P.F. Universal heteroplasmy of human mitochondrial DNA. *Hum. Mol. Genet.* **2013**, *22*, 384–390. [CrossRef] [PubMed]
6. Li, M.; Schröder, R.; Ni, S.; Madea, B.; Stoneking, M. Extensive tissue-related and allele-related MtDNA heteroplasmy suggests positive selection for somatic mutations. *Proc. Natl. Acad. Sci. USA* **2015**, *112*, 2491–2496. [CrossRef]
7. Ye, K.; Lu, J.; Ma, F.; Keinan, A.; Gu, Z. Extensive pathogenicity of mitochondrial heteroplasmy in healthy human individuals. *Proc. Natl. Acad. Sci. USA* **2014**, *111*, 10654–10659. [CrossRef]
8. Nunes, M.D.S.; Dolezal, M.; Schlötterer, C. Extensive paternal MtDNA leakage in natural populations of *Drosophila* melanogaster. *Mol. Ecol.* **2013**, *22*, 2106–2117. [CrossRef]
9. Robison, G.A.; Balvin, O.; Schal, C.; Vargo, E.L.; Booth, W. Extensive mitochondrial heteroplasmy in natural populations of a resurging human pest, the bed bug (Hemiptera: Cimicidae). *J. Med. Entomol.* **2015**, *52*, 734–738. [CrossRef]
10. Luo, S.; Valencia, C.A.; Zhang, J.; Lee, N.-C.; Slone, J.; Gui, B.; Wang, X.; Li, Z.; Dell, S.; Brown, J.; et al. Biparental Inheritance of Mitochondrial DNA in Humans. *Proc. Natl. Acad. Sci. USA* **2018**, *115*, 13039–13044. [CrossRef]

11. Rokas, A.; Ladoukakis, E.; Zouros, E. Animal mitochondrial DNA recombination revisited. *Trends Ecol. Evol.* **2003**, *18*, 411–417. [CrossRef]
12. Sato, M.; Sato, K. Maternal inheritance of mitochondrial DNA by diverse mechanisms to eliminate paternal mitochondrial DNA. *Biochim. Biophys. Acta Mol. Cell Res.* **2013**, *1833*, 1979–1984. [CrossRef] [PubMed]
13. DeLuca, S.Z.; O'Farrell, P.H. Barriers to male transmission of mitochondrial DNA in sperm development. *Dev. Cell* **2012**, *22*, 660–668. [CrossRef]
14. Politi, Y.; Gal, L.; Kalifa, Y.; Ravid, L.; Elazar, Z.; Arama, E. Paternal mitochondrial destruction after fertilization is mediated by a common endocytic and autophagic pathway in *Drosophila*. *Dev. Cell* **2014**. [CrossRef] [PubMed]
15. Li, M.; Rothwell, R.; Vermaat, M.; Wachsmuth, M.; Schröder, R.; Laros, J.F.J.; Van Oven, M.; De Bakker, P.I.W.; Bovenberg, J.A.; Van Duijn, C.M.; et al. Transmission of Human MtDNA heteroplasmy in the genome of the netherlands families: Support for a variable-size bottleneck. *Genome Res.* **2016**, *26*, 417–426. [CrossRef] [PubMed]
16. Schwartz, M.; Vissing, J. Paternal inheritance of mitochondrial DNA. *N. Engl. J. Med.* **2002**, *347*, 576–580. [CrossRef]
17. Doublet, V.; Souty-Grosset, C.; Bouchon, D.; Cordaux, R.; Marcadé, I. A thirty million year-old inherited heteroplasmy. *PLoS ONE* **2008**, *3*, e2938. [CrossRef] [PubMed]
18. Polovina, E.S.; Parakatselaki, M.E.; Ladoukakis, E.D. Paternal leakage of mitochondrial DNA and maternal inheritance of heteroplasmy in *Drosophila* hybrids. *Sci. Rep.* **2020**, *10*, 1–9. [CrossRef]
19. Skibinski, D.O.; Gallagher, C.; Beynon, C.M. Sex-limited mitochondrial DNA transmission in the marine mussel mytilus edulis. *Genetics* **1994**, *138*, 3. [CrossRef]
20. Zouros, E.; Oberhauser Ball, A.; Saavedra, C.; Freeman, K.R. An unusual type of mitochondrial DNA inheritance in the blue mussel mytilus. *Proc. Natl. Acad. Sci. USA* **1994**, *91*, 7463–7467. [CrossRef]
21. Venetis, C.; Theologidis, I.; Zouros, E.; Rodakis, G.C. No evidence for presence of maternal mitochondrial DNA in the sperm of *Mytilus galloprovincialis* males. *Proc. R. Soc. B Biol. Sci.* **2006**, *273*, 2483–2489. [CrossRef]
22. Zouros, E. Biparental inheritance through uniparental transmission: The Doubly Uniparental Inheritance (DUI) of mitochondrial DNA. *Evol. Biol.* **2013**, *40*, 1–31. [CrossRef]
23. Dokianakis, E.; Ladoukakis, E.D. Different degree of paternal MtDNA leakage between male and female progeny in interspecific *Drosophila* Crosses. *Ecol. Evol.* **2014**, *4*, 2633–2641. [CrossRef] [PubMed]
24. Naue, J.; Hörer, S.; Sänger, T.; Strobl, C.; Hatzer-Grubwieser, P.; Parson, W.; Lutz-Bonengel, S. Evidence for frequent and tissue-specific sequence heteroplasmy in human mitochondrial DNA. *Mitochondrion* **2015**, *20*, 82–94. [CrossRef] [PubMed]
25. Sharpley, M.S.; Marciniak, C.; Eckel-mahan, K.; Mcmanus, M.; Crimi, M.; Waymire, K.; Lin, C.S.; Masubuchi, S.; Friend, N.; Koike, M.; et al. Heteroplasmy of mouse MtDNA is genetically unstable and results in altered behavior and cognition. *Cell* **2012**, *151*, 333–343. [CrossRef] [PubMed]
26. Floros, V.I.; Pyle, A.; DIetmann, S.; Wei, W.; Tang, W.W.C.; Irie, N.; Payne, B.; Capalbo, A.; Noli, L.; Coxhead, J.; et al. Segregation of mitochondrial DNA heteroplasmy through a developmental genetic bottleneck in human embryos. *Nat. Cell Biol.* **2018**, *20*, 144–151. [CrossRef] [PubMed]
27. Hauswirth, W.; Laipis, P. Mitochondrial DNA polymorphism in a maternal lineage of holstein cows. *Proc. Natl. Acad. Sci. USA* **1982**, *79*, 4686–4690. [CrossRef] [PubMed]
28. Kang, E.; Wu, J.; Gutierrez, N.M.; Koski, A.; Tippner-hedges, R.; Agaronyan, K.; Platero-luengo, A.; Martinez-redondo, P.; Ma, H.; Lee, Y.; et al. Mitochondrial replacement in human oocytes carrying pathogenic mitochondrial DNA mutations. *Nature* **2016**, *540*, 270–275. [CrossRef]
29. Yang, L.; Long, Q.; Liu, J.; Tang, H.; Li, Y.; Bao, F.; Qin, D.; Pei, D.; Liu, X. Mitochondrial fusion provides an "initial metabolic complementation" controlled by MtDNA. *Cell. Mol. Life Sci.* **2015**, *72*, 2585–2598. [CrossRef]
30. Lareau, C.A.; Ludwig, L.S.; Muus, C.; Gohil, S.H.; Zhao, T.; Chiang, Z.; Pelka, K.; Verboon, J.M.; Luo, W.; Christian, E.; et al. Massively parallel single-cell mitochondrial DNA genotyping and chromatin profiling. *Nat. Biotechnol.* **2020**, 1–11. [CrossRef]
31. Maeda, R.; Kami, D.; Maeda, H.; Shikuma, A.; Gojo, S. High throughput single cell analysis of mitochondrial heteroplasmy in mitochondrial diseases. *Sci. Rep.* **2020**, *10*, 1–10. [CrossRef]
32. Jaberi, E.; Tresse, E.; Grønbæk, K.; Weischenfeldt, J.; Issazadeh-Navikas, S. Identification of unique and shared mitochondrial DNA mutations in neurodegeneration and cancer by single-cell mitochondrial DNA structural variation sequencing (MitoSV-Seq). *EBioMedicine* **2020**, *57*, 102868. [CrossRef]
33. Garrido-Ramos, M.A.; Stewart, D.T.; Sutherland, B.W.; Zouros, E. The distribution of male-transmitted and female-transmitted mitochondrial DNA types in somatic tissues of blue mussels: Implications for the operation of doubly uniparental inheritance of mitochondrial DNA. *Genome* **1998**, *41*, 818–824. [CrossRef]
34. Solignac, M.; Monnerott, M.; Mounolout, J. Mitochondrial heteroplasmy mauritiana. *Genetics* **1983**, *80*, 6942–6946.
35. Harrison, R.G.; Rand, D.M.; Wheeler, W.C. Mitochondrial DNA size variation within individual crickets. *Science* **1985**, *228*, 1446–1448. [CrossRef] [PubMed]
36. Bentzen, S.M.; Poulsen, H.S.; Kaae, S.; Myhre Jensen, O.; Johansen, H.; Mouridsen, H.T.; Daugaard, S.; Arnoldl, C. Prognostic factors in osteosarcomas: A regression analysis. *Cancer* **1988**, *62*, 194–202. [CrossRef]
37. Buroker, N.E.; Brown, J.R.; Gilbert, T.A.; O'Hara, P.J.; Beckenbach, A.T.; Thomas, W.K.; Smith, M.J. Length heteroplasmy of sturgeon mitochondrial DNA: An illegitimate elongation model. *Genetics* **1990**, *124*, 1. [CrossRef]

38. Boyce, T.M.; Zwick, M.E.; Aquadro, C.F. Mitochondrial DNA in the bark weevils: Size, structure and heteroplasmy. *Genetics* **1989**, *123*, 825–836. [CrossRef]
39. Maté, M.L.; Di Rocco, F.; Zambelli, A.; Vidal-Rioja, L. Mitochondrial heteroplasmy in control region DNA of South American Camelids. *Small Rumin. Res.* **2007**, *71*, 123–129. [CrossRef]
40. Tully, L.A.; Parsons, T.J.; Steighner, R.J.; Holland, M.M.; Marino, M.A.; Prenger, V.L. A sensitive denaturing gradient-gel electrophoresis assay reveals a high frequency of heteroplasmy in hypervariable region 1 of the human MtDNA control region. *Am. J. Hum. Genet.* **2000**, *67*, 432–443. [CrossRef]
41. El-Schahawi, M.; De López Munain, A.; Sarrazin, A.M.; Shanske, A.L.; Basirico, M.; Shanske, S.; DiMauro, S. Two large spanish pedigrees with nonsyndromic sensorineural deafness and the MtDNA mutation at Nt 1555 in the 12s RRNA gene: Evidence of heteroplasmy. *Neurology* **1997**, *48*, 453–456. [CrossRef] [PubMed]
42. Ladoukakis, E.D.; Saavedra, C.; Magoulas, A.; Zouros, E. Mitochondrial DNA variation in a species with two mitochondrial genomes: The case of *Mytilus galloprovincialis* from the atlantic, the mediterranean and the black sea. *Mol. Ecol.* **2002**, *11*, 755–769. [CrossRef] [PubMed]
43. Nesbø, C.L.; Arab, M.O.; Jakobsen, K.S. Heteroplasmy, length and sequence variation in the MtDNA control regions of three percid fish species (*Perca Fluviatilis, Acerina Cernua, Stizostedion Lucioperca*). *Genetics* **1998**, *148*, 1907–1919. [CrossRef]
44. Ladoukakis, E.D.; Zouros, E. Direct evidence for homologous recombination in mussel (*Mytilus galloprovincialis*) Mitochondrial DNA. *Mol. Biol. Evol.* **2001**, *18*, 1168–1175. [CrossRef]
45. Rodríguez-Pena, E.; Verísimo, P.; Fernández, L.; González-Tizón, A.; Bárcena, C.; Martínez-Lage, A. High incidence of heteroplasmy in the MtDNA of a natural population of the spider crab *Maja brachydactyla*. *PLoS ONE* **2020**, *15*, e0230243. [CrossRef]
46. Lutz-Bonengel, S.; Sänger, T.; Pollak, S.; Szibor, R. Different methods to determine length heteroplasmy within the mitochondrial control region. *Int. J. Leg. Med.* **2004**, *118*, 274–281. [CrossRef]
47. Poe, B.G.; Navratil, M.; Arriaga, E.A. Absolute quantitation of a heteroplasmic mitochondrial DNA deletion using a multiplex three-primer real-time PCR assay. *Anal. Biochem.* **2007**, *362*, 193–200. [CrossRef]
48. Bai, R.K.; Wong, L.J.C. Detection and quantification of heteroplasmic mutant mitochondrial DNA by real-time amplification refractory mutation system quantitative PCR analysis: A single-step approach. *Clin. Chem.* **2004**, *50*, 996–1001. [CrossRef]
49. Rong, E.; Wang, H.; Hao, S.; Fu, Y.; Ma, Y.; Wang, T. Heteroplasmy detection of mitochondrial DNA A3243G mutation using quantitative real-time PCR assay based on TaqMan-MGB probes. *BioMed Res. Int.* **2018**, *2018*, 1–8. [CrossRef]
50. Duan, M.; Tu, J.; Lu, Z. Recent advances in detecting mitochondrial DNA heteroplasmic variations. *Molecules* **2018**, *23*, 323. [CrossRef]
51. Chin, R.M.; Panavas, T.; Brown, J.M.; Johnson, K.K. Patient-derived lymphoblastoid cell lines harboring mitochondrial DNA Mutations as tool for small molecule drug discovery. *BMC Res. Notes* **2018**, *11*, 205. [CrossRef]
52. Hindson, B.J.; Ness, K.D.; Masquelier, D.A.; Belgrader, P.; Heredia, N.J.; Makarewicz, A.J.; Bright, I.J.; Lucero, M.Y.; Hiddessen, A.L.; Legler, T.C.; et al. High-throughput droplet digital PCR system for absolute quantitation of DNA copy number. *Anal. Chem.* **2011**, *83*, 8604–8610. [CrossRef] [PubMed]
53. Trifunov, S.; Pyle, A.; Valentino, M.L.; Liguori, R.; Yu-Wai-Man, P.; Burté, F.; Duff, J.; Kleinle, S.; Diebold, I.; Rugolo, M.; et al. Clonal expansion of MtDNA deletions: Different disease models assessed by digital droplet PCR in single muscle cells. *Sci. Rep.* **2018**, *8*, 1682. [CrossRef]
54. Fazzini, F.; Fendt, L.; Schönherr, S.; Forer, L.; Schöpf, B.; Streiter, G.; Losso, J.L.; Kloss-Brandstätter, A.; Kronenberg, F.; Weissensteiner, H. Analyzing low-level MtDNA heteroplasmy—Pitfalls and challenges from bench to benchmarking. *Int. J. Mol. Sci.* **2021**, *22*, 935. [CrossRef]
55. Marquis, J.; Lefebvre, G.; Kourmpetis, Y.A.I.; Kassam, M.; Ronga, F.; De Marchi, U.; Wiederkehr, A.; Descombes, P. MitoRS, a method for high throughput, sensitive, and accurate detection of mitochondrial DNA heteroplasmy. *BMC Genom.* **2017**, *18*, 326. [CrossRef] [PubMed]
56. Calabrese, C.; Simone, D.; Diroma, M.A.; Santorsola, M.; Guttà, C.; Gasparre, G.; Picardi, E.; Pesole, G.; Attimonelli, M. MToolBox: A highly automated pipeline for heteroplasmy annotation and prioritization analysis of human mitochondrial variants in high-throughput sequencing. *Bioinformatics* **2014**, *30*, 3115–3117. [CrossRef] [PubMed]
57. Dierckxsens, N.; Mardulyn, P.; Smits, G. Unraveling heteroplasmy patterns with NOVOPlasty. *NAR Genom. Bioinf.* **2020**, *2*. [CrossRef]
58. Weissensteiner, H.; Forer, L.; Fuchsberger, C.; Schöpf, B.; Kloss-Brandstätter, A.; Specht, G.; Kronenberg, F.; Schönherr, S. MtDNA-server: Next-generation sequencing data analysis of human mitochondrial DNA in the cloud. *Nucleic Acids Res.* **2016**, *44*, W64–W69. [CrossRef]
59. Zhou, K.; Mo, Q.; Guo, S.; Liu, Y.; Yin, C.; Ji, X.; Guo, X.; Xing, J. A novel next-generation sequencing-based approach for concurrent detection of mitochondrial DNA copy number and mutation. *J. Mol. Diagn.* **2020**, *22*, 1408–1418. [CrossRef]
60. He, Y.; Wu, J.; Dressman, D.C.; Iacobuzio-Donahue, C.; Markowitz, S.D.; Velculescu, V.E.; Diaz, L.A.; Kinzler, K.W.; Vogelstein, B.; Papadopoulos, N. Heteroplasmic mitochondrial DNA mutations in normal and tumour cells. *Nature* **2010**, *464*, 610–614. [CrossRef]
61. Liu, Y.; Guo, S.; Yin, C.; Guo, X.; Liu, M.; Yuan, Z.; Zhao, Z.; Jia, Y.; Xing, J. Optimized PCR-based enrichment improves coverage uniformity and mutation detection in mitochondrial DNA next-generation sequencing. *J. Mol. Diagn.* **2020**, *22*, 503–512. [CrossRef] [PubMed]

62. Kelly, P.S.; Clarke, C.; Costello, A.; Monger, C.; Meiller, J.; Dhiman, H.; Borth, N.; Betenbaugh, M.J.; Clynes, M.; Barron, N. Ultra-deep next generation mitochondrial genome sequencing reveals widespread heteroplasmy in Chinese hamster ovary cells. *Metab. Eng.* **2017**, *41*, 11–22. [CrossRef] [PubMed]
63. Radojičić, J.M.; Kristoffersen, J.B.; Polovina, E.-S.; Pavlidis, P.; Ladoukakis, E.D. Pervasive non-random MtDNA heteroplasmy in a natural hybrid water frog population.
64. Duan, M.; Chen, L.; Ge, Q.; Lu, N.; Li, J.; Pan, X.; Qiao, Y.; Tu, J.; Lu, Z. Evaluating heteroplasmic variations of the mitochondrial genome from whole genome sequencing data. *Gene* **2019**, *699*, 145–154. [CrossRef] [PubMed]
65. Grandhi, S.; Bosworth, C.; Maddox, W.; Sensiba, C.; Akhavanfard, S.; Ni, Y.; LaFramboise, T. Heteroplasmic shifts in tumor mitochondrial genomes reveal tissue-specific signals of relaxed and positive selection. *Hum. Mol. Genet.* **2017**, *26*, 2912–2922. [CrossRef]
66. Xu, J.; Nuno, K.; Litzenburger, U.M.; Qi, Y.; Corces, M.R.; Majeti, R.; Chang, H.Y. Single-cell lineage tracing by endogenous mutations enriched in transposase accessible mitochondrial DNA. *Elife* **2019**, *8*, 1–14. [CrossRef]
67. Bensasson, D.; Zhang, D.X.; Hartl, D.L.; Hewitt, G.M. Mitochondrial pseudogenes: Evolution's misplaced witnesses. *Trends Ecol. Evol.* **2001**, 314–321. [CrossRef]
68. Albayrak, L.; Khanipov, K.; Pimenova, M.; Golovko, G.; Rojas, M.; Pavlidis, I.; Chumakov, S.; Aguilar, G.; Chávez, A.; Widger, W.R.; et al. The ability of human nuclear DNA to cause false positive low-abundance heteroplasmy calls varies across the mitochondrial genome. *BMC Genom.* **2016**, *17*, 1017. [CrossRef]
69. Richly, E.; Leister, D. NUMTs in sequenced eukaryotic genomes. *Mol. Biol. Evol.* **2004**, *21*, 1081–1084. [CrossRef]
70. Mourier, T.; Hansen, A.J.; Willerslev, E.; Arctander, P. The human genome project reveals a continuous transfer of large mitochondrial fragments to the nucleus. *Mol. Biol. Evol.* **2001**, *18*, 1833–1837. [CrossRef]
71. Shi, H.; Xing, Y.; Mao, X. The little brown bat nuclear genome contains an entire mitochondrial genome: Real or artifact? *Gene* **2017**, *629*, 64–67. [CrossRef]
72. Wei, W.; Pagnamenta, A.T.; Gleadall, N.; Sanchis-Juan, A.; Stephens, J.; Broxholme, J.; Tuna, S.; Odhams, C.A.; Ambrose, J.C.; Baple, E.L.; et al. Nuclear-mitochondrial DNA segments resemble paternally inherited mitochondrial DNA in humans. *Nat. Commun.* **2020**, *11*, 1740. [CrossRef] [PubMed]
73. Lutz-Bonengel, S.; Niederstätter, H.; Naue, J.; Koziel, R.; Yang, F.; Sänger, T.; Huber, G.; Berger, C.; Pflugradt, R.; Strobl, C.; et al. Evidence for multi-copy Mega-NUMT s in the human genome. *Nucleic Acids Res.* **2021**, *49*, 1517–1531. [CrossRef] [PubMed]
74. Tu, J.; Guo, J.; Li, J.; Gao, S.; Yao, B.; Lu, Z. Systematic characteristic exploration of the chimeras generated in multiple displacement amplification through next generation sequencing data reanalysis. *PLoS ONE* **2015**, *10*, e0139857. [CrossRef]
75. Balciuniene, J.; Balciunas, D. A Nuclear MtDNA concatemer (Mega-NUMT) could mimic paternal inheritance of mitochondrial genome. *Front. Genet.* **2019**, *10*, 2018–2020. [CrossRef]
76. Salas, A.; Schönherr, S.; Bandelt, H.J.; Gómez-Carballa, A.; Weissensteiner, H. Extraordinary claims require extraordinary evidence in asserted MtDNA biparental inheritance. *Forensic Sci. Int. Genet.* **2020**, *47*, 102274. [CrossRef] [PubMed]
77. Bris, C.; Goudenege, D.; Desquiret-Dumas, V.; Charif, M.; Colin, E.; Bonneau, D.; Amati-Bonneau, P.; Lenaers, G.; Reynier, P.; Procaccio, V. Bioinformatics tools and databases to assess the pathogenicity of mitochondrial DNA variants in the field of next generation sequencing. *Front. Genet.* **2018**, *11*, 632. [CrossRef] [PubMed]
78. Santibanez-Koref, M.; Griffin, H.; Turnbull, D.M.; Chinnery, P.F.; Herbert, M.; Hudson, G. Assessing mitochondrial heteroplasmy using next generation sequencing: A note of caution. *Mitochondrion* **2018**, *46*, 302–306. [CrossRef]
79. Ring, J.D.; Sturk-Andreaggi, K.; Alyse Peck, M.; Marshall, C. Bioinformatic removal of NUMT-associated variants in mitotiling next-generation sequencing data from whole blood samples. *Electrophoresis* **2018**, *39*, 2785–2797. [CrossRef]
80. Cihlar, J.C.; Strobl, C.; Lagacé, R.; Muenzler, M.; Parson, W.; Budowle, B. Distinguishing mitochondrial DNA and NUMT sequences amplified with the precision ID MtDNA whole genome panel. *Mitochondrion* **2020**, *55*, 122–133. [CrossRef]
81. Gorman, G.S.; Chinnery, P.F.; DiMauro, S.; Hirano, M.; Koga, Y.; McFarland, R.; Suomalainen, A.; Thorburn, D.R.; Zeviani, M.; Turnbull, D.M. Mitochondrial diseases. *Nat. Rev. Dis. Prim.* **2016**, *2*, 16080. [CrossRef] [PubMed]
82. Stewart, J.B.; Chinnery, P.F. The dynamics of mitochondrial DNA heteroplasmy: Implications for human health and disease. *Nat. Rev. Genet.* **2015**, *16*, 530–542. [CrossRef]
83. Machado, T.S.; Macabelli, C.H.; Del Collado, M.; Meirelles, F.V.; Guimarães, F.E.G.; Chiaratti, M.R. Evidence of selection against damaged mitochondria during early embryogenesis in the mouse. *Front. Genet.* **2020**, *11*, 1762. [CrossRef] [PubMed]
84. Rossignol, R.; Faustin, B.; Rocher, C.; Malgat, M.; Mazat, J.-P.; Letellier, T. Mitochondrial threshold effects. *Biochem. J.* **2003**, *370*, 751–762. [CrossRef] [PubMed]
85. Schon, E.A.; DiMauro, S.; Hirano, M. Human mitochondrial DNA: Roles of inherited and somatic mutations. *Nat. Rev. Genet.* **2012**, *13*, 878–890. [CrossRef] [PubMed]
86. Burr, S.P.; Pezet, M.; Chinnery, P.F. Mitochondrial DNA heteroplasmy and purifying selection in the mammalian female germ line. *Dev. Growth Differ.* **2018**, *60*, 21–32. [CrossRef] [PubMed]
87. Gorman, G.S.; Schaefer, A.M.; Ng, Y.; Gomez, N.; Blakely, E.L.; Alston, C.L.; Feeney, C.; Horvath, R.; Yu-Wai-Man, P.; Chinnery, P.F.; et al. Prevalence of nuclear and mitochondrial DNA mutations related to adult mitochondrial disease. *Ann. Neurol.* **2015**, *77*, 753–759. [CrossRef]
88. Mustafa, M.F.; Fakurazi, S.; Abdullah, M.A.; Maniam, S. Pathogenic mitochondria DNA mutations: Current detection tools and interventions. *Nature* **2020**, *11*, 192. [CrossRef]

89. Zascavage, R.R.; Hall, C.L.; Thorson, K.; Mahmoud, M.; Sedlazeck, F.J.; Planz, J.V. Approaches to whole mitochondrial genome sequencing on the oxford nanopore MinION. *Curr. Protoc. Hum. Genet.* **2019**, *104*. [CrossRef] [PubMed]
90. Mok, B.Y.; de Moraes, M.H.; Zeng, J.; Bosch, D.E.; Kotrys, A.V.; Raguram, A.; Hsu, F.S.; Radey, M.C.; Peterson, S.B.; Mootha, V.K.; et al. A bacterial cytidine deaminase toxin enables CRISPR-free mitochondrial base editing. *Nature* **2020**, *583*, 631–637. [CrossRef]
91. Jackson, C.B.; Turnbull, D.M.; Minczuk, M.; Gammage, P.A. Therapeutic manipulation of MtDNA heteroplasmy: A shifting perspective. *Trends Mol. Med.* **2020**, *26*, 698–709. [CrossRef] [PubMed]
92. Craven, L.; Tuppen, H.A.; Greggains, G.D.; Harbottle, S.J.; Murphy, J.L.; Cree, L.M.; Murdoch, A.P.; Chinnery, P.F.; Taylor, R.W.; Lightowlers, R.N.; et al. Pronuclear transfer in human embryos to prevent transmission of mitochondrial DNA disease. *Nature* **2010**, *465*, 82–85. [CrossRef]
93. Tachibana, M.; Sparman, M.; Sritanaudomchai, H.; Ma, H.; Clepper, L.; Woodward, J.; Li, Y.; Ramsey, C.; Kolotushkina, O.; Mitalipov, S. Mitochondrial gene replacement in primate offspring and embryonic stem cells. *Nature* **2009**, *461*, 367–372. [CrossRef]
94. Flood, J.T.; Chillik, C.F.; van Uem, J.F.; Iritani, A.; Hodgen, G.D. Ooplasmic transfusion: Prophase germinal vesicle oocytes made developmentally competent by microinjection of metaphase II egg cytoplasm. *Fertil. Steril.* **1990**, *53*, 1049–1054. [CrossRef]
95. Reznichenko, A.S.; Huyser, C.; Pepper, M.S. Mitochondrial transfer: Implications for assisted reproductive technologies. *Appl. Translat. Genom.* **2016**, *11*, 40–47. [CrossRef]
96. Yamada, M.; Emmanuele, V.; Sanchez-Quintero, M.J.; Sun, B.; Lallos, G.; Paull, D.; Zimmer, M.; Pagett, S.; Prosser, R.W.; Sauer, M.V.; et al. Genetic drift can compromise mitochondrial replacement by nuclear transfer in human oocytes. *Cell Stem Cell* **2016**, *18*, 749–754. [CrossRef] [PubMed]
97. Tachibana, M.; Kuno, T.; Yaegashi, N. Mitochondrial replacement therapy and assisted reproductive technology: A paradigm shift toward treatment of genetic diseases in gametes or in early embryos. *Reprod. Med. Biol.* **2018**, *17*, 421–433. [CrossRef] [PubMed]
98. Dunbar, D.R.; Moonie, P.A.; Jacobs, H.T.; Holt, I.J. Different cellular backgrounds confer a marked advantage to either mutant or wild-type mitochondrial genomes. *Proc. Natl. Acad. Sci. USA* **1995**, *92*, 6562–6566. [CrossRef] [PubMed]
99. Avise, J.C. Introduction. In *Molecular Markers, Natural History and Evolution*; Springer: New York, NY, USA, 1994; pp. 3–15. [CrossRef]
100. Schierup, M.H.; Hein, J. Consequences of recombination on traditional phylogenetic analysis. *Genetics* **2000**, *156*, 879–891. [CrossRef] [PubMed]
101. Tsaousis, A.D.; Martin, D.P.; Ladoukakis, E.D.; Posada, D.; Zouros, E. Widespread recombination in published animal MtDNA sequences. *Mol. Biol. Evol.* **2005**, *22*, 925–933. [CrossRef] [PubMed]
102. Piganeau, G.; Eyre-Walker, A. A reanalysis of the indirect evidence for recombination in human mitochondrial DNA. *Heredity* **2004**, *92*, 282–288. [CrossRef]
103. Hagström, E.; Freyer, C.; Battersby, B.J.; Stewart, J.B.; Larsson, N.G. No recombination of MtDNA after heteroplasmy for 50 generations in the mouse maternal germline. *Nucleic Acids Res.* **2014**, *42*, 1111–1116. [CrossRef]
104. Radzvilavicius, A.L.; Lane, N.; Pomiankowski, A. Sexual conflict explains the extraordinary diversity of mechanisms regulating mitochondrial inheritance. *BMC Biol.* **2017**, *15*, 94. [CrossRef]
105. Greiner, S.; Sobanski, J.; Bock, R. Why are most organelle genomes transmitted maternally? *Bioessays* **2015**, *37*, 80–94. [CrossRef] [PubMed]
106. Wallace, D.C. Why do we still have a maternally inherited mitochondrial DNA? Insights from evolutionary medicine. *Annu. Rev. Biochem.* **2007**, *76*, 781–821. [CrossRef]
107. Freyer, C.; Cree, L.M.; Mourier, A.; Stewart, J.B.; Koolmeister, C.; Milenkovic, D.; Wai, T.; Floros, V.I.; Hagström, E.; Chatzidaki, E.E.; et al. Variation in germline MtDNA heteroplasmy is determined prenatally but modified during subsequent transmission. *Nat. Genet.* **2012**, *44*, 1282–1285. [CrossRef] [PubMed]
108. Wolff, J.N.; White, D.J.; Woodhams, M.; White, H.E.; Gemmell, N.J. The strength and timing of the mitochondrial bottleneck in salmon suggests a conserved mechanism in vertebrates. *PLoS ONE* **2011**, *6*, e20522. [CrossRef]
109. Otten, A.B.C.; Stassen, A.P.M.; Adriaens, M.; Gerards, M.; Dohmen, R.G.J.; Timmer, A.J.; Vanherle, S.J.V.; Kamps, R.; Boesten, I.B.W.; Vanoevelen, J.M.; et al. Replication errors made during oogenesis lead to detectable de novo MtDNA mutations in zebrafish oocytes with a low MtDNA copy number. *Genetics* **2016**, *204*, 1423–1431. [CrossRef]
110. Rebolledo-Jaramillo, B.; Su, M.S.W.; Stoler, N.; McElhoe, J.A.; Dickins, B.; Blankenberg, D.; Korneliussen, T.S.; Chiaromonte, F.; Nielsen, R.; Holland, M.M.; et al. Maternal age effect and severe germ-line bottleneck in the inheritance of human mitochondrial DNA. *Proc. Natl. Acad. Sci. USA* **2014**, *111*, 15474–15479. [CrossRef]
111. Fan, W.; Waymire, K.G.; Narula, N.; Li, P.; Rocher, C.; Coskun, P.E.; Vannan, M.A.; Narula, J.; Macgregor, G.R.; Wallace, D.C. A mouse model of mitochondrial disease reveals germline selection against severe MtDNA mutations. *Science* **2008**, *319*, 958–962. [CrossRef] [PubMed]
112. Stewart, J.B.; Freyer, C.; Elson, J.L.; Larsson, N.G. Purifying selection of MtDNA and its implications for understanding evolution and mitochondrial disease. *Nat. Rev. Genet.* **2008**, *9*, 657–662. [CrossRef] [PubMed]
113. Ma, H.; Xu, H.; O'Farrell, P.H. Transmission of mitochondrial mutations and action of purifying selection in *Drosophila melanogaster*. *Nat. Genet.* **2014**, *46*, 393–397. [CrossRef]
114. Hill, J.H.; Chen, Z.; Xu, H. Selective propagation of functional mitochondrial DNA during oogenesis restricts the transmission of a deleterious mitochondrial variant. *Nat. Genet.* **2014**, *46*, 389–392. [CrossRef] [PubMed]

115. De Fanti, S.; Vicario, S.; Lang, M.; Simone, D.; Magli, C.; Luiselli, D.; Gianaroli, L.; Romeo, G. Intra-individual purifying selection on mitochondrial DNA variants during human oogenesis. *Hum. Reprod.* **2017**, *32*, 1100–1107. [CrossRef]
116. Rand, D.M. The Units of Selection of Mitochondrial DNA. *Annu. Rev. Ecol. Syst.* **2001**, *32*, 415–448. [CrossRef]
117. Latorre-Pellicer, A.; Lechuga-Vieco, A.V.; Johnston, I.G.; Hämäläinen, R.H.; Pellico, J.; Justo-Méndez, R.; Fernández-Toro, J.M.; Clavería, C.; Guaras, A.; Sierra, R.; et al. Regulation of mother-to-offspring transmission of MtDNA heteroplasmy. *Cell Metab.* **2019**, *30*, 1120–1130.e5. [CrossRef] [PubMed]
118. Lieber, T.; Jeedigunta, S.P.; Palozzi, J.M.; Lehmann, R.; Hurd, T.R. Mitochondrial fragmentation drives selective removal of deleterious MtDNA in the germline. *Nature* **2019**, *570*, 380–384. [CrossRef] [PubMed]
119. Palozzi, J.M.; Jeedigunta, S.P.; Hurd, T.R. Mitochondrial DNA purifying selection in mammals and invertebrates. *J. Mol. Biol.* **2018**, *430*, 4834–4848. [CrossRef] [PubMed]
120. Wilton, P.R.; Zaidi, A.; Makova, K.; Nielsen, R. A population phylogenetic view of mitochondrial heteroplasmy. *Genetics* **2018**, *208*, 1261–1274. [CrossRef]
121. Zaidi, A.A.; Wilton, P.R.; Su, M.S.W.; Paul, I.M.; Arbeithuber, B.; Anthony, K.; Nekrutenko, A.; Nielsen, R.; Makova, K.D. Bottleneck and selection in the germline and maternal age influence transmission of mitochondrial DNA in human pedigrees. *Proc. Natl. Acad. Sci. USA* **2019**, *116*, 25172–25178. [CrossRef]
122. Barrett, A.; Arbeithuber, B.; Zaidi, A.; Wilton, P.; Paul, I.M.; Nielsen, R.; Makova, K.D. Pronounced somatic bottleneck in mitochondrial DNA of human hair. *Philos. Trans. R. Soc. B Biol. Sci.* **2020**, *375*. [CrossRef]
123. Rajasimha, H.K.; Chinnery, P.F.; Samuels, D.C. Selection against pathogenic mtDNA mutations in a stem cell population leads to the loss of the 3243A→G mutation in blood. *Am. J. Hum. Genet.* **2008**, *82*, 333–343. [CrossRef]
124. Su, T.; Grady, J.P.; Afshar, S.; McDonald, S.A.C.; Taylor, R.W.; Turnbull, D.M.; Greaves, L.C. Inherited pathogenic mitochondrial DNA mutations and gastrointestinal stem cell populations. *J. Pathol.* **2018**, *246*, 427–432. [CrossRef] [PubMed]
125. Szczepanowska, K.; Trifunovic, A. Origins of MtDNA mutations in ageing. *Essays Biochem.* **2017**, *61*, 325–337. [CrossRef] [PubMed]
126. Durham, S.E.; Samuels, D.C.; Chinnery, P.F. Is selection required for the accumulation of somatic mitochondrial DNA mutations in post-mitotic cells? *Neuromuscul. Disord.* **2006**, *16*, 381–386. [CrossRef] [PubMed]
127. Elson, J.L.; Andrews, R.M.; Chinnery, P.F.; Lightowlers, R.N.; Turnbull, D.M.; Howell, N. Analysis of European MtDNAs for recombination. *Am. J. Hum. Genet.* **2001**, *68*, 145–153. [CrossRef]
128. Baines, H.L.; Stewart, J.B.; Stamp, C.; Zupanic, A.; Kirkwood, T.B.L.; Larsson, N.G.; Turnbull, D.M.; Greaves, L.C. Similar patterns of clonally expanded somatic MtDNA mutations in the colon of heterozygous MtDNA mutator mice and ageing humans. *Mech. Ageing Dev.* **2014**, *139*, 22–30. [CrossRef]
129. Lechuga-Vieco, A.V.; Latorre-Pellicer, A.; Johnston, I.G.; Prota, G.; Gileadi, U.; Justo-Méndez, R.; Acín-Pérez, R.; Martínez-De-Mena, R.; Fernández-Toro, J.M.; Jimenez-Blasco, D.; et al. Cell identity and nucleo-mitochondrial genetic context modulate OXPHOS performance and determine somatic heteroplasmy dynamics. *Sci. Adv.* **2020**, *6*, eaba5345. [CrossRef]
130. Muller, H.J. The relation of recombination to mutational advance. *Mutat. Res. Mol. Mech. Mutagen.* **1964**, *1*, 2–9. [CrossRef]
131. Andersson, G.E.; Karlberg, O.; Canbäck, B.; Kurland, C.G. On the origin of mitochondria: A *Genom.* perspective. *Philos. Trans. R. Soc. Lond. Ser. B Biol. Sci.* **2003**, *358*. [CrossRef]
132. Soares, P.; Abrantes, D.; Rito, T.; Thomson, N.; Radivojac, P.; Li, B.; Macaulay, V.; Samuels, D.C.; Pereira, L. Evaluating purifying selection in the mitochondrial DNA of various mammalian species. *PLoS ONE* **2013**, *8*, e58993. [CrossRef]
133. Loewe, L. Quantifying the genomic decay paradox due to Muller's ratchet in human mitochondrial DNA quantifying the genomic decay paradox due to Muller's ratchet in human mitochondrial DNA. *Genet. Res.* **2006**, *87*, 133–159. [CrossRef]
134. Radzvilavicius, A.L.; Hadjivasiliou, Z.; Pomiankowski, A.; Lane, N. Selection for mitochondrial quality drives evolution of the germline. *PLoS Biol.* **2016**, *14*, e2000410. [CrossRef] [PubMed]
135. Fisher, R.A. *The Genetical Theory of Natural Selection*, 2nd ed.; Dover: New York, NY, USA, 1958.
136. Edwards, D.M.; Røyrvik, E.C.; Chustecki, J.M.; Giannakis, K.; Glastad, R.C.; Radzvilavicius, A.L.; Johnston, I.G. Avoiding organelle mutational meltdown across eukaryotes with or without a germline bottleneck. *PLoS Biol.* **2021**, *19*, e3001153. [CrossRef] [PubMed]
137. Hoekstra, R.F. Evolutionary origin and consequences of uniparental mitochondrial inheritance. *Hum. Reprod.* **2000**, *15*, 102–111. [CrossRef]
138. Gordo, I.; Charlesworth, B. The degeneration of asexual haploid populations and the speed of Muller's ratchet. *Genetics* **2000**, *154*, 1379–1387. [CrossRef]
139. Frank, S.A.; Hurst, L.D. Mitochondria and male disease. *Nature* **1996**, 224. [CrossRef] [PubMed]
140. Gemmell, N.J.; Metcalf, V.J.; Allendorf, F.W. Mother's curse: The effect of MtDNA on Individual fitness and population viability. *Trends Ecol. Evol.* **2004**, *19*, 238–244. [CrossRef]
141. Kuijper, B.; Lane, N.; Pomiankowski, A. Can paternal leakage maintain sexually antagonistic polymorphism in the cytoplasm? *J. Evol. Biol.* **2015**, *28*, 468–480. [CrossRef]
142. Garcia, L.E.; Zubko, M.K.; Zubko, E.I.; Sanchez-Puerta, M.V. Elucidating genomic patterns and recombination events in plant cybrid mitochondria. *Plant Mol. Biol.* **2019**, *100*, 433–450. [CrossRef]
143. Sato, A.; Nakada, K.; Akimoto, M.; Ishikawa, K.; Ono, T.; Shitara, H.; Yonekawa, H.; Hayashi, J.-I. Rare creation of recombinant MtDNA haplotypes in mammalian tissues. *Proc. Nat. Acad. Sci. USA* **2005**, *102*, 6057–6062. [CrossRef]

144. Guo, X.; Liu, S.; Liu, Y. Evidence for recombination of mitochondrial DNA in triploid crucian carp. *Genetics* **2006**, *172*, 1745–1749. [CrossRef] [PubMed]
145. Gantenbein, B.; Fet, V.; Gantenbein-Ritter, I.A.; Balloux, F. Evidence for recombination in scorpion mitochondrial DNA (Scorpiones: Buthidae). *Proc. Biol. Sci.* **2005**, *272*, 697–704. [CrossRef] [PubMed]
146. Ma, H.; O'Farrell, P.H. Selections that isolate recombinant mitochondrial genomes in animals. *Elife* **2015**, *4*, 1–16. [CrossRef] [PubMed]
147. Kraytsberg, Y.; Schwartz, M.; Brown, T.A.; Ebralidse, K.; Kunz, W.S.; Clayton, D.A.; Vissing, J.; Khrapko, K. Recombination of human mitochondrial DNA. *Science* **2004**, *304*, 981. [CrossRef] [PubMed]
148. Bandelt, H.J.; Kong, Q.P.; Parson, W.; Salas, A. More evidence for non-maternal inheritance of mitochondrial DNA? *J. Med. Genet.* **2005**, *42*, 957–960. [CrossRef] [PubMed]
149. Piganeau, G.; Gardner, M.; Eyre-Walker, A. A broad survey of recombination in animal mitochondria. *Mol. Biol. Evol.* **2004**, *21*, 2319–2325. [CrossRef] [PubMed]
150. Smith, J.M. A Short-term advantage for sex and recombination through sib-competition. *J. Theor. Biol.* **1976**, *63*, 245–258. [CrossRef]

MDPI
St. Alban-Anlage 66
4052 Basel
Switzerland
Tel. +41 61 683 77 34
Fax +41 61 302 89 18
www.mdpi.com

Life Editorial Office
E-mail: life@mdpi.com
www.mdpi.com/journal/life

www.ingramcontent.com/pod-product-compliance
Lightning Source LLC
LaVergne TN
LVHW070416100526
838202LV00014B/1465